Principles of Brazing

David M. Jacobson

Giles Humpston

ASM International®
Materials Park, Ohio 44073-0002
www.asminternational.org

First printing, August 2005
Digital printing, October 2010

Prepared under the direction of the ASM International Technical Book Committee (2004–2005), Yip-Wah Chung, Chair.

ASM International staff who worked on this project include Charles Moosbrugger, Acquisitions Editor; Bonnie Sanders, Manager of Production; Madrid Tramble, Production Editor; Kathryn Muldoon, Production Assistant; and Scott Henry, Assistant Director of Reference Publications.

Library of Congress Cataloging-in-Publication Data
Jacobson, David M.
Principles of brazing / David M. Jacobson, Giles Humpston.
p. cm.
Includes bibliographical references and index.
ISBN-10: 0-87170-812-4
ISBN-13: 978-0-87170-812-0
1. Brazing. I. Humpston, Giles. II. Title

TT267.J33 2005
671.5'6—cd22 2005042106
SAN: 204-7586

ASM International®
Materials Park, OH 44073-0002
www.asminternational.org

Printed in the United States of America

Contents

Preface

Principles of Brazing primarily aims at presenting the subject in a form that is readily accessible to practitioners of this joining technology, while at the same time offering a scientific perspective of brazing. It focuses on fundamental principles rather than providing recipes for producing brazed joints. Based on *Principles of Soldering and Brazing,* published in 1993, this volume includes much new material on brazing, covering progress over the past decade.

The largely artificial distinctions between brazing and soldering are preserved because, despite the many commonalities, it has been found that practicing engineers are concerned with either soldering or brazing and seldom are involved with both simultaneously. The companion volume, *Principles of Soldering,* addresses this complementary need. A large proportion of the literature on brazing and soldering may be charged with being heavy on description and light on critical analysis. We have endeavored to redress the balance, while striving to avoid being unduly simplistic or overly mathematical in our approach. Admittedly, we may not always have succeeded in this aim.

As in *Principles of Soldering and Brazing,* we have striven to maintain the focus on the fundamental aspects of brazing and have deliberately avoided entering into specific joining technologies in detail. Therefore, it is inevitable that some topics are not accorded due consideration, although it is hoped that sufficient references are provided to enable the reader to pursue these further.

The authors recognize that the range and extent of the knowledge base of metal joining is not immediately obvious, and mastery of the subject requires a fairly deep understanding of materials. This point is particularly apposite to the wetting of ceramics by metals, discussed in Chapter 7. The behavior of self-fluxing copper-phosphorus brazes and the poor mechanical integrity of joints made with these brazes to steel unless nickel is present represents another example of the interplay of factors that need to be considered and understood (see Chapter 3).

Two areas of brazing that have benefited from significant research efforts in recent years are active brazing and diffusion brazing. These are also areas in which the authors have been directly involved, and accordingly, individual chapters are devoted to each of these topics in this new edition (Chapters 6 and 7).

No attempt has been made to gather a comprehensive list of publications in our bibliography. Those that are included have been selected because they are useful basic texts, cover important subject matter, or relate to exemplary pieces of work, whether in respect of methodology, technique, or other noteworthy feature. It was felt that if the value of the book depended on its bibliography, it would rapidly become dated. The advent of computer search facilities and databases of scientific journal and conference abstracts should enable the reader who wishes to chase up references on specific topics to obtain further information without too much difficulty.

The reader should note that all compositions given in this book are expressed in weight percentage in accordance with the standard industrial practice, and these have, for the most part, been rounded to the nearest integer. The ratio of elements in intermetallic compounds, again by convention, refers to the atomic weight of the respective constituents. The general convention used for referring to braze alloy families is that adopted by the alloy phase diagram community, namely, in the alphabetical order of the elements, by chemical symbol. However, when referring to specific braze compositions, for the most part we have listed the elements in order of concentration. The alternative, which is widely accepted in the metal joining literature, is to express compositions in alphabetical order of the chemical symbols of the major constituents, followed by the minor constituents. The authors prefer the system based on concentration because, otherwise, lesser ingredients can punctuate the composition order, which would be illogical and avoids debate as to whether a constituent is a major or minor element. Thus, for example, we refer to an alloy of composition Ag-22Zn-21Cu-2Sn-0.01Ce as a silver-copper-zinc braze.

Specific references are given with each chapter. For those wishing to read more generally on particular topics, the authors recommend the texts listed as Selected References at the end of this preface.

Many phase diagrams are subject to ongoing research, resulting in continued improvement in the accuracy and detail of the information. The most recent version of a diagram may be identified by consulting the latest cumulative index of phase diagrams, published in the Cumulative Index of the periodical *Journal of Phase Equilibria* (ASM International). This index refers to the source of the thermodynamically assessed diagram of interest. The reader is advised that the four compendia of binary phase diagrams published in the 1960s, 1970s, and 1980s (colloquially referred to as Hansen, Elliott, and Shunk) are now known to contain many errors and omissions.

Information on new developments in soldering and brazing is scattered throughout a wide range of periodicals, as reflected in the sources cited in the references appended to the individual chapters. To keep abreast of the literature, the authors have found especially useful the following abstract publications: *Metals Abstracts* and *Science Abstracts.* Technical libraries can provide automated searches against specified key words as a monthly service.

We wish to thank our many colleagues and ex-colleagues for their helpful advice and encouragement, particularly Chris Corti of the World Gold Council for sharing his insights into the brazing of jewelery.

David M. Jacobson
Giles Humpston

SELECTED REFERENCES

- Brandon, D.G., and Kaplan, W.D., 1997. *Joining Processes: An Introduction,* John Wiley & Sons
- Eustathopoulos, N., Nicholas, M.G., and Drevet, B., 1999. *Wettability at High Temperatures,* Pergamon Press
- International Organization for Standardization (IOS), 1990. *Welding, Brazing and Soldering Processes: Vocabulary,* (IOS/DIS 857-2) ISO (currently under revision)
- Liebermann, E., 1988. *Modern Soldering and Brazing Techniques,* Business News
- Nicholas, M.G., 1990. *Joining of Ceramics,* The Institute of Ceramics/Chapman and Hall
- Olson, D.L. et al., Eds., 1993. *Welding, Brazing and Soldering,* Vol 6, *ASM Handbook,* ASM International
- Schwartz, M.M., 2003. *Brazing,* 2nd ed., ASM International
- Schwartz, M.M., 1990. *Ceramic Joining,* American Society for Materials
- Thwaites, C.J., 1983. *Capillary Joining: Brazing and Soft-Soldering,* Books Demand UMI

About the Authors

David M. Jacobson graduated in physics from the University of Sussex in 1967 and obtained his doctorate in materials science there in 1972. Between 1972 and 1975 he lectured in materials engineering at the Ben Gurion University, Beer-Sheva, Israel, returning as Visiting Senior Lecturer in 1979–1980. Having gained experience in brazing development with Johnson Matthey Ltd., he extended his range of expertise to soldering at the Hirst Research Centre, GEC-Marconi Ltd., which he joined in 1980. Currently, he holds the position of Professor of Manufacturing Technology at the Centre for Rapid Design and Manufacture, Buckinghamshire Chilterns University College, in High Wycombe. He is the author of more than 80 scientific and technical publications in materials science and technology and more than a dozen patents. He has been awarded three prestigious awards for his work on brazing.

Professor Jacobson's principal outside interests are archaeology and architectural history, focusing on the Near East in the Graeco-Roman period. He has published extensively in these fields, which extend to the numismatics and early metallurgy of that region. He recently completed a Ph.D. thesis on Herodian architecture at King's College, London, and teaches part-time in this subject area at University College, London. Professor Jacobson is married with two grown-up children and lives in Wembley, England, close to the internationally famous football stadium.

Giles Humpston took a first degree in metallurgy at Brunel University in 1982, followed by a Ph.D. on the constitution of solder alloys in 1985. He has since been employed by several leading industrial companies, where he has been involved with determining alloy phase diagrams and developing processes and procedures for producing precise and high-integrity soldered, brazed, and diffusion-bonded joints to a wide variety of metallic and nonmetallic materials. His expertise extends to fine-pitch flip-chip; new materials development; and packaging and interconnection for electronics, radio frequency, and optical products. He is a cited inventor on over 75 patents, the author of more than 60 papers, and the recipient of six international awards for his work on soldering and brazing.

Dr. Humpston is a licensed amateur radio enthusiast and has published several articles and reviews on electronics, radio, and computing. His other interests include exploring vertical axis wind turbines, building grid-tie inverters, flying radio-controlled gliders, winemaking, and growing bonsai. He lives with his wife, Jacqueline, and their three children in a small village in Buckinghamshire, England and in San Jose (Silicon Valley), California.

David Jacobson and **Giles Humpston** are the coauthors of the companion volume, *Principles of Soldering,* and the predecessor to this two-volume set, *Principles of Soldering and Brazing,* which was published by ASM International in 1993, with more than 4000 copies sold.

History

Origins of Brazes and Brazing

Brazing is not a modern invention. Archaeological evidence shows that it has been practiced continuously since ancient times. Because the earliest forms of metalwork often required joints, and the ingredients of common brazes were either of noble metals or base metals extracted from readily winnable ores, brazing appears to predate soldering heritage by 2 to 3 millennia, despite the higher temperatures involved.

Brazing of gold and silver using foils of copper or copper alloyed to the more precious metals has been practiced earlier than 3000 B.C. One of the oldest known applications of brazing is the fabrication of a gold chalice and other objects found in the tomb of Queen Shab-Ad of Ur in the Euphrates valley, dated to about 3200 B.C. [Maryon 1936]. A Proto-Elamite figurine of this vintage, in silver, was clearly assembled by brazing with copper [Lang and Hughes 1991]. From about 2800 B.C., we have the copper fittings on the canopy of the sarcophagus of the Egyptian Queen Hetep-heres, which were joined using a copper-silver braze [Brooker, Beatson, and Roberts 1975]. Brazing is documented as a clearly recognizable process in the 4th century A.D. Leiden Papyrus X, recipes 31 and 33 [Caley 1926]. Abundant direct evidence of brazing being used as a joining method has also been obtained from objects dated from the Roman Imperial period. Thus, brazing in its most traditional form takes advantage of the reduction of the melting point of copper on alloying with gold or silver to below that of all three individual metals. The 12th century writer on craftsmanship, Theophilus, describes a brazing alloy made of two parts of silver and one part of copper, which is not far from the silver-copper eutectic (Ag-28Cu) [Hawthorne and Smith 1963, p 107].

Pliny the Elder (1st century A.D.) mentions another version of the same brazing method, which was used in antiquity to join gold granules to gold sheet [Pliny, *Natural History,* xxxiii 29 (see Rackham, 1952); Wolters 1983]. A powdered copper-bearing ore, such as malachite, is heated with a flux in a reducing environment, generally provided by the organic gum, which is used to hold the granules in place. This operation yielded finely divided copper, which on alloying with the noble metal makes the brazed joint. The same author provides the first documented account of a copper-zinc braze. This involved the addition of one part of white copper (brass) to two parts of white lead (tin). A joint in a gold ewer dating from the 4th century was found to contain both copper and zinc [Lang and Hughes 1991]. At this point, brazing appears to have halted its development. We know that verb braze derives partly from the Old French, *braser,* meaning to burn. In the 16th century, the English verb braze denoted "expose to the action of fire." By the late 17th century, it acquired the meaning "solder with an alloy of copper and zinc," i.e., essentially its present connotation, including the original precious metal origins.

An early application of diffusion brazing, known as Sheffield plate, was developed by Thomas Bolsover in 1743 [Bradbury 1912]. Sheffield plate is produced by rolling together a plate of copper sandwiched between two thin sheets of silver. The pressed assembly is heated in air using a small amount of flux at the edges to prevent oxidation. Above the silver-copper eutectic temperature of 780 °C (1436 °F), diffusion during heating produces a liquid phase, which fuses the silver sheets to the copper plate. This material is then rolled down to a thinner sheet, and this was exploited by the industrial pioneer Matthew Boulton in the second half of the 18th century as a cheap substitute for solid silver in domestic wares.

In the early 20th century, brazing alloys and techniques underwent further evolution with the development of low-temperature silver brazing alloys, which have found widespread application in industry.

REFERENCES

- Bradbury, F., 1912. *History of Old Sheffield Plate,* MacMillan
- Brooker, H.R., Beatson, E.V., and Roberts, P.M., 1975. *Industrial Brazing,* 2nd ed., Newnes-Butterworth
- Caley, E.R., 1926. The Leiden Papyrus X, *J. Chemical Education,* Vol 3, p 1149–1168
- Hawthorne, G. and Smith, C.S., (Ed. and Transl.), 1963. *On Divers Arts; The Treatise of Theophilus,* Chicago University Press
- Lang, J. and Hughes, M.J., 1991. *Joining Techniques,* in *Aspects of Early Metallurgy, British Museum Occasional Papers No. 17,* British Museum, p 169–177
- Maryon, H., 1936. Soldering and Welding in the Bronze and Early Iron Ages, *Technical Studies,* (No. 5), p 75–108
- Rackham, H. (Transl.), 1952. *Pliny: Natural History,* Vol 10, Harvard University Press
- Wolters, J., 1883. *Die Granulation: Geschichte und Technik einer alten Goldschmiedekunst,* Callwey [in German]

CHAPTER 1

Introduction

1.1 Joining Methods

BRAZING AND SOLDERING jointly represent one of several methods for joining solid materials. These methods may be classified as:

- Mechanical fastening
- Adhesive bonding
- Brazing and soldering
- Welding
- Solid state joining

Other methods, such as glass/metal sealing, electrostatic welding, and so forth, are dealt with elsewhere [Messler 1999; Nicholas 1998; O'Brien 1991; Bever 1986].

Schematics of these joining methods are given in Fig. 1.1. These methods have a number of common features but also certain significant differences. For example, brazing and soldering are the only joining methods that can produce smooth and rounded fillets at the periphery of the joints. The joining methods just listed are in the order in which they lead to fusion of the joint surfaces and tend toward a "seamless" joint.

Because brazing and soldering lie in the middle of this sequence, they share several of the features of the other methods. For example, brazed and soldered joints often can be endowed with the advantageous mechanical properties of welded and diffusion-bonded joints; at the same time, in most cases they can be disassembled readily, without detriment to the components, like mechanically fastened joints. These features make brazing and soldering highly versatile.

The principal characteristics of the various joining methods are summarized in the paragraphs that follow.

1.1.1 Mechanical Fastening

Mechanical fastening involves the clamping together of components without fusing the joint surfaces. This method often, but not always, relies on the use of clamping members such as screws and rivets. In crimping, the components are keyed together by mechanical deformation.

Characteristic features of mechanical fastening include:

- A heating cycle is generally not applied to the components being joined. A notable exception is riveting, where the rivets used for clamping are heated immediately prior to the fastening operation. On subsequent cooling the rivets shrink, causing the components to be clamped tightly together. Historically, riveting has been used in shipbuilding because this joining method can achieve watertight joints between large plates of steel that make up the hull of a ship.
- The reliance on local stressing to effect joining requires thickening or some other means of reinforcement of the components in the joint region. This places a severe restriction on the joint geometries that may be used and imposes a weight penalty on the assembly. Another constraint on permissible joint configurations is the need for access to insert the clamping member.
- The method usually requires special mechanical preparation, such as drilling holes, machining screw threads, or perhaps chamfering of abutting surfaces, in the case of components to be crimped.
- The choice of suitable joint configurations is highly dependent on service conditions, for example, whether leak-tightness is required. Joints may be designed to accommodate thermal expansion mismatch between the components in the assembly. In the extreme case, joints can be made to permit complete freedom of movement in the plane perpendicular to the clamping member, as applied to the joint bars (fishplates) used to couple train rails.

- The electrical and thermal conductance across the joint is a function of the effective area that is in contact. This depends on many other parameters, such as the clamping force and the materials used; in service, the conductance is unlikely to be constant.

Mechanical fastening is used widely in conjunction with brazing processes, particularly as a method of holding preforms of brazing alloy in place during the heating cycle. Figure 1.2 shows one such application where multiple strips of brazing foil are held in place by twist-tag fasteners. In the assembly of aircraft structures made of light aluminum alloys, riveted joints are complemented and reinforced by adhesive bonding.

1.1.2 Adhesive Bonding

Adhesive bonding involves the use of a polymeric material, often containing various additives, to "stick" the components together. The process involves a chemical reaction, which may simply occur through exposure of the adhesive to air, leading to the formation of a hydrogen-type bond between the cured adhesive and the respective components. The original interfaces of the joint are preserved in this type of bonding process.

Characteristic features of adhesive bonding include:

- It is inherently a low-stress joining method because it is carried out at relatively low

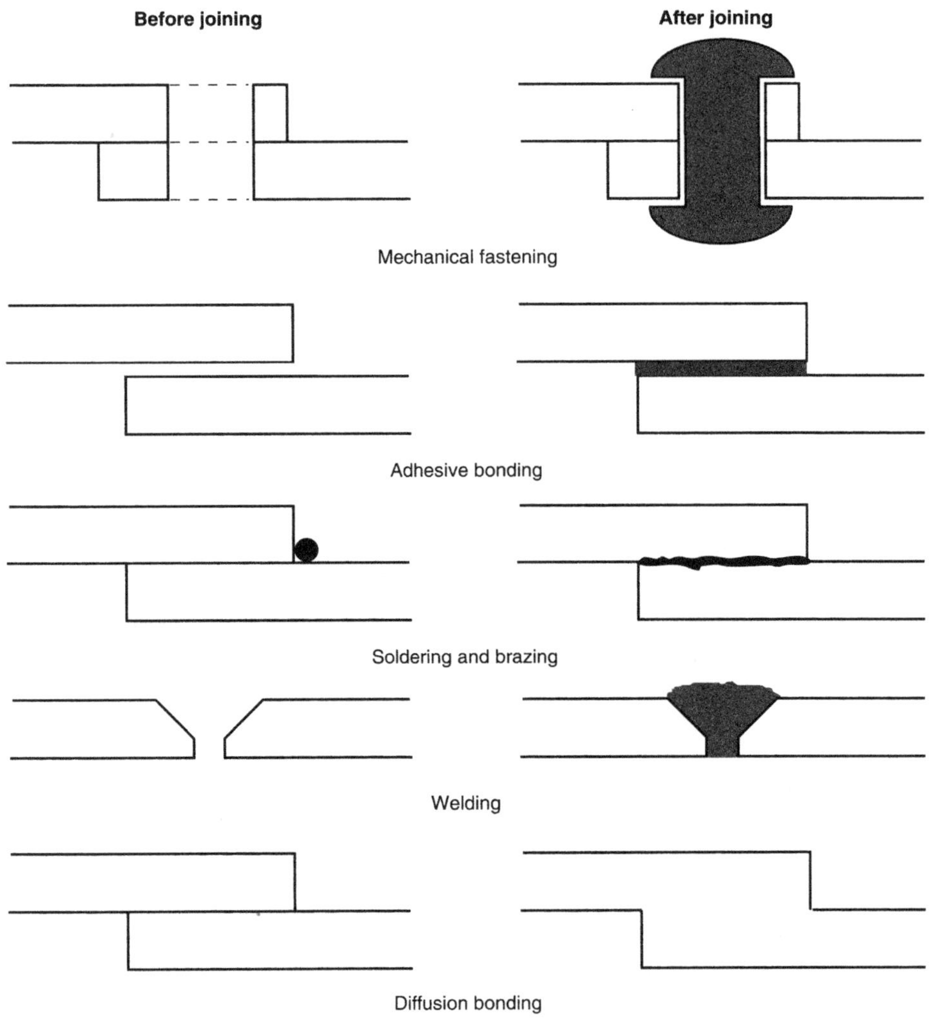

Fig. 1.1 Principal methods for joining engineering materials

temperatures, and most adhesives have high compliance.

- A diverse range of methods is available for curing adhesives.
- The geometry of the components tends not to be critical.
- Constraints apply to the geometry of the actual joint; in particular, large areas and very narrow gaps are necessary to ensure mechanical integrity.
- Joints tend to be weak when subject to forces that cause peeling. For this reason, adhesive joints are used frequently in combination with mechanical fastening, for example, in air-frame assemblies.
- Joint integrity tends to be sensitive to the state of cleanliness of the mating surfaces and to the atmosphere of the service environment.
- The service temperature range of adhesively bonded joints is usually limited, owing to the restricted temperature range over which they are stable, as is their compatibility with organic and aqueous media.
- Joints usually possess poor electrical and thermal conductance, although by loading the organic adhesive with metal particles, moderate conductance can be achieved, which approaches that of some brazing alloys. Such loading, however, is often at the expense of adhesive strength.

Polymer chemistry is a rapidly evolving science and some very advanced adhesives have appeared on the market with properties highly tailored for particular functions. However, as adhesives do not have high temperature stability, they are never used in combination with brazes.

1.1.3 Brazing and Soldering

Brazing and soldering use a molten filler metal to wet the mating surfaces of a joint, with or without the aid of a fluxing agent, leading to the formation of metallurgical bonds between the filler and the respective components. In these processes, the original surfaces of the components are "eroded" by virtue of the reaction occurring between the molten filler metal and the solid components, but the extent of this "erosion" is usually at the microscopic level (<100 μm, or 4000 μin.). Joining processes of this type, by convention, are defined as brazing if the filler melts above 450 °C (840 °F) and as soldering if it melts below this temperature.

Characteristic features of brazing and soldering include:

- All brazing operations and most soldering operations involve heating the filler metal and joint surfaces above ambient temperature.
- In most cases, the service temperature of the assembly must be lower than the melting temperature of the filler metal.
- It is not always necessary to clean the surfaces of components prior to the joining operation because fluxes that are capable of

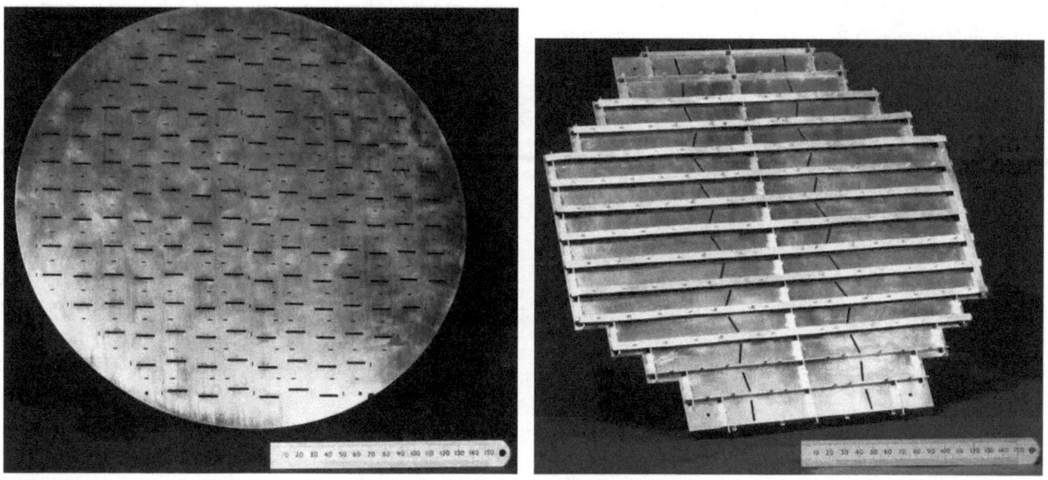

(a) (b)

Fig. 1.2 A phased array radar antenna prepared for brazing. The (a) front face and (b) interior show strips of brazing foil held in place by twist-tag mechanical fasteners. Courtesy of BAE Systems Ltd.

removing common oxides and organic films are available. However, there are penalties associated with the use of fluxes; for example, they leave behind residues, which are often corrosive and can be difficult to remove.
- The appropriate joint and component geometries are governed by the filler/component material combination and by service requirements (need for hermeticity, stress loading, positional tolerances, and so forth). Complex geometries and combinations of thick and thin sections can usually be brazed or soldered together.
- Intricate assemblies can be produced with low distortion, high fatigue resistance, and good resistance to thermal shock.
- Joints tend to be strong if well filled, unless embrittling phases are produced by reaction between the filler metal and the components.
- Brazed and soldered joints can be endowed with physical and chemical properties that approximately match and, in some cases, even exceed those of the components, but solders and some brazes usually have limited elevated-temperature service capability and stability.
- Fillets are formed under favorable conditions. These fillets can act as stress reducers at the edges of joints that benefit the overall mechanical properties of the joined assembly.

Brazing and soldering can be applied to a wide variety of materials, including metals, glasses, ceramics, plastics, and composite materials. For many materials, and plastics in particular, it is necessary to apply a surface metallization prior to joining.

1.1.4 Welding

Welding involves the fusion of the joint surfaces by controlled melting through heat being directed specifically toward the joint. Commonly used heating sources are plasma arcs, electron beams, lasers, and electrical current that is passed through the components and across the joints (electrical resistance) [Messler 1999; O'Brien 1991]. Filler metals may be used to supplement the fusion process for components of similar composition, as, for example, when the joint gap is wide and, possibly, of variable width. In that situation, the filler is often chosen to have a marginally lower melting point than the components in order to help ensure that it melts completely. The joining process then has many similarities with wide joint gap brazing.

Characteristic features of welding include:

- Welding invariably involves a heating cycle, which tends to be rapid. A very wide variety of welding processes are available.
- Welding cannot be used to join metals to nonmetals or materials of greatly differing melting points. There are exceptions, but these are generally limited to precise combinations of materials and highly specific welding methods.
- Joint geometries are limited by the requirement that all joint surfaces are accessible to the concentrated heat source.
- Welded joints may approach the physical integrity of the components but are often inferior in their mechanical properties, particularly fatigue resistance. This is due to stress concentrations produced by the high thermal gradients developed during joining and the relatively rough surface texture of welds.
- The heating cycle usually affects the microstructure and hence the properties of the components over a macroscopic region around the joint, called the *heat-affected zone* (HAZ). The HAZ is often influential in determining the properties of welded joints.
- Welding tends to distort the components in the region of the HAZ. This is associated with the thermal gradients developed through the use of a concentrated heat source to fuse the joint surfaces.

Where the brazing cycle is particularly rapid and conducted using a high-intensity heat source, as, for example, in brazing of refractory metals, the process has many similarities to welding. Likewise, there exist hybrid processes, such as braze welding. This joining method is of particular value for joining advanced multiphase materials such as metal matrix composites. Although the metal component of such materials can often be brazed readily, the nonmetal species can usually be wetted using only very different, and hence often incompatible, filler metals and processing conditions. Braze-welding overcomes these limitations. The method combines heating, usually achieved by the passage of electric current, combined with compression of the joint, and only a small quantity of additional, preplaced, filler metal. The heating cycle is so rapid that dewetting, porosity, and other phenomena associated with liquid phase joining, which would normally be prevalent when attempting to braze such materials in a conventional manner, are suppressed. Advanced, ultralightweight bicycle frames have been formed from tubular members of long-fiber-reinforced aluminum-boron composites using this method [Schwartz 2003, Zvolinskii 1995].

1.1.5 Solid-State Joining

The term *solid-state joining* covers a wide range of joining processes, most notably pressure welding, friction welding, and diffusion bonding [Messler 1999; Nicholas 1998]. Pressure welding, at its simplest, involves deforming physically two abutting surfaces to disrupt any intervening surface films and enable direct metal-to-metal contact. This process can be performed either hot or cold. In friction welding, the heat and disruption of surface films needed to achieve a sound bond are achieved by subjecting the components to relative movement while applying a compressive force. The relative movement can be rotary or involve angular or linear reciprocation. Diffusion bonding, in its purest form, requires placing two faying surfaces in contact and heating the assembly until the voids at the interface have been removed by diffusion. Messler [1999] includes wet plating (electrolytic and electroless) and vapor deposition (evaporation, sputtering, etc.) as solid-state joining processes because the deposit forms as a solid, and chemical bonds are usually formed when a metal is deposited on another material. Deposition requires that the interface is atomically clean for the deposit to be joined effectively to the substrate.

Solid-state joining constitutes a subject in its own right, quite separate from brazing and soldering, which rely on liquid state metal joining. However, the development of friction brazing and diffusion brazing processes, which are both hybrids between solid-state joining and brazing, require some consideration of the solid-state joining (see section 1.1.7). Pressure welding is sometimes used to prepare filler metals in various geometries and for tacking preforms in position.

Characteristic features of solid-state joining are:

- This method generally involves heating of the joint to a temperature below the melting point of the components.
- Pressure welding (especially in the form of explosive welding) and friction welding are often much faster processes than soldering or brazing ($<$1 s), while diffusion bonding is much slower ($>$10 min).
- The joints have no fillets.
- The service temperature of joined assemblies can be higher than the joining temperature and tend toward the melting point of the components.
- Solid-state joining is limited in application to specific combinations of materials that provide specific combinations of mechanical or diffusion characteristics.
- Of all the joining methods, solid-state joining is the least tolerant to poor mating of the joint surfaces.
- Joint surfaces need to be scrupulously clean because solid-state joining is a fluxless process.
- The properties of solid-state joints can approach those of the parent materials.

Further details on pressure welding are given in section 1.1.7.1, friction welding in section 1.1.7.2, and diffusion bonding in section 1.1.7.3 of this chapter.

1.1.6 Comparison between Brazes and Solders

In many respects, it is fruitful to consider brazes together with solders. This integrated treatment can be justified on metallurgical grounds. These two classes of filler cannot simply be demarcated by the 450 °C (840 °F) temperature boundary as is habitually done. This distinction has a historical origin: the earliest brazes were based on alloys of copper, while solders were based on alloys of tin (see the introductory section "History: Origins of Brazes and Brazing" in this volume and the similar historical summary on solders and soldering in the companion volume *Principles of Soldering*).

The type of metallurgical reaction that occurs between a molten filler metal and parent metal is sometimes used to differentiate brazing from soldering. Brazes mostly alloy with the parent materials to form solid solutions, which are mixtures of the constituents on an atomic scale. By contrast, solders usually react to form intermetallic phases, that is, compounds of the constituent elements that have different atomic arrangements from the elements in solid form.

However, this distinction does not have universal validity. For example, silver-copper-phosphorous brazes react with steels to form the interfacial phase of Fe_3P in a similar manner to the reaction of tin-base solders with iron and steels to form $FeSn_2$. As for the temperature convention used to differentiate brazes from solders, there exist brazes for aluminum that melt below 450 °C (840 °F) and gold-indium solders above this temperature.

Brazing and soldering involve essentially the same bonding mechanism: that is, reaction with the parent material, usually alloying, so as to form metallic bonds at the interface. In both situations, good wetting promotes the formation of fillets that serve to enhance the strength of the joints. Similar processing conditions are required and the physical properties of both classes of filler metal are fairly comparable, provided the same homologous temperature (the temperature at which the properties are measured as a fraction of the melting temperature, expressed in degrees Kelvin) is used for the comparison.

The perpetuation of the distinction of brazes from solders based on the temperature convention of a 450 °C (840 °F) demarcation has arisen from the significant gap that exists between the melting points of available brazes and solder alloys. The highest melting point commercially produced solder is Au-3Si that melts at 363 °C (685 °F), and the lowest temperature standard braze is the Al-10Si-4Cu alloy, which melts at 524 °C (975 °F) but, being a noneutectic alloy, is fully liquid only above 585 °C (1085 °F). Eutectic alloys are defined in Chapter 2, section 2.3. For now, it shall suffice to state that eutectic alloys are akin to pure metals in melting and freezing at a unique temperature. The temperature ranges of the principal braze and solder alloy families are shown in Fig. 1.3 and 1.4, respectively.

For most purposes, the temperature gap between brazes and solders is substantially wider than 160 °C (290 °F). This is because the gold-base solders are very expensive and are largely limited in use to high added-value manufacturing applications in the electronics and photonics industries. Removing the high-gold-content alloys from consideration, the highest melting point solders are the lead-rich alloys, which melt at about 300 °C (570 °F). On the other hand, the lowest melting point brazes that are used commercially in significant quantities are the reasonably ductile aluminum-silicon-base alloys, which melt at 577 °C (1071 °F). However, these have limited compatibility with many components and "low" melting point brazes are generally considered to be silver-base alloys that melt at about 600 °C (1112 °F). The practical

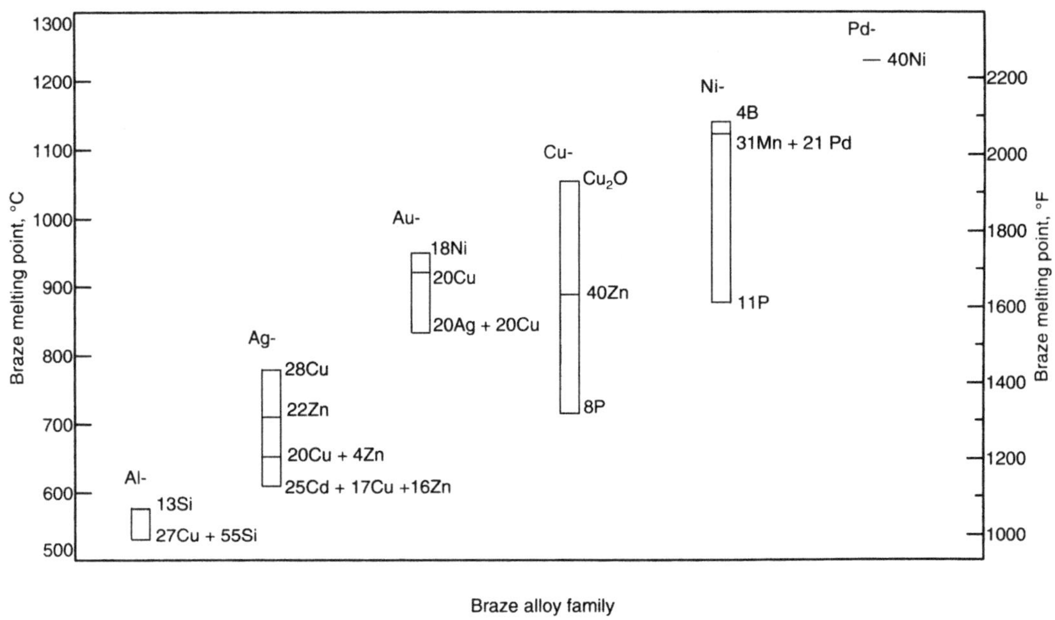

Fig. 1.3 Principal braze alloy families and their melting ranges

gap in temperature between brazes and solders is therefore closer to 300 °C (540 °F).

The dearth of filler metals with melting points in the range 300 to 600 °C (570 to 1112 °F) is not necessarily a handicap. Techniques are available for making joints using molten filler metals with effective melting points in this temperature range. Transient-liquid-phase diffusion bonding (otherwise known as *diffusion soldering* when carried out below 450 °C, or 840 °F), or *diffusion brazing* when this temperature threshold is exceeded) is one such example and is discussed in Chapter 6 of this volume and in Chapter 5, section 5.9 of the companion volume *Principles of Soldering*.

From the "maps" of brazes and solders in Fig. 1.3 and 1.4, it might appear that there are far fewer brazes than solders. In fact, the contrary is true. The alloys that are specifically indicated in these figures are mostly eutectic compositions or those characterized by minimum melting ranges. Most commercially used solders are included because these solders are almost all of (different) eutectic composition, but whole families of brazes have been omitted because there is no eutectic in the alloy system. Instead, these particular alloys exhibit complete intersolubility so that a continuum of alloy compositions exists that are suitable as brazes. Examples are the copper-nickel, silver-gold, silver-palladium, and silver-gold-palladium alloys. Alloys in such systems melt over a temperature range that varies with the composition.

Solders are usually referred to directly by composition in weight percentage, e.g., Pb-63Sn. Brazes are often denoted by commercial (or common) names. Some of these are coded such that they indicate the principal constituents. Some examples are listed in Table 1.1.

Because more thermal activation energy is present, the higher process temperatures needed to make a brazed joint have important consequences:

- More extensive metallurgical reaction between the filler metal and the substrate. Solders typically do not dissolve more than a few microns of the component surfaces, whereas brazes often dissolve tens of microns. Larger changes in the composition of the filler metal therefore occur during brazing, which in turn significantly affects the fluidity of and wetting by the molten filler as well as the properties of the joint.
- Greater reactivity with the atmosphere surrounding the workpiece. All other factors being equal, brazes are less tolerant to oxidizing atmospheres than solders but, for the same reasons, are also better suited to cleaning by reducing atmospheres. When joints are made in air with the aid of a flux, the greater reactivity of brazes means that a

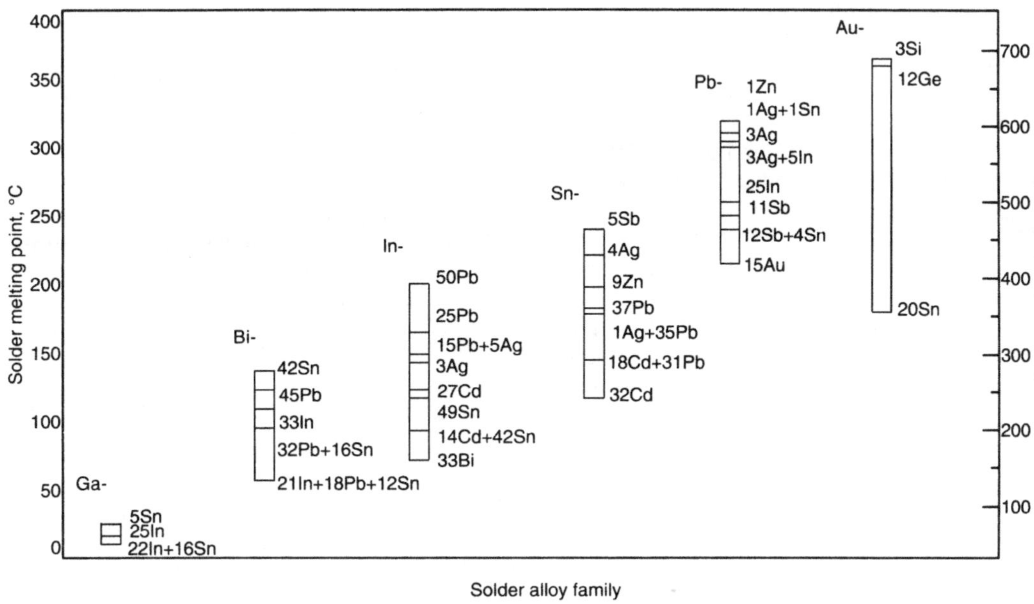

Fig. 1.4 Principal solder alloy families and their melting ranges

higher ratio of flux to filler metal is generally required. In consequence, flux-cored solders are adequate for use in air, while brazing rods intended for use in ambient atmosphere must be provided with a higher volume of flux, achieved by an external coating on the braze preform. Fluxes are discussed in Chapter 3, section 3.2.

Several general features distinguish the majority of common brazes from solders:

- Most brazes possess mutual solid solubility between their constituents and are therefore offered with a wide range of compositions and melting ranges.
- Most commercial solders, by comparison, are of eutectic composition because there is usually a need to minimize the processing temperature while maintaining reasonable fluidity of the molten filler. Also, solders are generally soft, even at room temperature, and must be conferred with optimum mechanical properties; generally, these are achieved by having a fine-grained microstructure, which is a characteristic feature of a true eutectic alloy. The low degree of intersolubility and the propensity to form intermetallic compounds possessed by solder alloys is related to their low melting-point constituent elements, principally tin and indium, having a noncubic crystal symmetry.
- Brazes tend to be used at temperatures that are usually below half their melting point in degrees Kelvin. The principal failure modes of brazed joints are traditional metallurgical processes such as fatigue, stress overload, and corrosion. Solders find application at temperatures at a fraction of between 50 and 90% of their melting point in degrees Kelvin, under strain levels that often exceed 10%. Under these conditions, the alloys are not metallurgically stable and the joint microstructure tends to change with time.
- Brazes are predominantly used for structural applications, while the major use of solders is for making electrical connections in electronic circuits.

These points are discussed in further detail in Chapters 2 and 3 and reference should also be made to the companion volume *Principles of Soldering*. Notwithstanding the differences, brazes and solders operate on similar principles, and hence the frequent use of the collective term *filler metal* throughout this book and the companion volume *Principles of Soldering* has some justification.

1.1.7 Pressure Welding, Friction Welding, and Diffusion Bonding

Solid-state joining methods are not new and examples of gold-base artifacts fabricated using pressure welds have been dated to between 1400 and 1000 BC [Tylecote 1968]. A gold ribbon used as a torch or neck ornament, found in a Celtic grave, was joined by the same method. Other ancient examples of welding from Egypt and the Black Sea region are cited by Tylecote [1967]. Although more recent interest in welding has been almost totally dominated by fusion welding processes, pressure welding, friction welding, and diffusion bonding continue to satisfy niche applications because of the unique combination of process and joint parameters they offer. Among the principal advantages of these nonfusion welding processes are:

- Melting of the components either does not occur or is very slight and highly localized so that changes to their microstructure are minor.
- Because there is very little alloying between components joined in these processes, formation of brittle intermetallic compounds is minimized so that a wider range of materials are amenable to joining by nonfusion welding than by fusion welding.

Some solid state joining procedures are a combination of pressure welding, friction welding, and diffusion bonding, as evidenced by the fundamental characteristics of each.

Table 1.1 Examples of brazing alloys with commercial (now common) names that indicate the principal constituents

The name may be augmented by a number if there are several brazes in the family in which the proportion of one element varies while the ratio of the other constituents is largely constant.

Commercial/common name	Typical composition
Cusil	Ag-28Cu
Cusiltin 5	Ag-27Cu-5Sn
Gapsil	Ag-9Ga-9Pd
Incuro 60	Au-37Cu-3In
Nicuman 23	Cu-23.5Mn-9Ni
Palco	Pd-35Co
Palcusil 10	Ag-32Cu-10Pd
Palnicusil	Ag-22.5Pd-18.9Cu-10Ni
Silcoro 60	Au-20Ag-20Cu
Ticusil	Ag-26.7Cu-4.5Ti

1.1.7.1 Pressure Welding

Pressure welding utilizes pressure to rupture surface films at the joint interface and also to extrude virgin parent metal between islands of surface contamination so that metallic bonding can take place. Thus, the process is characterized by high pressures, applied for short periods of time, on metals that may be either cold or hot. By necessity, bulk plastic deformation of the metals will occur. The pressure can be exerted through the application of a uniaxial force, alternatively isostatically, or by rolling, or explosively. Of these, explosion welding is the most "exotic" form of pressure welding; controlled detonation is used to force metal workpieces together under a rapid impulse. During this event, air between the parts is expelled in the form of a supersonic jet, which strips away surface oxides and causes localized heating, to promote bonding. The deformation at high strain rate that results from the impact of the components supplies additional heating in explosion welding. The different types of pressure welding are succinctly described by Messler [1999]. Possibly the most common examples of pressure welding that are pertinent to brazing are butt-welding to join lengths of wires, roll-bonding, and indentation welding.

In pressure welding, it is generally accepted that bond formation is controlled by the extent of deformation of the faying surfaces. The term *threshold deformation* is used extensively in the literature on this subject and is defined as the minimum deformation needed to achieve any bonding, although the strength of a bond at this level of deformation is generally much less than that of the parent metal (Fig. 1.5). The pressure that must be exerted to achieve flow by a ductile member between two effectively rigid components increases with the width-to-thickness ratio of the interface. In accordance with the von Mises criterion, the pressure is $(2/\sqrt{3})\sigma_s$ for a width-to-thickness ratio of 1:1, and nearly six times this value for a ratio of 20:1, where σ_s is the yield stress of the ductile member. Consequently it is preferable that pressure welded joints are small and radially symmetric.

The bonding process can be described as involving four consecutive stages:

- Removal of surface contamination and breakup of brittle surface layers, in particular, oxides. This is frequently assisted by mechanically abrading the surface immediately prior to bonding. Adsorbed water is believed to be the main surface contaminant and responsible for preventing bonding if the plastic deformation is less than 8%. Typically, 40% deformation is required to affect a sound joint when bonding base metals in atmospheres other than vacuum.
- Establishing physical contact between regions of uncontaminated metal as virgin metal extrudes between gaps in the ruptured surface films
- Activation of contacting atoms to form a metallic bond. The contact area determines the extent of bonding.
- Subsequent atom rearrangement as a consequence of post-weld heat treatment and/or stress relaxation

Pressure welding is particularly effective when joining dissimilar metals. For good weldability, the softer metal of the two abutting components should have the more brittle and stronger oxide film and vice versa. The hard oxide layer on the softer metal component can then promote and assist in the breakup of the surface layers on the harder metal but is itself easily ruptured by yielding of the metal supporting it. For example, the oxide on aluminum fulfills the requirements of hardness in relation to the oxides of most other metals, while the metal itself is relatively soft. Therefore, pressure welding of aluminum to harder metals occurs at lower deformations than when autogenously welded to itself. Also, the different deformation characteristics of dissimilar metals may result in interfacial movement that will enhance bonding com-

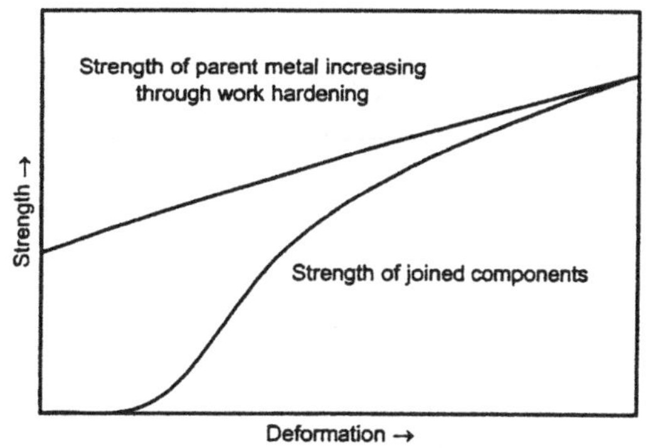

Fig. 1.5 The strength of pressure-welded joints as a function of the deformation induced during the bonding process. No joining occurs below the threshold deformation level. With increasing deformation the joint strength also increases eventually up to that of the parent materials. Note that the joining process modifies the properties of the parent material as it will work-harden when mechanically deformed.

pared with autogenous welding. The use of pressure welding to fabricate ductile preforms of brittle alloys, by partitioning of their constituents, is discussed further in Chapter 4, section 4.1.5.

1.1.7.2 Friction Welding

Friction welding is a thermally activated process in which mechanical energy is converted into heat at the interface between two impacting parts through their rapid, relative movement. Metal contact is achieved under the combination of pressure and rubbing together of the parts, which generates the heat of the process [Elmer and Kautz 1993; Messler 1999]. Material transport occurs through plastic deformation of the parts and thermally activated diffusion.

Frictional welding processes have proved difficult to model quantitatively, although they are fairly well understood at a qualitative level. The quality of a friction weld depends on five conditions:

- The pressure applied to the area forming the joint
- The relative velocity of the impacting surfaces
- The temperature at the interface
- Specific bulk material characteristics
- Condition of the surfaces to be joined together

The first three parameters are self-evident, being also common to other nonfusion welding processes, relating to pressure and temperature. The relative velocity of the impacting surfaces relates to the kinetic energy available for conversion to heat at the interface. The relevant bulk material characteristics involve greater complexity. However, from a pragmatic point of view, intrinsic property criteria that favor good friction welds between materials are:

- The materials have good forging characteristics
- The materials generate sufficient frictional resistance when rubbed together

The first criterion excludes pairs of brittle materials, such as ceramics; cemented carbides; and hard, brittle materials in general, although it is occasionally possible to friction weld a ductile material to a hard material. The second criterion excludes materials that contain constituents that provide dry lubrication, such as cast iron and other graphite-containing alloys. Lead-bearing bronzes are also unsuitable candidates for friction welding.

The relative movement most commonly involves:

- Rotation
- Angular or linear reciprocation
- Ultrasonic agitation
- Frictional stirring

Rotational friction welding equipment operates by direct drive or inertia drive. The first of these arrangements makes use of a motor running at constant speed to rotate one of the parts as it is driven into contact with the second, stationary part. The resulting friction and abrasion heats up the two surfaces (the friction phase), and a point is reached when the rotation is stopped and pressure is applied to join the parts together (the forging phase) (Fig. 1.6). The conditions used for the process cycle need to be optimized for the material combinations and dimensions of the parts. In inertia-drive friction welding, energy stored in a flywheel is imparted to the joint. Axial pressure is applied to force the part connected to the freely rotating flywheel against the stationary part, and this pressure is maintained until the welding operation is completed. Here, the moment of inertia of the flywheel and its initial speed are two important parameters of the process because they govern the energy available to the joining process. Typical spindle speeds in rotational friction welding are

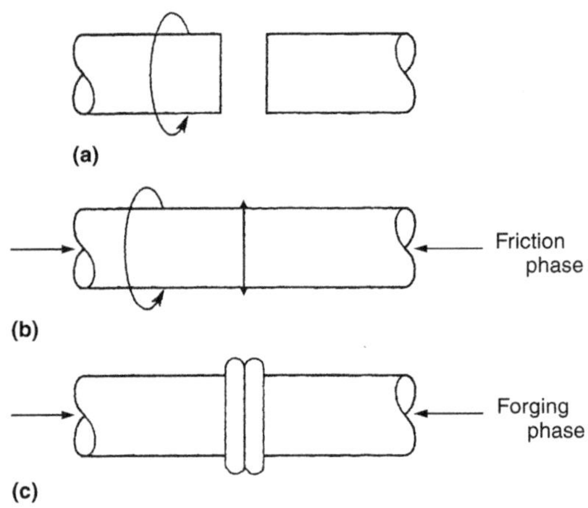

Fig. 1.6 Schematic showing fundamental steps in the frictional welding process (involving rotational movement). (a) One part or workpiece is rotated and the other part is held stationary. (b) Both parts are brought together, and axial stress is applied to begin frictional welding. (c) Rotation is stopped and the welding operation is completed. Source: Elmer and Kautz [1993]

in the range 1000 to 10,000 rpm, and the total process time is under 3 s, making this process relatively fast. Other procedures for frictional welding by rotation and other means such as friction stir welding are described in the literature [Messler 1999].

Similar materials that have been joined by friction welding, and reported in the literature, include low-alloy steels, austenitic stainless steels, and various aluminum and titanium alloys. This nonfusion welding technique has also proved successful for joining copper, tool steels, nickel alloys, and austenitic stainless steels to low-alloy steels, aluminum to austenitic stainless steel, and copper to aluminum alloys. It has also been possible to join high-carbon steel to low-alloy steel by friction welding [Elmer and Kautz 1993]. However, when joining dissimilar materials together, the quality of the joint is more sensitive to surface preparation of the joint surfaces than for friction welds made to similar materials. This is particularly the case where aluminum and its alloys constitute one of the dissimilar materials. Then, the alumina on the surface of the former tends to attract contaminants such as water and hydrocarbons, which results in mechanical weakness at the interface.

1.1.7.3 Diffusion Bonding

Diffusion bonding relies on a combination of temperature, pressure, and time to remove voids from the free interfaces between two abutting metal parts [Messler 1999]. Fundamentally, the process is defined as one in which no plastic deformation of the components being joined takes place, although it is normal to apply some pressure to ensure that the nominally flat faying surfaces are indeed in intimate contact. Typically, diffusion bonding requires process times of up to several hours at temperatures that may be as high as two-thirds of the melting point of the least thermally stable metal in the bonded couple. The use of long times at relatively high temperatures necessitates some form of atmosphere control to preserve surface cleanliness. Soft (roughing) vacuum and controlled atmospheres are equally suitable.

Since diffusion processes are the main mechanisms for bonding by this process, with no means for the physical displacement of any intervening nonmetallic surface films, two related requirements need to be satisfied. The first of these is that these films must be prevented from constituting a barrier to atom migration. Secondly, in bimetallic or higher order systems, the formation of intermetallic compounds and porosity resulting from an imbalance of the diffusion rate of different atomic species (Kirkendall porosity) must be controlled. Table 1.2 presents some of the better-known direct diffusion bonding combinations of metals and metalloids.

In a simplified model of diffusion bonding, it is assumed that the process occurs in three sequential stages [Messler 1999]:

- Contact between asperities in the abutting or overlapping parts, i.e. intermittent contact. The application of pressure causes the asperities to deform plastically, resulting in the formation of metallic interfaces between the parts in these contact regions.
- Under the combined effect of pressure and temperature, migration of grain boundaries and of material occurs through creep, and gaps between the abutting parts are progressively reduced to leave isolated pores.
- Surface and volume diffusion of metallic constituents occurs and the pores are eliminated.

In reality, one or more of these mechanisms is dominant and they may operate in parallel rather than sequentially. Each mechanism results in material (or void) transport so that the surface energy associated with the interface is progressively reduced as joining proceeds.

A detailed theoretical treatment of solid-state diffusion bonding is provided by Hill and Wallach [1989]. In practice, the extent of bonding and the rate at which it is achieved is governed both by materials properties (such as surface, grain boundary and volume diffusion coefficients, creep and yield strength, etc.) and process parameters of which the four main variables are:

- *Pressure:* Adequate pressure is required to achieve contact on an atomic scale by localized deformation of asperities on the nominally flat surfaces being joined and also to allow creep mechanisms to contribute to bonding.
- *Temperature:* Thermal energy promotes faster bonding because plastic deformation, creep, and all diffusion mechanisms are temperature dependent. Typically, temperatures around 0.7 T_m are used, where T_m is the absolute melting temperature of the lowest melting point component, in order to decrease the yield stress of the metals and en-

Table 1.2 Matrix of metals and metalloids that can be diffusion bonded

X indicates potentially workable combinations

	Ag	Al	Au	Be	Cr	Co	Cu	Fe	Nb	Mg	Mo	Ni	Pd	Pt	Ta	Ti	U	V	W	Stainless steel	Cast iron	Carbides	Graphite
Ag	X	X	...	X	...	...	X	...	...	...	...	...	...	...	...	X	...	...	...	X	...	...	...
Al	X	X	...	X	...	...	X	X	...	X	...	X	...	...	...	X	...	X	...	X	...	...	...
Au	...	...	X	X	...	...	X	...	...	...	...	X	...	...	...	...	...	...	...	X	...	...	...
Be	X	...	X	X	...	...	X	...	...	...	...	...	...	...	...	...	...	...	...	...	...	...	...
Cr	...	...	...	...	...	...	...	...	...	...	...	...	...	...	...	...	...	...	X	...	...	...	...
Co	...	...	...	...	...	X	...	...	...	...	...	X	X	...	...	...	...	...	...	...	...	...	...
Cu	X	X	X	X	...	...	X	X	X	...	X	X	...	...	...	X	...	X	...	X	...	...	...
Fe	X	...	...	...	...	...	X	...	...	...	X	X	X	X	...	X	...	...	X	...	...	...	...
Nb	...	...	...	...	...	...	X	...	X	...	X	X	...	...	X	X	...	X	X	X	...	X	...
Mg	...	X	...	...	...	...	...	...	...	X	...	X	...	...	...	...	...	...	...	...	...	...	...
Mo	...	...	...	...	...	...	X	...	X	...	X	X	X	X	X	X	...	...	X	...	...	...	...
Ni	...	X	X	...	...	X	X	X	X	X	X	X	X	X	...	X	X	...	X	X	X	X	...
Pd	...	...	...	...	...	...	X	X	...	...	X	X	X	...	...	...	...	...	X	X	...	...	...
Pt	...	...	...	...	...	...	...	...	...	...	X	X	...	X	...	...	...	...	X	...	...	...	...
Ta	...	...	...	...	...	...	...	...	X	...	X	...	...	...	X	X	...	X	...	...	...	...	...
Ti	X	X	...	...	...	...	X	X	X	...	X	X	...	...	X	X	X	...	...	X	...	...	X
U	...	...	...	...	...	...	...	...	...	...	...	X	...	...	...	X	X	...	...	...	X	...	...
V	...	X	...	...	...	...	X	...	X	...	...	...	...	...	X	...	...	X	...	...	...	...	...
W	...	...	...	...	X	...	...	X	X	...	X	X	X	X	...	...	...	...	X	...	X	...	...
Stainless steel	X	X	...	X	...	...	...	X	...	...	X	X	X	...	...	X	...	...	...	X	...	...	X
Cast iron	...	...	...	...	...	...	...	X	...	...	...	X	...	...	...	...	...	...	X	...	X	...	...
Carbides	...	...	...	...	...	...	...	...	X	...	...	X	...	...	...	...	...	...	...	...	...	X	...
Graphite	...	...	...	...	...	...	...	...	...	...	...	...	...	...	...	X	...	...	...	X	...	...	...

Adapted from [Feature 1976]

sure that diffusion occurs at a useful rate. Heating rates are not critical.

- *Time:* Creep and diffusion mechanisms are also strongly time dependent and there must be a sufficient interval afforded to allow for void closure by material transfer. As the temperature increases, so the time required for bonding decreases.
- *Surface condition:* The height and frequency of surface asperities defining the joint control the extent of initial surface contact and thus influence the bonding rate. Generally, flatter and more highly polished surfaces are easiest to bond. The removal of surface contamination and thick oxides prior to bonding is essential because these will either persist at the joint line or must be removed by solution in the parent material as bonding proceeds. It, therefore, takes higher relative temperatures and pressures to bond aluminum-base alloys than copper-base alloys or gold, as can be seen from Fig. 1.7.

Of these four process variables, temperature is the most important, followed in order by pressure, time, and surface condition. The trade-off between temperature and pressure can be seen in Fig. 1.8. Changing the process time by almost an order of magnitude affects the bonded area by less than 10%, which is within the limits of experimental error. The reason for the order of the variables is that temperature plus pressure cause creep, which operates at both the macrolevel and microlevel to remove interfacial gaps. The combination of temperature and time promotes diffusion, which is usually a much slower mechanism for material transport. Without raising the temperature, the combination of pressure and time generally results in inelastic deformation, i.e., pressure welding. Reducing the surface roughness from several microns to one micron has little effect on the extent of bonding because this process in itself does not ensure significant metal-to-metal contact between the abutting parts—pressure is also needed. However, when the surface roughness is reduced further to an optically polished surface, the bond quality tends to improve measurably, with the other variables being unchanged.

Diffusion-bonded joints normally exhibit 80 to 100% of the strength of the parent materials. One perceived problem with diffusion bonding is the relatively long process cycle time, particularly in comparison with fusion welding. However, a complex welding operation may require several hours to prepare and jig the components, in which case diffusion bonding might offer overall advantages. Unlike most welding processes, the process time curve for diffusion bonding is almost flat in relation to the size of the joint because the process time is essentially independent of joint area, provided adequate compressive stress is applied.

Titanium and some of its alloys (Ti-6Al-4V, Ti-6Al-2Sn-2V, etc.) are of particular industrial interest because they can be diffusion bonded and super-plastically shaped in one processing operation (Fig. 1.9). This is made possible by the fact that, above temperatures of about 925 °C (1700 °F), titanium and certain of its alloys exhibit creep and superplasticity (both congenial for shaping and forming), and titanium can dissolve oxygen into its bulk as fast as a surface scale can form, thereby facilitating diffusion bonding. Indeed, at these temperatures, material

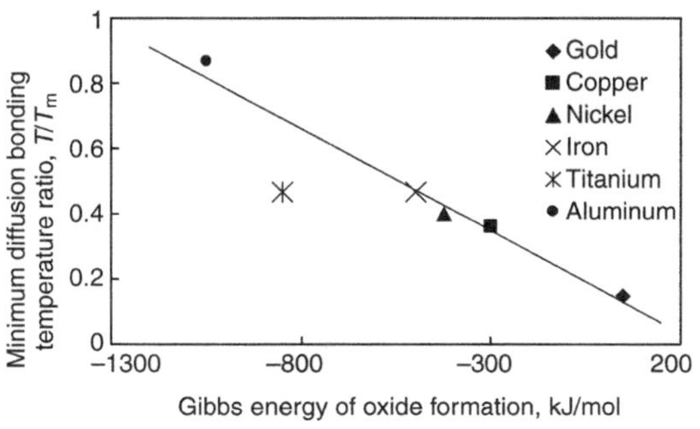

Fig. 1.7 Relationship between the stability of metal oxides (in terms of the Gibbs free energy of formation) and the ratio of the process temperature (T) to the melting point of the metal (T_m)

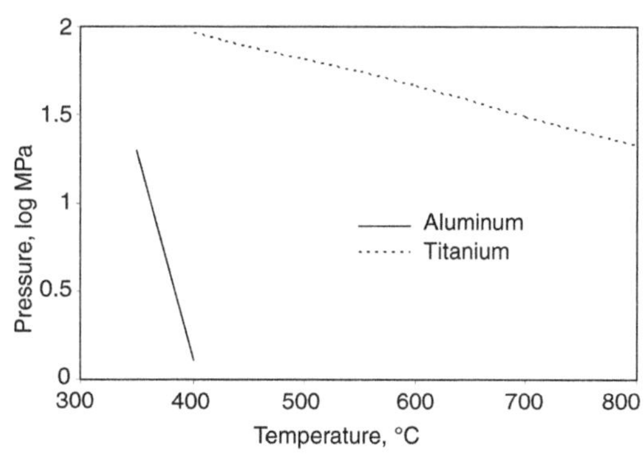

Fig. 1.8 Effect of temperature on the pressure necessary to produce good quality diffusion bonds in titanium and aluminum by diffusion bonding

flow by creep or superplasticity occurs at such low applied stress that gas pressure at just a few MPa (~1 ksi) can be used to form and weld these materials in any chosen order. Diffusion bonding occurs so readily that inert oxides need to be applied to areas to prevent bonding where it is not required. This conjunction explains the anomalously low position of titanium in the representation of metals shown in Fig. 1.7.

Interlayers are often used to facilitate diffusion bonding dissimilar metals. For example, silver foil is used for bonding steel to titanium and nickel foil is often used to bond high-carbon steel to itself and other materials. A gold flash applied to a precleaned surface permits diffusion bonding of nickel and copper components. An obvious extension to this approach is making use of interlayers that melt, thereby increasing diffusion rates, helping to fill the joint gap and to disrupt surface films. These benefits are characteristic of diffusion brazing and diffusion soldering processes. Diffusion brazing is discussed in Chapter 6 of this volume and diffusion soldering in Chapter 5, section 5.9 of the companion volume *Principles of Soldering*.

1.2 Key Parameters of Brazing

The quality of brazed joints depends strongly on the combination of filler and component materials and also on the processing conditions that are used. It is precisely for this reason that a sound understanding of the metallurgical changes accompanying the sequence of events that occur in making brazed joints is so vital for developing reliable joining processes.

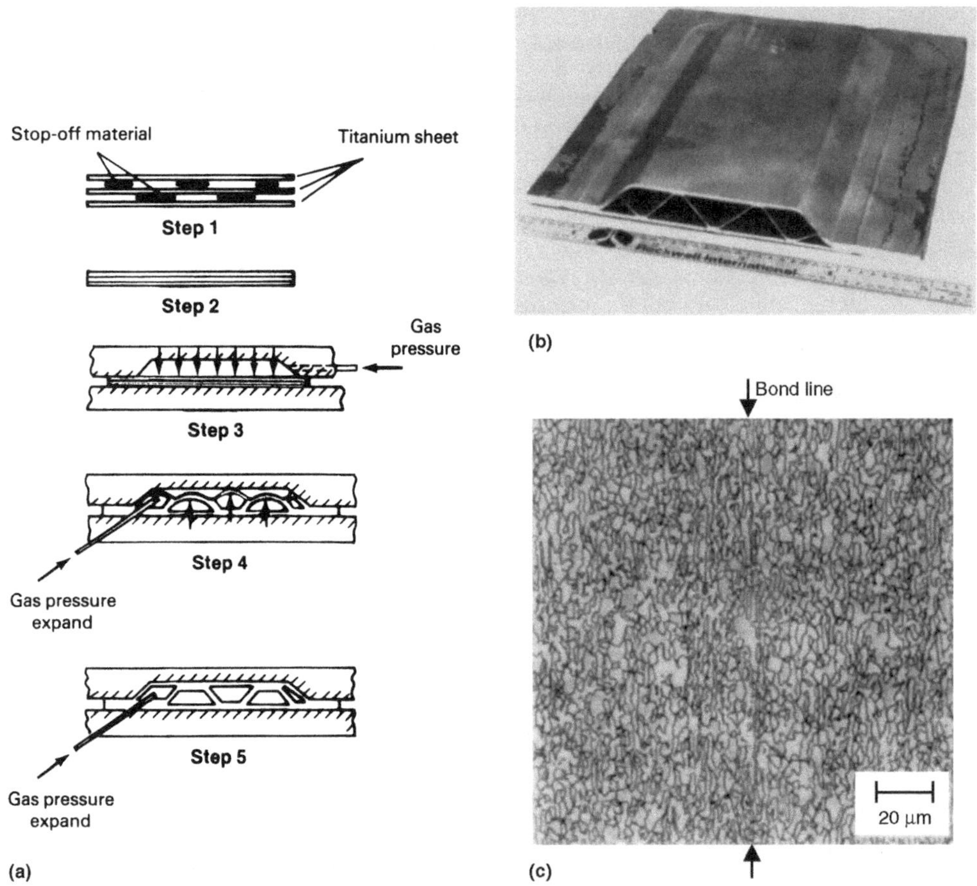

Fig. 1.9 Superplastic forming and diffusion bonding of titanium. (a) Schematic of the steps involved. (b) Typical three-sheet titanium alloy component formed superplastically and diffusion bonded. (c) Cross section through a diffusion-bonded joint in titanium alloy, made at 980 °C (1795 °F) for 2 h, under an applied pressure of 10 MPa (1.5 ksi), and in a vacuum of 0.1 mPa

Brazing technology has generally evolved in an empirical manner, largely by trial and error. Theoretical principles have helped to furnish insights, guidelines, and qualitative explanations for this technology but have rarely provided reliable data for use in the design of joining processes. The basic difficulty is that the real situation is highly complex because it brings into play a large number of variables, some of which may not be easy to recognize. Among the relevant factors are the condition of the solid surfaces (i.e., the nature of any oxides or other coatings, surface roughness, etc.); the temperature gradients that develop during the joining operation, not to mention the metallurgical reactions involving the filler and parent materials and also the chemical reactions with fluxes, where these are used; and the process atmosphere.

Key aspects of joining with brazes are the manner and extent of flow of the molten filler into the joint. These aspects are influenced by:

- Dimensions of the joint
- Surface condition of the components
- Spreading characteristics of the filler metal
- Alloying between the filler metal and components

The limitation of theory in accounting for observed behavior is well illustrated by the classical model of wetting and spreading. In this model, the surface of the solid is taken to be invariant as a liquid droplet spreads over it. That is, the reaction between the liquid and the solid components across their common interface is considered negligible. It is also assumed that the composition and other characteristics of the solid and liquid components, likewise, do not change with time. These assumptions are not generally valid, as will be shown. The model nevertheless does provide useful guidance so the principles on which it is founded, namely surface energy and surface tension, are worth reviewing.

1.2.1 Surface Energy and Surface Tension

A simplified representation of the atomic structure of a solid close to one of its free surfaces is provided in Fig. 1.10. The atom at position A, in the bulk of the solid, has a balanced array of neighboring atoms, whereas atom B at the surface of the solid is lacking in neighbors above it, apart from the occasional vapor molecule and, therefore, it has some unsaturated bonds.

The potential energy of atoms at the free surface, such as B, is higher than the energy of atoms within the bulk of the solid, such as A, by the energy of the unsaturated bonds. The aggregate of this excess energy that is possessed by atoms in the vicinity of the free surface constitutes the surface energy of the solid. In a similar manner, a liquid also possesses a surface energy, which is directly manifested in the tendency to draw up into drops. If small, the droplets are perfect spheres. Because a sphere has the smallest surface-to-volume ratio, it is clear that the surface energy of a liquid is greater than its volume energy. In the classical model, when a liquid spreads over a surface, the volume remains constant because evaporation and reaction with the substrate are excluded. Therefore, only surface energy changes must be considered.

The surface of a liquid acts like an elastic skin covering the volume; in other words, the surface is in a state of tension. The tensile force (F), known as surface tension (γ), is defined as the force acting at right angles to a line of unit length drawn in the surface. The relationship between surface tension and surface energy under specific conditions can be seen as follows.

Consider a liquid film of length L and width W. Apply a force F at a barrier AB, as shown in Fig. 1.11, parallel to one surface of the film, so as to extend the liquid film a distance x. The increase in area of the film is $x \cdot L$. The work done in obtaining this increase is the mathematical product of the force applied times the distance moved, or $F \cdot x$.

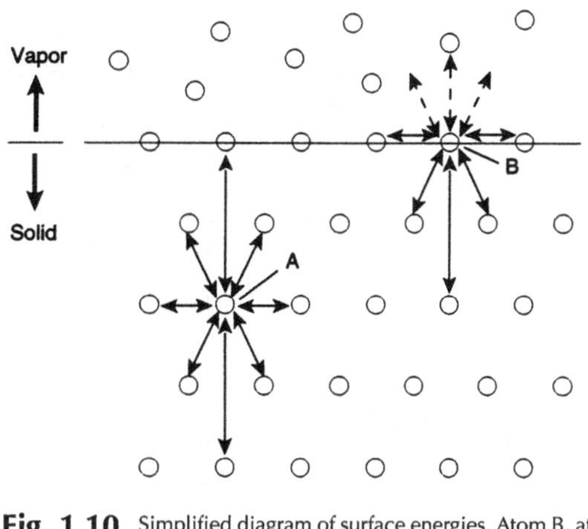

Fig. 1.10 Simplified diagram of surface energies. Atom B, at the surface, has unsaturated bonds and thus a higher energy than atom A. This difference in energy is the origin of surface energy γ_{SV}.

The work done by the liquid film in opposing the increase in area, under isothermal conditions (constant temperature), is $2 \cdot \gamma \cdot x \cdot L$, where γ is the surface tension force acting on each surface at the prescribed temperature.

At a fixed temperature (under isothermal conditions):

$$Fx = 2\gamma xL$$

Rearranging, $F/L = 2\gamma$ or $F/L = \gamma$ for each surface.

Thus, surface energy is equivalent to surface tension under isothermal conditions. In the modern metric or International System of Units (SI), the unit of surface energy is joule per square meter (J/m^2) and that of surface tension is newton per meter (N/m). Because these parameters are properties of an interface (e.g., between liquid and air), surface energy and tension must be defined with reference to the appropriate pair of materials that meet at the interface, and the test conditions, such as temperature and atmosphere, also must be specified.

Modeling of liquid droplets in contact with surfaces can be done with a program called "Surface Evolver" [Brakke 2003]. A search of the World Wide Web using this as the keyword should identify a site from which the latest version of the software can be obtained.

1.2.2 Wetting and Contact Angle

The classical model of wetting is based on the behavior of a liquid drop on a solid with a flat surface, which is rigid and is inert with respect to the liquid material. Accordingly, the liquid will spread over a solid surface until the three surface tensions—between the liquid droplet and the solid substrate, the liquid droplet and the atmosphere, and the substrate and the atmosphere—are in balance as shown in Fig. 1.12.

According to the balance of forces:

$$\gamma_{SL} = \gamma_{SV} - \gamma_{LV}\cos\theta \qquad \text{(Eq 1.1)}$$

where, γ_{SL} is the surface tension between the solid and liquid, γ_{LV} is the surface tension between the liquid and vapor, γ_{SV} is the surface tension between solid and vapor, and θ is the contact angle of the liquid droplet on the solid surface.

Equation 1.1, known as the wetting or Young's equation, shows that $\theta < 90°$ corresponds to the condition $\gamma_{SV} > \gamma_{SL}$. This imbalance in surface tension (i.e., surface energy) provides the driving force for the spreading of liquid over the solid surface and diminution of the unwetted surface area.

The contact angle θ provides a measure of the quality of wetting. Thus, if $90° < \theta < 180°$, some wetting is said to occur, but a liquid droplet will not spread on the surface with which it is in contact. On the other hand, if $\theta < 90°$, a liquid droplet will wet the substrate and also spread over an area defined by the contact angle θ. Clearly, the area of spreading will increase with decreasing contact angle. For further details of the interrelationship between these two parameters, refer to Appendix A1.1. The relationship between contact angle spreading can be demonstrated by a numerical model. Figures 1.13(a), (b) and (c) show a droplet of constant volume wetted onto a planar surface at three different contact angles. The boundary between spreading and diminution of the wetted area occurs for $\theta = 90°$.

Rewriting Eq 1.1 in terms of $\cos\theta$:

$$\cos\theta = \frac{\gamma_{SV} - \gamma_{SL}}{\gamma_{LV}}$$

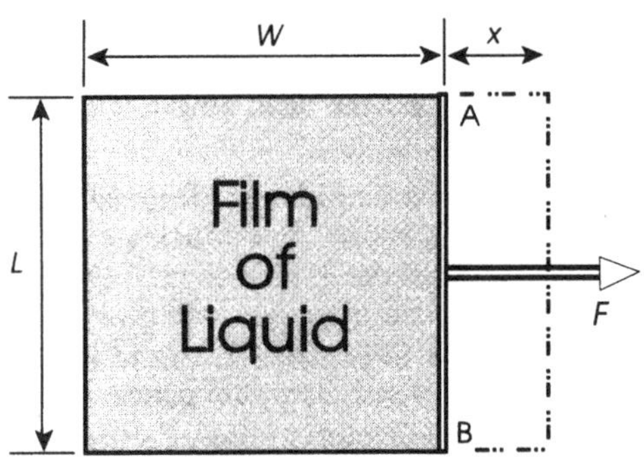

Fig. 1.11 Relationship between surface energy and surface tension

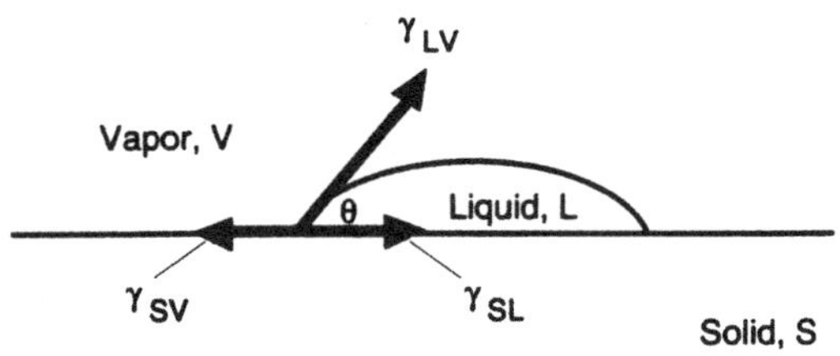

Fig. 1.12 Surface tension forces acting when a liquid droplet wets a solid surface, according to the classical model

Thus, wetting is improved by decreasing θ, i.e., as cosθ increases, i.e., as θ approaches zero, and cosθ may be maximized by:

- Increasing γ_{SV}
- Decreasing γ_{SL}
- Decreasing γ_{LV}

The term γ_{SV} can be maximized for a given solid by cleaning the surfaces. The presence of adsorbed material such as water vapor, dust, and other nonmetallic surface films on a metal surface markedly reduces γ_{SV} and correspondingly raises the contact angle θ. Therefore, it is important in brazing operations that joint surfaces are clean and metallic, hence the need for fluxes or protective atmospheres to achieve and then sustain this condition. This is exemplified by molten silver spreading on a solid nickel surface at 970 °C (1780 °F). The intersolubility of nickel in molten silver is extremely low and so the nickel may be considered as essentially unreactive toward silver. After 30 min of exposure to air, the contact angle of the molten droplet is 90°, whereas in a helium atmosphere it is close to 10°. The difference between these two cases is the degree of oxidation of the nickel substrate because silver oxide is unstable at typical brazing temperatures.

The term γ_{SL} is a constant at a fixed temperature for a particular solid-liquid combination, according to the classical model of wetting. This parameter can be reduced by changing the composition of the materials system, but this is not usually easy to achieve in practice because component materials are specified to fulfill certain functional requirements. Fortunately, γ_{SL} is temperature dependent, declining in the range 0.1 to 1%/K, and thereby providing a ready means of controlling spreading.

According to the classical model, the term γ_{LV} is constant at a fixed temperature and pressure for a particular liquid-vapor combination but can be varied by altering the composition of the atmosphere. Although the composition of the atmosphere used for the joining operation is known to affect the contact angle, in practice it is often easier to promote spreading (at constant temperature) by reducing the pressure of the atmosphere. This is one of the reasons for the popularity of vacuum-based joining processes, especially when chemical fluxes need to be excluded for various reasons.

In general, the relative magnitudes of the surface energies are $\gamma_{SV} > \gamma_{SL} > \gamma_{LV}$. The surface energies of pure metals correlate quite well with their melting points [Howe 1993]. This relationship, which is illustrated in Fig. 1.14, is to be expected because the temperature stability of metals reflects the strength of the bonds between adjacent atoms in the lattice, and the difference

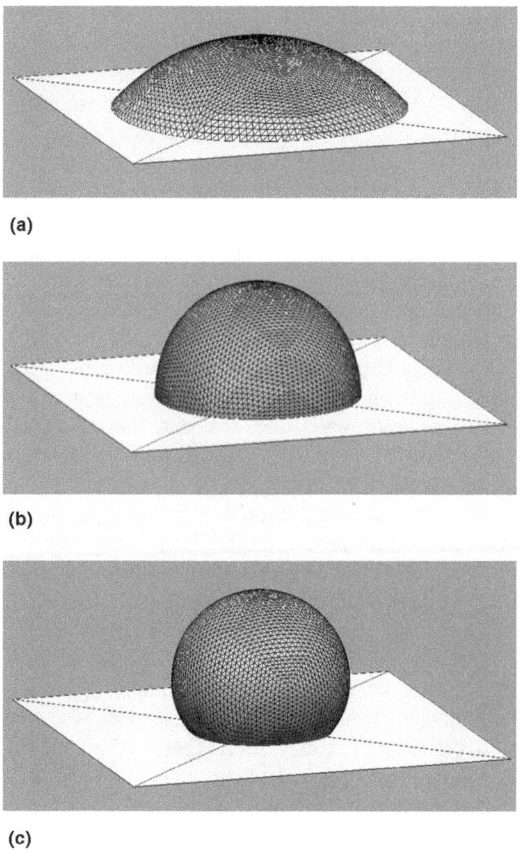

Fig. 1.13 Numerical model of a liquid droplet of constant volume wetted on a plane at contact angles of (a) 45°, (b) 90°, and (c) 135°. Spreading occurs if the contact angle is less than 90°.

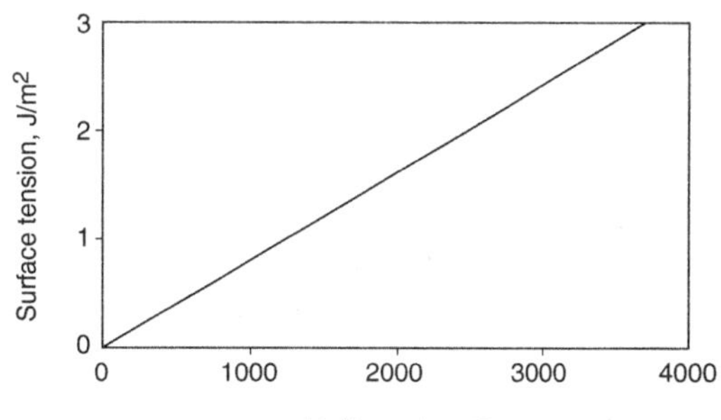

Fig. 1.14 Surface tension, γ_{LV}, of liquid elemental metals at their melting points [Howe 1993]

between the potential energy of atoms within the bulk of a solid metal and that of atoms of the same metal in the vapor is responsible for surface energy and tension. It is possible to calculate the surface tension of brazes from thermodynamic principles using data for the pure metals. To a first approximation it varies as an essentially linear relationship between the values for the constituent pure metals, with appropriate corrections for temperature. Thus, the surface tension of molten silver at its melting point is 0.903 N/m and molten copper at its melting point is 1.285 N/m, giving a calculated surface tension for silver-copper braze at the eutectic temperature of 0.967 N/m, which is reasonably close to the measured value of 0.952 N/m at 781 °C (1438 °F).

As noted previously, the contact angle, θ, is an important measure of wetting, but from a thermodynamic perspective, the work of adhesion, W_a, is the appropriate parameter used to characterize the extent of wetting of a liquid (i.e., molten) braze on a solid substrate material and also provides an index of the mechanical adhesion.

The work of adhesion, W_a, may be expressed in terms of γ_{SV}, γ_{SL}, and γ_{LV}, according to the Dupré equation:

$$W_a = \gamma_{LV} + \gamma_{SV} - \gamma_{SL} \quad \text{(Eq 1.2)}$$

That is, the work of adhesion equals the work required to incrementally increase liquid-vapor and solid-vapor interfaces from a liquid-solid interface.

Combining Eq 1.1 and 1.2 yields the Young-Dupré equation:

$$W_a = \gamma_{LV}\,(1 + \cos\theta)$$

The Young-Dupré equation permits W_a to be calculated from values of γ_{LV} and θ.

As just mentioned, the wetting equation (Eq 1.1) applies when the liquid is practically insoluble in the solid over which it spreads (i.e., the solubility is less than 0.1%). For binary metal systems where this condition is satisfied (e.g., molten silver on iron or on nickel), it has been shown that the wetting equation can be reduced to:

$$\cos\theta = 1 + k\left[\frac{T_m^s}{T_m^l} - 1\right]$$

where k is a constant equal to approximately 0.3, T_m^s is the melting point of the solid metal, and T_m^l is the melting point of the liquid metal. This expression has been verified experimentally [Eustathopoulos and Coudurier 1979]. Higher-order metal systems (ternary, quaternary, etc.) are considerably more complex, and the wetting equation cannot be truncated to such a simple form. A more sophisticated analysis of wetting that takes into account the influence of certain microscopic features, including the influence of local defects and van der Waal forces, is provided by de Gennes [1985]. However, this is still a continuum analysis and does not consider the local atomic environment. Indeed, it has been suggested that Young's equation is valid only under certain special cases and there are some difficulties with the theoretical definition of solid-surface tension [Xian 2000].

Another assumption of the classical model of wetting that has recently been challenged is the static equilibrium of surface tensions and surface energies. Recent studies examining the wetting of molten aluminum on sapphire (α-Al_2O_3) in ultrahigh vacuum have shown that the measured contact angle of the molten drop for these combinations declines from values above 90° to below 90° over time, regardless of temperature above the melting point of the metal involved [Levi and Kaplan 2003a; 2003b]. Experimental investigations, using a variety of surface analysis techniques, have revealed that wetting in these cases occurs through a nonequilibrium mechanism, characterized by dissolution of the sapphire at the boundary of the molten drop, or triple junction where the liquid, solid, and vapor are in contact with one another, and epitaxial deposition of sapphire at the liquid-solid interface beneath the drop.

Further academic endeavor is clearly needed to advance our understanding of wetting of molten metals. A review of recent theories of wetting is provided by Asthana and Sobczak [2000].

So far this chapter has considered filler spread over a single surface. In a joint there are always two facing surfaces. If both contact angles are less than 90°, the surface energies will give rise to a positive capillary force that will act to fill the joint. For a pair of vertical parallel plates D mm apart and partly immersed in a liquid, the capillary force per mm length of joint is equal to $2\gamma_{LV}\cos\theta$. Under this force, the liquid will rise to an equilibrium height h at which the capillary force balances the hydrostatic force (as shown schematically in Fig. 1.15, such that:

$$h = \frac{2\gamma_{LV}\cos\theta}{\rho \cdot g \cdot D} \qquad \text{(Eq 1.3)}$$

where ρ is the density of the liquid and g is the acceleration due to gravity.

As might be expected, experimental assessment of capillary rise of brazes reveals that capillary rise is less than predicted by theory, although the general principles of Eq 1.3 are substantiated. Meniscus rise is usually greatest for brazes that exhibit the lowest contact angle and surface tension and in the narrowest gaps. However, the correlation with gap width is generally weak, with other practical issues becoming manifest when the joint gap is narrow. (see Chapter 4, section 4.3.3.1 in this book).

The actual situation in brazing is much more complex than that represented by Eq 1.3 and the classical wetting model, as has already been shown. The irreversible nature of spreading and the time dependence of contact angle that is commonly observed are at variance with this equilibrium model. These and other departures from the classical model occur because the joining process almost invariably involves a degree of chemical reaction between the filler metal and the solid surface.

Reactions between a filler metal and the substrate often result in dissolution of the surface of the substrate; this process usually leads to a change in composition and sometimes the growth of new phases. These changes occur only because it is energetically favorable to do so. The energy of formation considered is the thermodynamic function known as the Gibbs free energy. This function and its properties are briefly explained in the Appendix A3.1.

Calculations made by Yost and Romig [1988] and Wang and Conrad [1995] have shown that the free energy of formation of new phases between a molten filler and a substrate is approximately two orders of magnitude larger than the energy release created by the surface energy imbalance during the advance of a spreading droplet, that is exclusively considered in the classical model. Therefore, in these cases, and probably more generally in brazing processes, the Gibbs free energy change that occurs on reaction by a filler with a substrate is demonstrably the dominant driving force for wetting. Empirical evidence for this is provided, for example, by the fact that the measured contact angle of molten germanium on silicon carbide at 1430 °C (2600 °F) is approximately 120°, whereas that of molten silicon on this ceramic at the same temperature is 38° [Li and Hausner 1991]. The substantial difference in these two contact angles cannot be accounted for by the difference in γ_{LV} in the wetting equation (Eq 1.1). It can be due only to the greater intersolubility of silicon with silicon carbide. This example, as that of molten aluminum and nickel on sapphire described previously, clearly demonstrates that the simple classical wetting equation cannot be relied on for a quantitative description of wetting, contact angle, or spreading. Another example is provided in Fig. 1.16. In this simple system of copper-silicon braze wetted on to vitreous carbon substrates, the final contact angle is insensitive to alloy composition, but the rate of attainment of steady-state wetting is directly related to the concentration of silicon, which is the active ingredient in the braze.

Modifications have been proposed to incorporate the Gibbs free energy change accompanying metallurgical reaction into the classical wetting equation by adding additional terms. In particular, the following equation has been developed for the contact angle in reactive wetting [Kritsalis, Coudurier, and Eustathopoulos 1991; Laurent, Chatain, and Eustathopoulos 1991]:

$$\cos\theta = \cos\theta^0 + \frac{\gamma_{SL} + \gamma_{SL}^0}{\gamma_{LV}} - \frac{\Delta G_r}{\gamma_{LV}} \qquad \text{(Eq 1.4)}$$

where γ_{SL} is the solid-liquid interfacial energy after reaction, γ_{SL}^0 is the interfacial energy before reaction, θ^0 is the contact angle before reaction, and ΔG_r is the Gibbs free energy of the reaction. Equation 1.4 is probably more of theo-

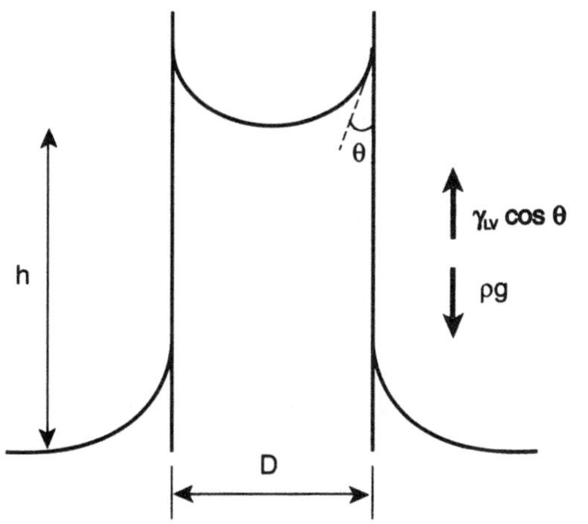

Fig. 1.15 Rise of a liquid between two parallel plates by capillary force

retical interest than practical value because its use presupposes knowledge not only of the Gibbs free energy of reaction but also values of the before-and-after contact angle, or the interfacial energy.

The effect of metallurgical interaction between a braze and the component (or parent) materials in promoting wetting is exploited in active brazes: the addition of a small fraction of a reactive metal such as titanium, hafnium, or zirconium to conventional brazes enables them to wet and spread over ceramic materials. In this instance, wetting of and reaction with the ceramic are inextricably linked. Activated filler alloys are discussed briefly in Chapter 4, section 4.1.2.2 and in detail in Chapter 7.

Although a low-contact angle is used as an index for judging the quality of wetting, there are situations where higher contact angles are preferred. This can be illustrated with reference to Fig. 1.17, which shows two joints, one between two component surfaces of unequal area and the other between component surfaces that correspond entirely. In the first case, a low-contact angle serves to form a gentle concave fillet, which enhances the mechanical properties of the joint. In the other configuration, a low-contact angle encourages the formation of a neck in the joint, which can be a source of weakness. A contact angle close to 90° and an interference fit joint gap will eliminate this problem.

A further point to be aware of in connection with wetting is that a situation can arise where the molten filler is physically prevented from achieving its equilibrium contact angle, as, for example, when a "stop-off" compound is used to confine the braze to a defined area. The enforced wetting angle is then not representative of the true wettability.

From a purely theoretical perspective, spreading driven by alloying should occur only when there is solution of an element of the liquid phase in the solid. Dissolution of an element from the solid phase into the liquid, or melt, requires the input of energy and hence this process cannot drive spreading. Rather, it has been observed that brazes spread preferentially along the exposed grain boundaries of substrates. Grain boundaries are disordered regions that have surface energy, γ_{SV}, up to 30% larger than the free surface, depending on the angular mismatch of adjacent grains.

1.2.3 Fluid Flow

The wetting equation does not provide information on the rate of wetting. Knowledge of the contact angle(s) enables the surface energy to be determined, assuming that the classical wetting model applies, and hence the force that acts to fill the joint gap with liquid may be estimated. The liquid will flow into the joint under this

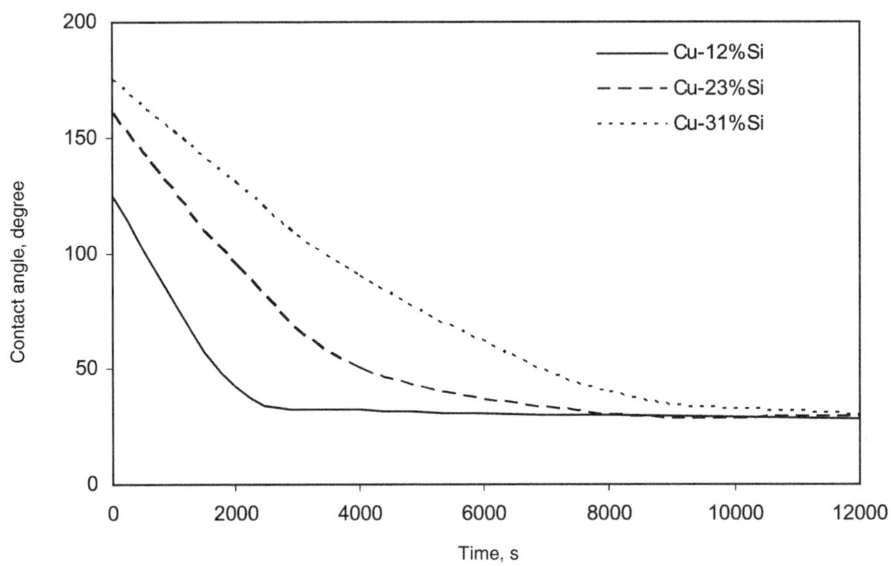

Fig. 1.16 Contact angle of copper-silicon brazes of different composition on vitreous carbon substrates demonstrating the effect of driving force of alloying on wetting rate and the dependence of the equilibrium wetting angle on the reaction product, which is the same in the three cases represented [Landry, Rado, and Eustathopoulos 1996]

force at a rate governed by its viscosity. Simple fluid flow theory assumes that:

- There is no interaction between the liquid and the solid surfaces with which it is in contact.
- All surfaces are smooth and perfectly clean.
- Flow is laminar, not turbulent.

For a detailed treatment of this subject, the reader is referred to the classic paper by Milner [1958]. Here, we shall merely quote the expression (given as Eq 8 in Milner's paper) for the volume rate of liquid flow, dV/dt, between a pair of horizontal parallel plates, length l, separated a distance D, under a pressure P per unit area transverse to the plates. The viscosity of the liquid is η.

$$\frac{dV}{dt} = \frac{PD^3}{12\eta l}$$

It is assumed that the liquid front will advance at a rate (dl/dt) equal to the mean velocity of flow, that is:

$$\frac{dl}{dt} = \left(\frac{1}{D}\right)\left(\frac{dV}{dt}\right) = \frac{PD^2}{12\eta l}$$

From the wetting equation (Eq 1.1), under isothermal conditions the change in surface energy as a unit area of a surface becomes wetted by the liquid is:

$$\gamma_{SV} - \gamma_{SL} = \gamma_{LV}\cos\theta$$

Therefore, the change in surface energy when the pair of parallel plates becomes wetted is:

$$2l\,(\gamma_{SV} - \gamma_{SL}) = 2l\,\gamma_{LV}\cos\theta$$

It follows that the force acting on the liquid to cause it to wet the plates is:

$$F = \frac{2l\,\gamma_{LV}\cos\theta}{l}$$

so that the pressure is:

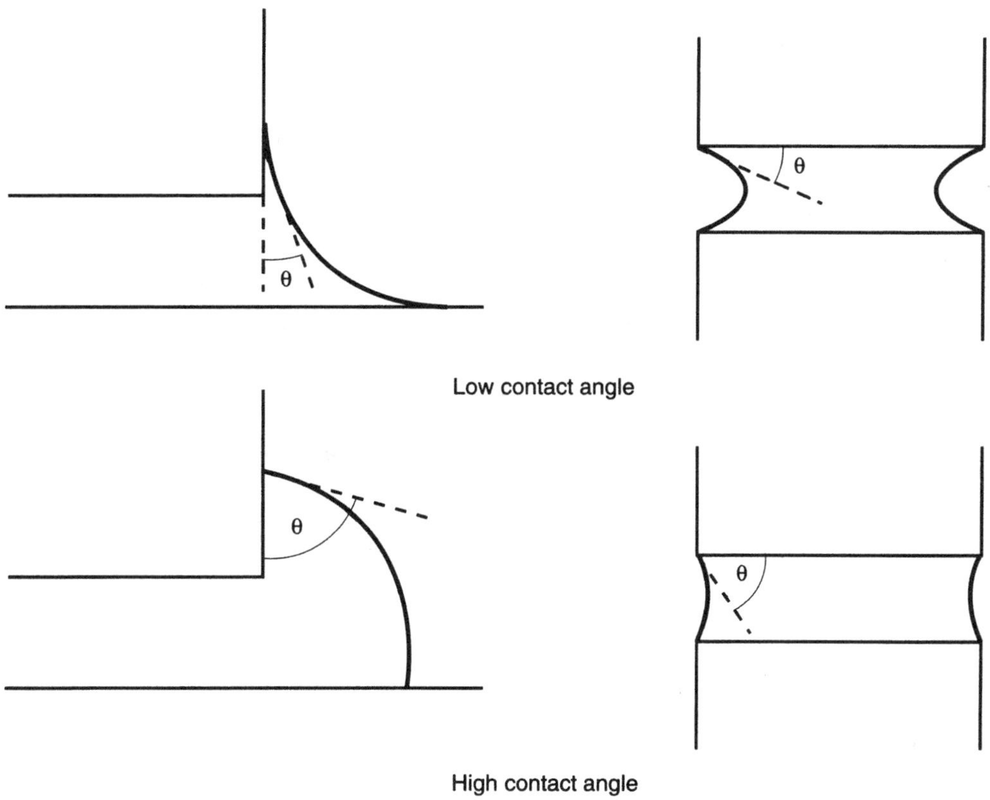

Fig. 1.17 Effect of contact angle on fillet formation and joint filling. Low contact angles tend to be preferred when external fillets can form. In other geometries, higher contact angles result in lower stress concentrations.

$$P = \frac{2\gamma_{LV}\cos\theta}{D}$$

and the velocity of flow of the liquid into the space between two parallel surfaces, of separation D, according to this simple model is given by:

$$\frac{dl}{dt} = \frac{\gamma_{LV} D \cos\theta}{6 \cdot \eta \cdot l} \quad \text{(Eq 1.5)}$$

Equation 1.5 shows that the rate of liquid flow increases when:

- The liquid-vapor surface tension, γ_{LV}, increases.
- The joint gap, D, increases.
- The contact angle θ decreases.
- Filler metal viscosity is low

Andrade [1952] derived an empirical formula relating viscosity, when molten, to the molecular weight of metals (in SI units):

$$\eta_m = \frac{1.65 \times 10^{-7}\, T_m^{0.5} A^{0.5}}{V^{2/3}}$$

where η_m is the viscosity at the melting point of the metal, T_m is the absolute melting point, V is the molar volume, and A is the atomic weight of the metal. By assuming limited solubility between the constituents in an alloy and applying the rule of mixtures, it is thereby possible to provide an estimate of the theoretical viscosity of a braze.

Rates of flow calculated from Eq 1.5 for molten brazes in joints 50 μm (2000 μin.) wide are typically 0.3 to 0.7 m/s (1 to 2.3 ft/s). In other words, a joint 5 mm (0.2 in.) long will be filled in a time of the order of 0.01 s. This implies that joint filling by the molten braze occurs virtually instantaneously and that transient effects associated with fluid flow can generally be neglected in joining processes. It should be noted that, although the rate of filling is proportional to the joint gap D, the driving force for filling, according to the classical model, is inversely proportional to D; that is, these two aspects of filling act in opposition.

This simple model needs to be modified in situations where interfacial reaction occurs while liquid spreading is proceeding. Models that have been tentatively proposed for this situation have been reviewed by Meier, Javernick, and Edwards [1999]. Currently, the lack of relevant data on reaction-rate kinetics, interfacial energy before and after reaction, and diffusion hampers a more complete understanding of spreading of molten brazes, especially where interfacial reaction with solid components is significant. However, much can be learned from empirical observations, as shown in the following section.

1.2.4 Filler Spreading Characteristics

All molten filler metals do not have the same spreading characteristics, although, with few exceptions, the degree of spread over an "ideal" substrate increases as the temperature is raised and the environment is made more reducing. In this context, an "ideal" substrate, suitable for reference purposes, needs to be defined. This is understood to possess a perfectly clean metal surface, which is highly wettable by the braze under consideration, but with which it does not significantly alloy. Any alloying reactions will be highly specific to the combination of materials in question so that the substrate will lose its ideal characteristics. This is particularly true for brazes where extensive reaction with the substrate is the norm rather than the exception. Thus, while it is often possible to draw comparisons within an alloy family (e.g., McDonald 1989), it is difficult to obtain true comparative data between different families of brazing alloys.

Alloys with narrow melting ranges, ideally of eutectic composition, are often regarded as having the best spreading characteristics, and this is frequently one of the reasons cited for their selection in preference to filler metals with wide melting ranges. The superior spreading of brazes with small melting ranges, in comparison with those in the same alloy system, but having wide melting ranges, as frequently observed, can be explained by the different melting characteristics in the two cases. In the ideal case, a pure metal or alloy of eutectic composition melts instantly. Spreading of the molten alloy is then driven by interaction with the substrate [Ambrose, Nicholas, and Stoneham 1992]. In the case of a noneutectic filler metal, wetting commences before the alloy is entirely molten and when its flow is relatively sluggish. By the time the alloy is completely molten, the filler will have partly alloyed with the substrate, and the driving force for spreading will have been diminished.

Whether the filler alloy has a narrow melting range is of much less importance to the phenomenon of spreading than the composition per se.

The spreading of a filler metal depends greatly on the elemental constituents present and the composition of the substrate. For example, Fig. 1.18 shows the spreading characteristics of a selection of alloys on stainless steel substrates, as a function of temperature. The data clearly show that the incorporation of palladium has a major beneficial effect, despite the associated widening of the melting range.

Although high fluidity of a filler metal is a desirable property when it is required to flow into the joint gap of a heated assembly by capillary action, it is not quite so important when the preferred method of applying the filler is to sandwich a thin foil preform between the components, which are then joined together in an appropriate heating cycle. For this type of configuration, a high degree of spreading is detrimental to joint filling, because then the filler tends to flow out of the joint. Placement of the filler metal and its influence on joint filling is discussed in Chapter 4, section 4.3.1.1.

Detailed investigation reveals that even an ostensibly simple characteristic such as spreading exhibits somewhat complex behavior. Figure 1.19 shows the spread area of silver-copper braze on nickel at 820 °C (1510 °F) as a function of the wetting time [Weirauch, Jr., and Krafick 1996]. The results indicate that there are at least four distinct stages of wetting and spreading. During the first tens of seconds of melting, the braze forms a spherical cap and spreads rapidly with a corresponding decline in the contact angle. The contact angle then temporarily stabilizes with the formation of a halo at the periphery of the braze pool. This region of the braze has a very different composition to the bulk and is formed by a selection of elements in the braze reacting preferentially with the substrate [Yunchen, Xiaoming, and Hongying 1992]. During this stage, the braze pool spreads by wetting up to an interface, demarcated by the halo. This situation persists for about 100 seconds. Thereafter, a further reduction in contact angle, accompanied by additional spreading, occurs. This stage is thought to be associated with a progressive alteration in the composition of the braze, which produces a sudden change in the phases formed in the halo at the edge of the pool of the molten braze. This, in turn, gives rise to a further reduction in the wetting angle. Finally, after many minutes, the contact angle reaches a settled value as the braze pool becomes saturated with the substrate metal, which is nickel in the reported case study [Weinrauch, Jr., and Krafick 1996]. The molten liquid then commences to freeze isothermally as the solidus temperature progressively increases, owing to the change of composition, caused by alloying with the substrate metal.

Depending on the braze/substrate combination, one or other of these four stages of interaction may be more dominant and visually obvious than the rest. Thus, while molten silver spreads readily on copper and forms an extensive halo, molten copper barely spreads at all on a titanium substrate but reacts strongly with it in the direction perpendicular to the interfacial surface. Molten aluminum wetted on copper represents a situation somewhere intermediate be-

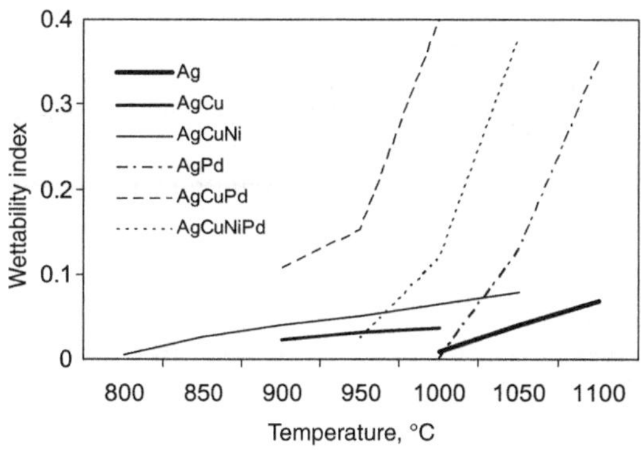

Fig. 1.18 Wettability index (defined as the product of the contact angle and spread area [Feduska 1959]) of silver-base brazes on 316L stainless steel, heated in vacuum for 5 min. Palladium additions clearly have a beneficial effect on wetting and spreading by the braze, despite widening of the melting range of the filler metal. Note: 316L stainless steel is sensitive to liquid metal embrittlement by copper-base brazing alloys. Adapted from Keller et al. [1990]

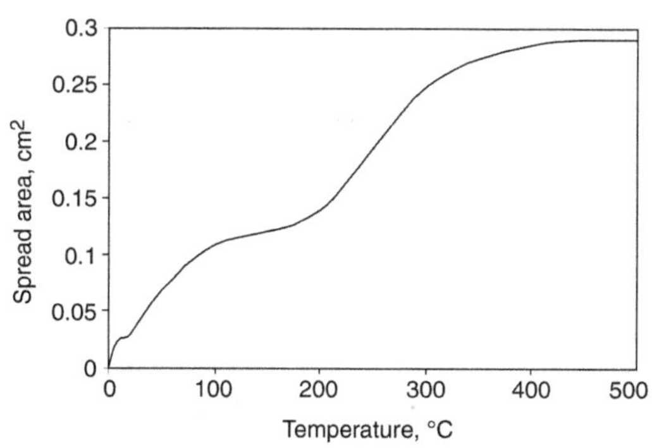

Fig. 1.19 Spread area of 0.5 mg spheres silver-copper eutectic braze on a nickel substrate in a nitrogen-10% hydrogen atmosphere as a function of holding time at 820 °C (1500 °F). Four distinct stages are observed, demonstrating the complexity of the wetting process. [Weirauch, Jr., and Krafick 1996]

tween these extremes, spreading quite well, with penetration into the substrate being most pronounced where the braze and substrate are in contact for longest; i.e., the reaction zone extends outward from the center of the pool of molten braze.

Some attempt has been made to undertake a theoretical analysis of the kinetics of spreading of a molten metal over a wettable solid surface. The current theoretical approach considers the spreading of an inert sessile drop on a smooth and perfectly wetted substrate as a balance between surface energy driving it forward and viscosity, which acts to impede spreading [de Gennes 1985]. However, comparison of this somewhat reductive theoretical model with practical experience discloses that it contains a number of flaws, not least that measured flow rates are about four orders of magnitude slower than predicted by theory. These discrepancies may be ascribed largely to the added metallurgical and physical complexity of the wetting and spreading of a filler metal, as discussed previously, which are not adequately taken into account in this analytical model.

Nevertheless, the de Gennes model does predict some interesting dependencies of spreading. First, the initial spreading of molten filler metal is described by the imbalance between surface tension forces and viscous damping. This model also predicts a relative insensitivity of spreading to excess temperature in filler/substrate combinations where wetting is good, as can be seen in Fig. 1.18. Continued research in this area may eventually achieve a more complete mathematical description of wetting and spreading by filler metals that takes into account the physical and chemical states of the surface and also the situation where isothermal solidification occurs in the course of spreading and interalloying.

1.2.5 Surface Roughness of Components

The roughness of joint surfaces can have a significant effect on both the wetting and spreading behavior of a braze. It is well known that for each parent material there is an optimum surface roughness for maximizing the spreading of a filler metal. For example, when brazing aluminum alloys with the Al-12Si filler alloy in high vacuum and in the absence of fluxes, the best results in terms of spreading of the molten filler metal and fillet formation have been obtained when the surface of the components were prepared by dry grinding with silicon carbide papers of between 400 and 600 grit size [Okamoto, Takemoto and Den 1976]. Table 1.3 indicates values of surface roughness that can be obtained by abrasion of copper by various means.

Surface roughness reduces the effective contact angle θ^*, where θ^* is related to θ, the contact angle for a perfectly flat surface through the relation:

$$\cos\theta^* = r \cos\theta$$

where

$$r = \frac{\text{Actual area of rough surface}}{\text{Plan area}}$$

R_a is the average roughness measured as the average deviation from the center line of the surface profile. In the equation, it is expressed as a ratio of surface areas.

At the same time, by producing a network of fine channels, the texturing may increase the capillary force acting between the filler and the component surfaces. Both phenomena will tend to aid spreading. A directionally orientated surface texture promotes preferential flow parallel to the channeling [Nicholas and Crispin 1986].

It is possible to show from surface energy calculations that if the instantaneous contact angle of the molten filler is less than the surface angle (i.e., the root angle of v-shaped valleys), then profuse wetting tends to occur along the valleys. This is a frequent observation and, indeed, represents a problem when brazing to rough machined surfaces in that the filler does not spread uniformly in all directions. Another factor that should be considered in connection with texturing is the extent of alloying between the filler and the parent material because the roughness will increase the interaction for a given area of spread. As wetting of the parent metals usually

Table 1.3 Surface roughness (R_a) of cold-rolled copper after sanding with wet silicon carbide paper or polishing with a colloidal suspension of alumina in water

Abrasive	Nominal particle size, μm	R_a obtained on cold-rolled copper, μm
80 grit	200	2.2
240 grit	63	0.95
400 grit	23	0.51
1200 grit	5	0.23
Polishing alumina	0.05	0.012

For comparison, copper surfaces on electronic component leads usually have an orientated R_a of approximately 0.1 μm (4 μin.).

serves to increase the melting point and stifle spreading, a rough surface can prove to be detrimental.

1.2.6 Dissolution of Parent Materials and New Phase Formation

It is frequently observed that a filler metal will continue to spread beyond an initially wetted surface area over an extended period of time (>10 s), which would not be expected from classical fluid flow theory. Clearly, classical expressions for fluid flow, exemplified by Eq 1.5, which assume that the solid substrate is inert toward the liquid braze, do not strictly apply in such cases. Indeed, this type of flow can usually be associated with solid-liquid interfacial reactions, which have been seen in section 1.2.4 to play a major role in the spreading behavior, but are neglected in the model described in Milner's paper [1958]. Where joint filling is sluggish because of reactions occurring between the filler and the solid surface, increasing the temperature to reduce the viscosity of the molten filler is unlikely to enhance filling because the reactions that are occurring transverse to the flow directions will accelerate [Tunca, Delamore, and Smith, 1990]. The alternative of widening the joint gap is not usually an option because this is likely to lead to a reduction in joint filling and/or joint strength, as discussed in Chapter 4, section 4.3.4. The solution then is to change the materials system; several means by which this can be achieved without changing the parent materials are described in Chapter 4, section 4.1.

Dissolution of a substrate in a braze and growth of intermetallic compounds, where these occur, both follow Arrhenius type rate relationships, represented by the expression:

$$\text{Rate} = \exp\left[\frac{-Q}{kT}\right]$$

where Q is an activation energy that characterizes the reaction taking place at temperature T (in degrees Kelvin) and k is the Boltzmann constant.

Interfacial reactions are important, not only in determining the flow characteristics of the filler and its wetting behavior, but also the properties of the resulting joints. Evidence of intersolubility between a molten filler and the parent materials is provided by erosion of the surfaces of the parent materials in the joint region and the formation of new phases at either interface between the parent materials and the molten filler or within the filler itself when it solidifies. The effects of dissolution of the parent materials and compound formation on joints are discussed in detail in Chapter 2, section 2.3.

The following expression describes the rate of dissolution of a solid metal in a molten metal [Weeks and Gurinsky 1958; Tunca, Delamore, and Smith 1990]:

$$\frac{dC}{dt} = \frac{K\,A\,(C_s - C)}{V} \qquad \text{(Eq 1.6)}$$

where C is the instantaneous concentration of the dissolved metal in the melt, C_s is the concentration limit of the dissolved metal in the melt at the temperature of interest, t is the time, K is the dissolution rate constant, A is the wetted surface area, and V is the volume of the melt. This equation is known as both the Nernst-Shchukarev and the Berthoud equation. In the integral form, Eq 1.6 becomes:

$$C = C_s\left[1 - \exp\left(\frac{-KAt}{V}\right)\right] \qquad \text{(Eq 1.7)}$$

assuming initial conditions of $C = 0$ and $t = 0$.

Equation 1.7 reflects the fact that, in general, the concentration of dissolved metal in the molten filler increases in an inverse exponential manner with respect to time. That is, the dissolution rate is initially very fast but then slows as the concentration of the dissolved parent material tends toward its saturation limit (i.e., equilibrium), as shown in Fig. 1.20. Substituting measured values into Eq 1.7 shows that, for a brazed joint of typical geometry, the equilibrium condition is reached within seconds at the process temperature. Thus, it is possible to use an equilibrium phase diagram to predict the change in the composition of the filler metal that will occur in typical joining operations and the associated depth of erosion of the joint surfaces. Equilibrium phase diagrams and their use in brazing are considered more fully in Chapter 2, section 2.3.

In some materials systems, the product of reaction between a molten filler metal and the parent materials takes the form of a continuous layer of an intermetallic compound over the joint interface. Once this intermetallic layer is established, the rate of erosion greatly decreases because it is then governed by the rate at which atoms of the parent material can diffuse through the solid intermetallic compound. As a rough guide, solid-state diffusion processes are two orders of magnitude slower than solid-liquid reactions, and thus continued dissolution of the parent materials effectively ceases within the timescales of typical joining processes. Intermetallic growth will, however, continue throughout the life of the product, even if it is barely detectible. The practical implications of this phenomenon are discussed in Chapter 4, section 4.1.4.

1.2.7 Significance of the Joint Gap

The joint gap at the process temperature influences both the joint filling and the mechanical properties of the resulting joint. The relationship between joint dimensions and mechanical properties is discussed in Chapter 4, section 4.3. In summary, the thinner a joint is, the greater its load-bearing capability tends to be, until a limiting condition is reached when joint filling becomes unreliable.

Contact angle, surface tension, and viscosity all reduce with increasing temperature, making good joint filling in narrow joints more readily achievable as the joining temperature is raised above the melting point of the filler metal. On the other hand, spreading by the braze will generally decline as the temperature is raised on account of increasing interdiffusion between the braze and parent material, leading to premature solidification before joint filling is complete. Therefore, it is generally good practice not to exceed the liquidus temperature of the braze by more than about 50 °C (90 °F), particularly where joints of normal or narrow width (<100 μm, or 4 mils) are required.

A lower practical limit to the joint gap is imposed by three factors:

- *The need to provide a path for vapors to escape:* Vapors evolved within the joint and pockets of air must be allowed to escape if the formation of voids through gas entrapment is to be prevented (see Chapter 4, section 4.3.1.1). At the same time, any reducing agent needs to gain access to all joint surfaces and be present in sufficient concentration to work effectively. This requirement is not nearly as critical for brazing as it is for soldering. Especially for brazing carried out at temperatures above about 750 °C (1380 °F), the pressure exerted by pockets of vapor is usually high enough to result in their expulsion from a joint while the braze is molten (a consequence of the Gas Law), especially if the geometry is planar and there are no reentrant features.
- *Reaction with the components:* The metallurgical reaction that occurs between a molten filler metal and the surfaces of the components can take one of two forms.

  a. *The surface region of the workpiece has limited solubility in the molten braze:* This is the preferred situation. The dissolution of metal from the surface of the components can result in either compound formation at the interface, which may prevent further dissolution, or alloying with the filler, which will change its composition and hence its melting point.

  On the whole, brazes usually exhibit extensive interalloying with the parent materials. This can be explained partly by the fact that most brazes are based on elements with crystal structures that are similar to most engineering metals and alloy metals. Consequently, solid solutions tend to form in preference to intermetallic compounds.

  A reaction that depresses the melting point of the filler metal is desirable for narrow joints because the fluidity of the braze will be enhanced by such a reaction at a constant temperature. A reaction that raises the melting point of the filler metal will tend to increase its viscosity and

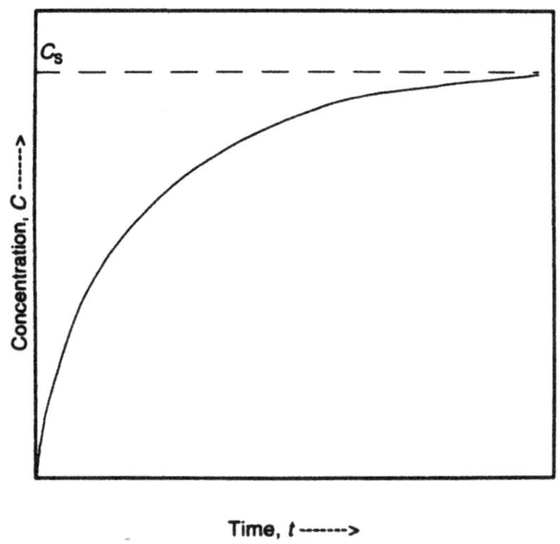

Fig. 1.20 The concentration of a solid metal in a liquid metal wetted by it changes in an inverse exponential manner with respect to time and is limited by the saturation concentration of the solid constituent in the liquid at that temperature.

holds out the danger that the filler will solidify isothermally at the process temperature before it has filled the entire joint. Wider joints mitigate this effect because the alloying will tend to be diluted.

b. *Dissolution of the filler in the parent metal:* In this situation, the volume of braze may shrink as the reaction progresses; therefore, a larger volume of filler metal accommodated in a wider joint gap is preferred. However, absorption of the filler is generally undesirable because its constituents will tend to penetrate into the parent materials, preferentially along grain boundaries, generally to the detriment of the mechanical properties of the assembly and sometimes resulting in embrittlement and/or hot shortness (sometimes also referred to as *liquid metal embrittlement*).

- *Control of the joint gap:* The width of the joint, i.e., the joint gap, should be predictable during the bonding cycle. The size of the gap will be influenced by the respective coefficients of thermal expansion of the components, and allowances need to be made for any differences. This is illustrated for tubular brazed joints between brass and steel members in Fig. 1.21. The coefficients of thermal expansion (CTEs) of a representative range of engineering materials at room temperature (25 °C, or 77 °F) are listed in Table 1.4. A joint gap that widens as the temperature is raised will cause the capillary forces to diminish, as noted in section 1.2.2. At the same time, the capillary forces are relied on to fill the joint with brazing alloy. This consideration is of added importance in relation to the widely used practice in brazing of drawing the filler into the joint from a preform placed adjacent to it.

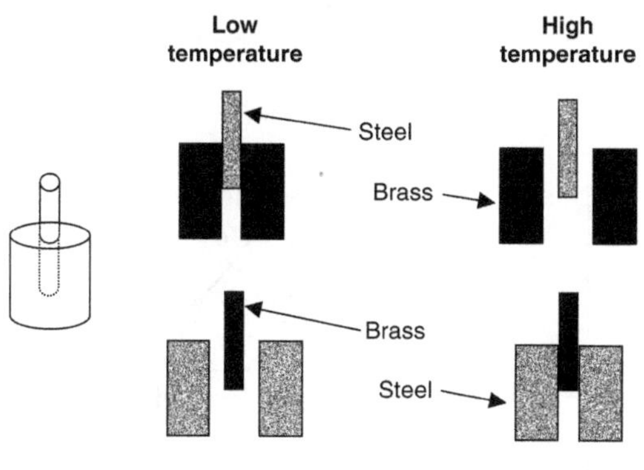

Fig. 1.21 Effect of temperature on the joint clearance between tubular brass and steel components arising from the difference in their thermal expansion coefficients

An upper practical limit to the joint gap is determined by two factors:

- *Mechanical properties of the joint:* As the gap is increased, the mechanical properties of the braze declines progressively to those of the bulk filler metal, which are usually inferior to most structural materials. This aspect is discussed further in Chapter 4, section 4.3.3.1.
- *Joint filling requirements:* Because the capillary force decreases as the joint gap increases, this will place a practical upper limit on the joint gap. At the same time, a sufficient quantity of filler must be supplied to the joint to fill it entirely. Hydrostatic forces will promote the flow of low-viscosity filler metals out of wide gap joints.

The optimal balance of these factors is achieved when the joint gap is about 10 to 100 μm (400 to 4000 μin.) wide, depending on the type of reaction that occurs between the braze

Table 1.4 Typical thermal expansivities of common engineering materials at normal ambient temperature

Material	Linear expansivity, 10^{-6}/K
Polymers	
Polymers, rubbers	150–300
Polymers, semicrystalline	100–200
Polymers, amorphous	50–100
Metals	
Zinc alloys	25–30
Aluminum alloys	20–23
Copper alloys	16–19
Stainless steels	15–17
Iron alloys	13–15
Nickel alloys	12–15
Cast irons	10–13
Titanium alloys	8–10
Tungsten/molybdenum alloys	4–7
Low expansion alloys (Fe-Ni-base)	1–5
Graphite	7–9
Ceramics	
Ceramics, glass	6–10
Ceramics, oxide	4–8
Ceramics, porcelain/clay	3–7
Ceramics, nitride/carbide	2–6
Diamond/silica/carbon fiber	−1 to 1

The values given are representative of the most widely used materials, rather than provide absolute limits for the different classes listed. The thermal expansivity will depend not only on elemental composition but also on microstructure and temper. Composite materials can have expansivities that effectively range between those of the constituents and depend on the relative proportions of the matrix and reinforcement phases. To convert to customary units of 10^{-6}/°F, multiply given values by 0.55556.

and the parent part, or component. This estimate is supported by theoretical calculations of capillary force and viscous drag of liquid flow (Fig. 1.22). A gap of this magnitude is usually readily achieved without resorting to expensive premachining operations. Generally, when components rest freely on one another and the assembly is heated until the filler is molten, the joint will tend to self-regulate to widths around 50 μm (2 mils), depending on the viscosity of the filler at the brazing temperature. Indeed, it has been demonstrated that, for a fixed combination of filler metal, component materials, and process conditions, the joint gap will tend to a fixed value specific for that combination. This value must be established by experiment. If there is insufficient brazing alloy to fill this gap, the joint will contain voids, or, if too much braze is made available, the excess will spill out and generate droplets on the free surfaces of the components [Bakulin, Shorshorov, and Shapiro 1992]. Where thinner or wider joints are required, it is necessary to insert spacers (such as wires) of the desired width between the components and, for thin joints, combine this measure with the application of a compressive loading on the joint during the bonding cycle to overcome the hydrostatic forces that will act to levitate the upper component.

Copper and silver are notable exceptions to this criteria governing the width of brazed joints. These metals possess the characteristic of exceptional fluidity, enabling the pure metals to be used as brazes to form exceedingly narrow (<1 μm, or 40 μin.) joints with steel components. Although copper and silver are soft and weak, the narrowness of the joints together with the high degree of joint filling that can be achieved and the negligible intersolubility with the parent material (i.e., steel), can result in joints that are stronger than the steel parts [Sloboda 1961]. The enhanced strength of extremely narrow, well-filled joints is explained in Chapter 4, section 4.3.3.1.

1.2.8 The Strength of Metals

The purpose of making brazed joints is usually to form a metallic bond between components. A fundamental question, therefore, is how strong is the interface between the parent material and the braze?

The cohesive strength of metals results from attractive forces between the constituent atoms. Normally, each atom will occupy a physical location where the net force upon it is zero. When the solid metal is strained by the application of an external load and the atoms move from their equilibrium positions, an opposing stress is set up in the crystal lattice of the metal. The attractive force between atoms that share the same electron cloud increases with the distance between them up to a maximum and thereafter decreases abruptly, when failure occurs. A perfect metal lattice (i.e., a defect-free crystal) will fail at this point by cleavage across the crystallographic plane because this is the region where the interatomic forces are weakest.

To a first approximation, the interatomic force per unit area varies with interatomic separation, x, according to a sine wave with wavelength λ, as shown in Fig. 1.23. The interatomic force per unit area may then be represented by a sine wave, as:

$$\sigma = \sigma_0 \sin 2\pi \frac{x}{\lambda} \qquad \text{(Eq 1.8a)}$$

where σ_0 is the maximum theoretical strength. The work done per unit area in completely separating neighboring planes of atoms, which are

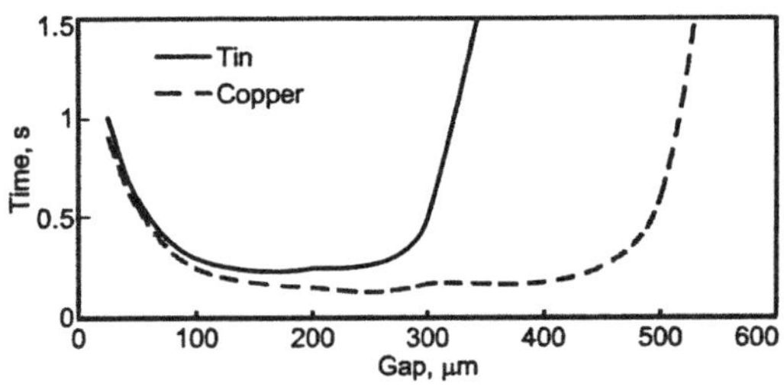

Fig. 1.22 Calculated time for molten tin and copper to flow up a perfectly wetted capillary [Nicholas 1989]

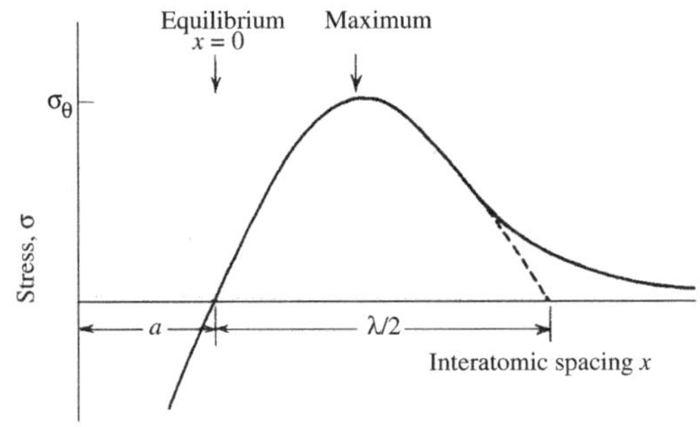

Fig. 1.23 Variation of interatomic force, per unit area, with distance

an equilibrium distance ($x = 0$; i.e., a $\sim \lambda/4$) apart, is then:

$$\int_0^{\lambda/2} \sigma \, dx = \int_0^{\lambda/2} \sigma_0 \sin 2\pi \frac{x}{\lambda} \, dx = \frac{\lambda \sigma_0}{\pi}$$

This work corresponds to the total surface energy of the two new surfaces created in the fracture, i.e., $2\gamma_{SV}$, where γ_{SV} is the surface energy per unit area of the solid.

Accordingly:

$$\sigma_0 = \frac{2\pi\gamma_{SV}}{\lambda} \qquad \text{(Eq 1.8b)}$$

Within the elastic range of strain, Hooke's law applies; that is:

$$\delta\sigma = E \frac{\delta x}{a}$$

Differentiating Eq 1.8(a) gives:

$$\frac{d\sigma}{dx} = \sigma_0 \frac{2\pi}{\lambda} \left(\cos 2\pi \frac{x}{\lambda} \right)$$

At zero strain; that is, $x = 0$:

$$\left(\frac{d\sigma}{dx} \right)_{x=0} = \sigma_0 \frac{2\pi}{a}$$

Hence:

$$\sigma_0 = \frac{E\lambda}{2\pi a} \qquad \text{(Eq 1.9)}$$

From Eq 1.8(b) and 1.9:

$$\lambda = \frac{2\pi\gamma_{SV}}{\sigma_0} = \sigma_0 \frac{2\pi a}{E}$$

so that:

$$\sigma_0 = \left(\frac{E\gamma_{SV}}{a} \right)^{1/2} \qquad \text{(Eq 1.10)}$$

The theoretical fracture stress is about $\sigma_0/10$ for metals, although, in practice, strengths of metals tend to be only one-tenth of this value (i.e., $\sigma_0/100$), owing to the presence of lattice defects and other discontinuities.

Possibly somewhat surprisingly, brazed joints subject to simple mechanical stress will often fail in a brittle manner. The reasons for this are elaborated in Chapter 4, section 4.3.3. In brittle materials, failure takes place by the propagation of cracks that either preexist in the structure or nucleate at lattice imperfections. The stress to cause fracture can be deduced from Eq 1.10 by replacing the denominator with c, where c is the crack length; thus:

$$\sigma_b = (E\gamma_{SV}/c)^{1/2}$$

Because c is much larger than a, the mechanical strength of a brittle material is low relative to its theoretical strength. Only in special materials, which are regularly ordered on an atomic scale, such as carbon fiber, are the two values remotely comparable.

In ductile metals, application of stress results in the movement of dislocations and other defects through the lattice of individual grains. The interfaces between grains represent another region where physical material transport and plastic flow takes place. Failure occurs when the rate of increase in strength of the material due to work hardening falls below the rate of decrease in the load-bearing cross section resulting from the plastic flow.

The preceding discussion pertained to bulk materials, i.e., the components and the filler metal, when considered in isolation. In reality, the joint interfaces will often be a source of voids, microcracks, local interfacial mismatch stresses, and brittle intermetallic phases. These features tend to be a common source of joint weakness and should be minimized through judicious choice of the filler/parent material and joining conditions.

1.3 The Design and Application of Brazing Processes

A brazed joint is usually required to satisfy a specific set of requirements. Most frequently, it must achieve a certain mechanical strength, which it must retain to the highest service temperature in the intended application. The joint must also endure a particular service environment, which may be corrosive, and it may have to provide good electrical and thermal conductance. In addition, the joint must be capable of being formed in a cost-effective manner without detriment to other parts of the assembly.

The principal aspects that need to be addressed can be divided as:

- The functional requirements of the application and the means of satisfying these through appropriate structural design.
- The achievement of the specified assembly through successful processing.

In the following sections, where these aspects are considered further, it is presumed, unless otherwise stated, that the faying surfaces of the components are metallic and, therefore, fundamentally wettable by a molten braze. Brazing of nonmetals is discussed in Chapter 7.

1.3.1 Functional Requirements and Design Criteria

All brazed joints used in manufactured products must remain solid in service and retain the associated components in fixed positions when subjected to stress. These requirements are usually satisfied by suitable design of the geometry and metallurgy of the joint, but there are also other aspects to consider. Several factors that affect the functional integrity of brazed joints are discussed next.

1.3.1.1 Metallurgical Stability

For a joint to remain solid, the melting point (solidus temperature) of the filler metal needs to exceed the peak temperature that the component is ever likely to experience in service. There are exceptions to this rule, which are discussed in Chapter 6. Because the strength of all metals decreases rapidly as the melting point is approached, in general, the peak operating temperature should not exceed about 70% of the melting point of the filler, in degrees Kelvin, if the joint is required to sustain a load.

The solidified filler metal and parent materials have different compositions and so the microstructural equilibrium is rarely achieved in practice. The composition of these materials and the phases that form on solidification in a brazed joint are frequently unstable or the relative proportions of these phases can change at elevated service temperatures. Instability of phases present in the joint at the service temperature may be undesirable. Such changes can result in a decrease in fracture toughness, if the new phases are brittle, or in a general loss of joint integrity should Kirkendall voids develop as some phases evolve at the expense of others. Therefore, the effects of continued reactions between the filler and the components also must be considered in relation to the application, as explained in Chapters 2 and 3.

1.3.1.2 Mechanical Integrity

The durability of engineering and consumer products often depends on joints maintaining their mechanical integrity for the duration of their expected service life. The mechanical integrity of a brazed joint depends on a number of factors:

- The mechanical properties of the bulk filler metal (Chapter 4, sections 4.3.3 and 4.3.4)
- The joint geometry, namely, area, width, and shape (Chapter 4, sections 4.3.1 and 4.3.3)
- The mechanical properties of any new phases formed in the joint by reaction between the filler and the components, either during the joining operation or subsequently in service (i.e., there is an interdependence with the microstructure) (Chapter 4, section 4.1.4)
- The number, size, shape, and distribution of voids within the joint (Chapter 4, section 4.3.1)
- The quality of fillets formed between the filler and the surface of the components at the edge of the joint (i.e., their radius of curvature and extent of continuity) (Chapter 4, section 4.2.5)

The mechanical properties of joints, taking into account the influence of joint geometry, are reviewed extensively elsewhere. The reader is particularly referred to Schwartz [2003] and Nicholas [1998]

1.3.1.3 Environmental Durability

Joints are normally expected to be robust in relation to the service environment. This may involve exposure to corrosive gases, including sulphur dioxide and other constituents of a polluted atmosphere; to moisture, perhaps laden with salt; and to variable temperature. The corrosion and stress-corrosion characteristics of the joint are then of relevance. Corrosion mechanisms are often somewhat complex and specific to a given combination of materials, chemical environment, and joint geometry. Therefore, each situation should be determined empirically. A typical example is provided by the case study of corrosion of brazed stainless steel pipes conveying drinking water, an environment in which extensive corrosion would not normally be expected [Jarman, Linekar, and Booker 1975].

The temperature of a joint can be shifted well beyond a normal ambient range, especially in aerospace applications and in situations where heat is generated within the assembly itself. Then, thermal fatigue and other changes to the metallurgical condition of the joint, such as the growth of phases, can occur, and these invariably affect the properties of the joint. In other words, there is interdependence between environmental stability and microstructural stability. An appropriate choice of the materials combination used should enable these changes to be constrained within predictable and acceptable limits. In other words, brazes should be used at service temperatures that are sufficiently below their melting points, with respect to the thermodynamic reference point of absolute zero temperature (−273 °C, or −459 °F) so that they are metallurgically stable and microstructural changes take place only very slowly.

1.3.1.4 Electrical and Thermal Conductivity

In certain applications, brazed joints are required to perform the function of providing electrical and/or thermal conductance between components. Generally, thin, well-filled brazed joints amply satisfy this requirement. Only in a few extreme situations are the thermal and electrical properties of such joints close to the allowed limits. A case in point is silicon high-power device assemblies, where the joint between the silicon device and the metal backing plate is required to conduct 1 W/mm^2, or more, of thermal power. Here, it is crucial to ensure that the brazed joints, which are conventionally made using an aluminum-silicon alloy, are kept thin (<30 μm, or 1200 μin.) and essentially void-free (<5% by volume), in order to meet the device performance specifications [Humpston et al. 1992].

1.3.2 Processing Aspects

An important aspect that must be considered when designing a joint is the practicality of the process involved. Among the relevant issues are:

- Jigging of the components
- Form of the filler metal
- Heating method
- Temperature measurement
- Joining atmosphere
- Coatings applied to surfaces of components (as necessary)
- Cleaning treatments
- Heat treatments prior to joining
- Heating cycle of the joining operation
- Post-joining treatments

Each of these aspects is considered in turn.

1.3.2.1 Jigging of the Components

Normally, the components being joined, and sometimes also the filler metal itself, must be held in the required configuration until the filler metal has solidified. Even if the components can be preplaced without a fixture, the use of some form of jig is still frequently beneficial to ensure that the components are not disturbed by capillary forces originating from the molten filler metal. On the other hand, it is also possible to exploit the capillary forces to make an assembly self-align during the brazing process. Self-alignment is widely practiced in the fabrication of electronic circuits using lightweight surface-mounted components and flip-chip assembly involving soldering (see the companion volume *Principles of Soldering* for details of these processes).

The jig can be used to fulfill more than a holding function. For example, it can serve as a heat spreader, as a heat sink, or as a heat source. A jig should be constructed from a nonporous material to prevent contamination of the atmosphere surrounding the workpiece. Moreover, the jig material should not be wettable by the molten filler metal, in case they come into contact through accidental spillage. Materials such as brass should be avoided as zinc readily volatilizes at elevated temperatures used for brazing. Care should be taken in designing jigs so as not to stress the constrained components through thermal expansion mismatch.

Graphite is often favored as a jig material. It is inexpensive, easy to machine to precise tolerances, is a good thermal conductor (making it an effective heat spreader), is an absorber of radiant heat, and is not wetted by the majority of molten filler metals. It can also be used as a heater by the passage of electric current. Graphite has the merit of "mopping-up" oxygen in an oxidizing atmosphere to form carbon monoxide and carbon dioxide, although the presence of these gases can result in carburization of some steel components. Care should be taken to ensure that the graphite used for jigs is of a dense grade so that it will be mechanically robust and have a low porosity to minimize outgasing. The desorption of water vapor, in particular, frequently determines the quality of a gas atmosphere in the vicinity of a workpiece.

Jigs are sometimes used to apply a controlled pressure to a joint in an assembly. One component can then be deliberately and elastically distorted to bring it into close and uniform contact with its mating part. This is an advantage when very narrow joints are required and when solid state diffusion constitutes an important part of the joining process. Compressive loading on the joint also aids expulsion of air and vapors from the joints, which are otherwise trapped in pockets and produce voids. An applied force also helps to puncture the oxide films on the surfaces of the filler and, when the filler metal melts, it acts against any dewetting capillary force of the liquid and ensures adequate spreading over the joint. The combination of these factors leads to improved joint filling. Typical pressures are in the range 1 to 5 kPa (0.14–0.72 psi).

Fixturing (i.e., jigging) represents an expensive add-on cost in a brazing operation, and many ingenious methods of self-supporting and integral jigs have been devised. They generally involve some means of mechanical fastening such as crimping, swaging, interlocking, peening, riveting, dimpling, knurling, and so forth, including spot welding. The jigging must be designed to provide the correct joint clearance at the brazing temperature, as discussed in section 1.2.7.

1.3.2.2 Form of the Filler Metal

Filler metals are available in many different forms. These forms include configurations that normally can be produced from an ingot by mechanical working, for example, wire, rings, and foil. Such geometries are not restricted to ductile alloys. If the constituents are individually ductile, the preform can be partitioned. This approach is discussed further in Chapter 4, section 4.1.5. The development of rapid-solidification processes has led to the availability of foils and wire of joining alloys that are inherently brittle. These foils are produced directly from the melt: the process involves forcing molten metal through a hole or slot onto a rapidly spinning, water-cooled, metal wheel. Fig. 1.24 shows such a strip casting process in operation, and Fig. 1.25 illustrates some typical foils produced by this route. The high rate of heat extraction that occurs in this process causes the molten metal to solidify almost immediately on striking the wheel, resulting in the formation of a strip of the alloy with a fine crystalline, or occasionally amorphous, microstructure. The dimensions of the cast material can be controlled by varying the nozzle dimensions, the ejection pressure, the speed of rotation of the wheel, and other parameters of the casting process [Otooni 1995; Srivatsan and Sudarshan 1993; Jones 1982]. The refined microstructure of the rapidly solidified alloys, together with their homogenous composition, generally improves strength and ductility compared with the same alloys produced by conventional casting and mechanical working (Fig. 1.26). A good example is the family of silver-copper-titanium brazes. These alloys are generally brittle in ingot form, but, when pre-

Fig. 1.24 Production of foil directly from a molten charge by strip casting. Source: Vacuumschelze GmbH, Germany

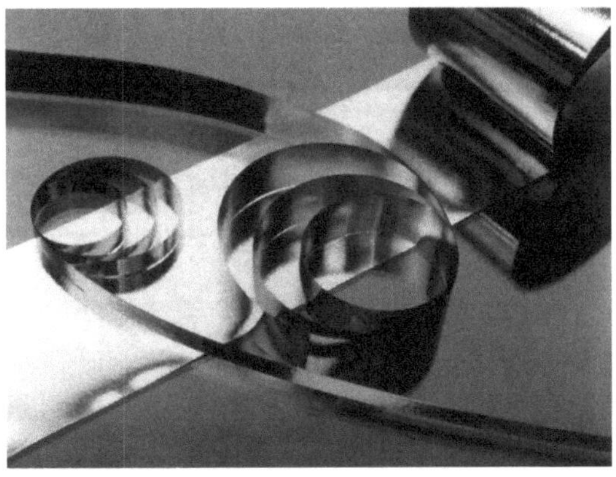

Fig. 1.25 Examples of foil strip produced by rapid-solidification casting technology. Source: Fleetwood et al. [1988]

pared as a foil by rapid solidification, their ductility is comparable to that of other brazes.

Brazes are also available as finely divided powders that can be mixed with a binder to form a paste capable of being screen printed onto a substrate or applied to the workpiece, via a dispenser, to suit an automated production line. Much industrial brazing is carried out in this manner. However, powders, and pastes containing powders, have an extremely high ratio of surface area to volume of braze, which generally results in high oxide fractions and, in the absence of suitable precautions, would be detrimental to the quality of the resulting joints. Braze paste manufacturers go to great lengths to produce spherical granules with smooth, clean surfaces specifically for this very reason. Powdered brazes can also be admitted to the joint gap in the form of tapes. These are powder compacts that are often supplied on an adhesive support. Preforms can thus be shaped with a sharp blade and are easily placed in the joint gap. The adhesive is selected to burn cleanly along with the other organic constituents.

In specialized joining processes, the braze can be deposited as a coating on the components by electroplating and by vapor deposition techniques such as evaporation. Where it is not possible to deposit the actual alloy, sequential layers of the constituent elements can be applied. The former is generally preferred because the melting point of an alloy is well defined, whereas there is no guarantee that melting will take place at the desired temperature in the case of the composite layers, unless significant solid-state diffusion has occurred first to form the appropriate low-melting-point phases.

The use of some form of preplaced filler metal has a number of advantages. Most particularly, because the thickness and area of filler metal are predetermined, the volume of molten braze may be carefully controlled. Also, the number of free surfaces is reduced from four (corresponding to a foil preform sandwiched in the joint) to just two, thereby considerably reducing the proportion of oxides and other impurities deriving from exposed surfaces. Piece-part inventory is reduced and jigging is simplified. Having the filler metal ready placed within the joint gap means there is no need for it to spread to effect joining and this arrangement eliminates operator error in placement or omission of brazing paste or preforms. Fine details and intricate parts are readily brazed and, generally, more consistent joints are obtained by this means. "Self-brazing metals," which offer these benefits, consist of a base metal that is clad on one or both sides with braze and are produced in many variants. A metallographic section through a commercial product of this type is shown in Fig. 1.27. The cladding is usually achieved by roll bonding, and typical cladding thicknesses range from 5 to 30%. Standard clad products include aluminum, copper, and ferrous alloys. Figure 1.28 shows two aluminum parts, one brazed with a preform and the other with a roll-clad substrate. Braze claddings can be made much thinner than brazing foils, a feature that generally benefits the mechanical properties of the joints. Once the brazing alloy is clad to the base metal, subsequent forming operations will not significantly alter the cladding ratio. The presence of a soft copper braze on medium-to-hard carbon steels is actually ben-

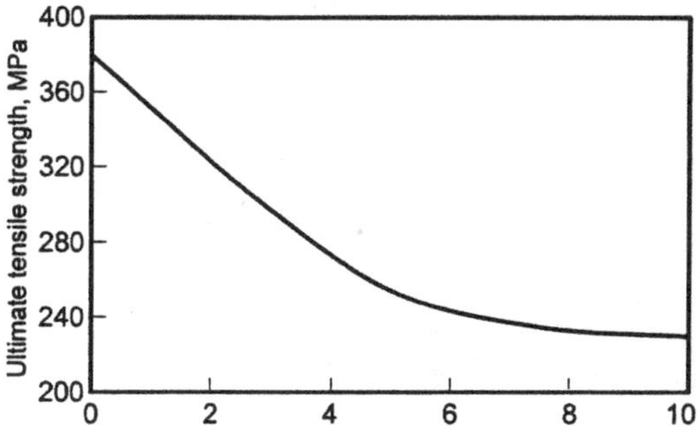

Fig. 1.26 Dendrite arm spacing decreases with increasing cooling rate and hence fine-grained microstructures have improved mechanical properties. The data pertain to hypereutectic cast iron. Adapted from Seah, Hemanth, and Sharma [1998]

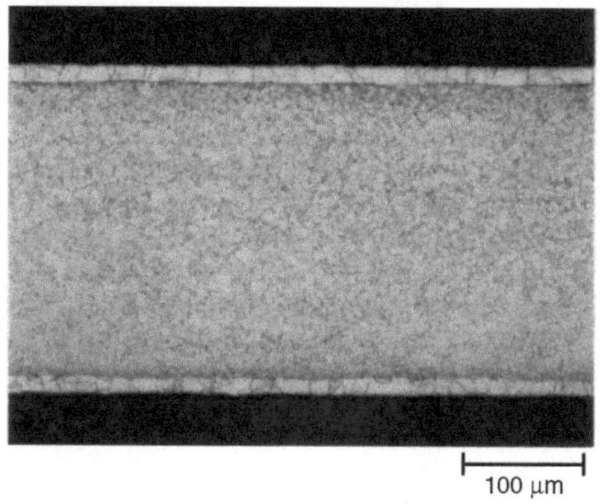

Fig. 1.27 Metallographic cross section of a stainless steel strip clad on both sides with copper braze. In this case, the ratio of braze cladding to core material is in the ratio 5/90/5.

eficial in reducing tool wear when parts are worked [Karavolis 1993].

1.3.2.3 Heating Methods

Heat must be supplied to the joint to raise the temperature of the filler metal and joint surfaces above the melting point of the filler. If the joint surfaces are maintained below the melting temperature of the braze, it will freeze on contact with the parent materials and "ball up." To prevent this situation, it is good practice always to heat the filler metal via the components to be joined and never vice versa.

The available methods of heating are:

- Local heating, in which only that part of the components in the immediate vicinity of the joint is heated to the desired temperature
- Diffuse heating, where the temperature of the entire assembly is raised

Common local heat sources include gas torches and resistance heating using the assembly as the resistive element and also heating elements external to the assembly. Other heating techniques that are now commonly used include induction heating and laser heating. Although some methods of local heating are applicable to joining in a controlled atmosphere, this is not usually the case with a gas torch and a flux must then be used.

In local heating, the rate of heat energy input must be high to swamp the heat conducted away by the components and jigging. A high rate of heat input can achieve the characteristic of fast heating and cooling of the joint. Fast heating coupled with short brazing cycles minimizes erosion of substrate surfaces and therefore restricts the formation of undesirable phases, while rapid cooling ensures a fine grain size in the solidified filler and thereby superior mechanical properties. However, these potential benefits can be offset by the generation of thermally induced stresses and distortion, and even cracking, in the components being joined (see also section 1.3.2.9). Local heating can be used to create specified temperature gradients that will restrict the flow of the molten filler metal to the immediate vicinity of the joint.

Diffuse heating sources include systems such as furnaces (both resistance and optical) and induction coils. The features of diffuse heating methods are the opposite of those of the local heating methods. For example, the total energy requirement is higher because the temperature of the entire assembly has to be raised, which also significantly increases the process cycle time. On the other hand, there is less risk of high-temperature gradients and thermal distortion, and accurate control of temperature at the joint is easier to achieve. Diffuse heating methods tend to impose fewer constraints on the atmosphere surrounding the workpiece because the source of heat is relatively remote from the components.

If diffuse heating is to be used in the fabrication of complex assemblies, the designer must ensure that all of the component parts are able to withstand the peak process temperature. With local heating, heat sinks can be used to protect

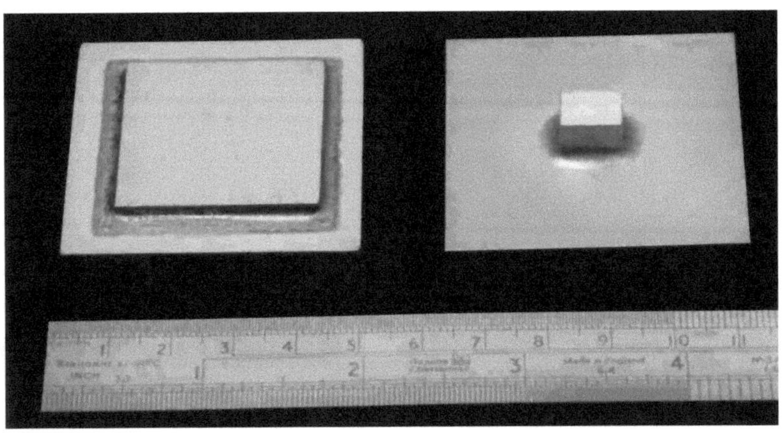

Fig. 1.28 Aluminum components brazed using a preform of (left) braze and (right) one component roll-clad with brazing alloy. In both cases, the brazing process was fluxless. Because roll-clad braze is generally thinner than a perform, the upper component needs to be smaller in area to achieve a comparably sized fillet. Courtesy of BAE Systems

sensitive areas from excessive thermal excursions. A related consideration when using diffuse heating in a situation where several joints must be made is that the melting point of the filler metal used for the preceding joining operations must be higher than the peak process temperature used in the current cycle. Several different filler metals will therefore be required to fabricate a multijointed product in a step joining process.

The oldest method of heating joints is by naked flame. The gases predominantly used now are acetylene and propane, burnt in oxygen. These gases are cheap, widely available, easy to use, and can be made oxidizing, reducing, or neutral by adjustment of the oxygen-to-gas ratio. These three combustion conditions are also readily discernible by eye allowing a skilled operator to adjust the torch to satisfy the requirements of the job in hand. The thermal characteristics of some common fuel gases burnt in oxygen are given in Table 1.5.

One of the more exotic methods of heating for making brazed joints is self-propagating high-temperature synthesis (SHS). The principle of the method is to place within the joint gap, or its vicinity, a mixture of two or more metals in the form of a powder compact or multilayer foil. The heating reaction is initiated by applying a spark, thermal energy (e.g., a lighted match), or passage of electric current (Fig. 1.29). The two metals are carefully chosen, for example a stoichiometric mix of palladium and aluminum, so that they react exothermically, in this case to form palladium aluminide with a large excess of heat ($\Delta G^0 = -94$ kJ/mol). An SHS heat source can be entirely internal to the joint. In brazing applications, either one or more of the reactants or products of the SHS acts as the filler metal itself, or the heat is used to melt a conventional filler metal [Hawk et al. 1993]. The intermetallic compounds formed as a result of SHS have relatively poor thermal conductivity (approximately 70 W/mK, or 1.2×10^{-2} Btu/h · ft · °F), compared with brazes that may need to be taken into account where this characteristic is important and the reactants constitute an appreciable volume fraction of the joint. Self-propagating high-temperature synthesis foils and powder mixtures are available commercially.

Table 1.5 Thermal characteristics of common fuel gases burnt in oxygen. In each case, the flame temperature is in the region of 3000 °C (5430 °F)

Fuel gas	Thermal output, kW/cm^3
Acetylene	15
Methane	7
Propane	6
Hydrogen	9

Ideally, during heating, the faying surfaces and brazing alloy should reach the brazing temperature at exactly the same time. In designing the brazing process it is therefore necessary to take into account the different thermal mass, conductivity, and heat capacity of the components together with their proximity to the heat source. Molten brazes flow preferentially toward the hottest region of the substrate. In recognition of this fact, it is best practice to place the filler metal in the joint gap, or, next best, to apply the filler metal to the edge of the joint and apply most heat to the central portion of the joint area.

1.3.2.4 Temperature Measurement

The liquid-solid metallurgical reactions that occur during brazing operations are highly temperature dependent. Therefore, reliable measurement of temperature is essential. Thermocouples and pyroelectric elements are the most common types of temperature sensors used in brazing operations.

A number of precautions should be taken when employing thermocouples. Regular in situ calibration checks should be made to determine whether the thermoelectric characteristics of the thermocouple materials have altered and to test for electrical interference affecting the display system. Correct temperature measurement requires good thermal contact between the ther-

Fig. 1.29 A quite remarkable photograph of a self-propagating high-temperature synthesis (SHS) reaction moments after ignition of a foil of a proprietary metal combination. The excess energy is sufficient to heat the foil to white heat. Courtesy of Maximilian Franz

mocouple and the object being monitored. This tends to present a problem in vacuum joining processes where thermal contact by mechanical means, namely, resting the thermocouple against a surface, tends to be inadequate. The thermal mass of the thermocouple and its protective sheath impedes the thermocouple junction from sensing the true temperature of the component surface. These effects can be minimized by embedding the thermocouple within the workpiece to improve thermal transfer.

Even when thermocouples are used for temperature measurement in gas atmospheres, where the thermal contact with the heated surface is better than it is in a vacuum, a change in the measured temperature will lag behind that actually occurring. This delay, which can be of the order of seconds, is difficult to measure accurately, but it must be taken into account if a thermocouple is being used to monitor the temperature of assemblies exposed to high heating and cooling rates.

Pyrometers have one important advantage over thermocouples: they are noncontacting sensors of temperature. Traditional pyrometers are designed primarily for operation above about 750 °C (1380 °F) and are widely used for monitoring brazing processes. Measurements may be made remotely from the workpiece, and the response time of the instrument can be determined accurately. However, it is necessary for the edge of the joint to be visible to the pyrometer and not obscured by jigging and other objects. Pyrometers and thermocouples provide only a localized measurement of temperature. As this might not be representative of the entire joint region, it is common to measure temperature at several places on an assembly, at least until a process is well established.

A new form of spot temperature measurement involves the use of lasers. A helium-neon laser is used to illuminate a surface and highly sensitive silicon photodiodes used to monitor the reflected light. Because reflectivity has a temperature dependence, known as *thermoreflectance,* this enables the temperature of the illuminated spot to be ascertained [Lee and Norris 1997]. The accuracy that can be achieved is about 1 °C (1.8 °F) with a low-cost, handheld instrument.

In manual brazing, the flux can function as an indicator of temperature. For example, many of the fluxes formulated for use with low melting point silver-base brazes transform from a white powder to a clear molten glass just as the temperature of the components approaches the solidus temperature of the braze.

1.3.2.5 Joining Atmosphere

For a molten filler metal to wet and bond to a metal surface, the latter must be free from nonmetallic surface films. Although it is possible to ensure that this condition is met at the beginning of the heating cycle, by prescribed cleaning treatments, significant oxidation will generally occur if the components are heated in air. Steps must therefore be taken to either prevent oxidation or remove the oxide film as fast as it forms.

The approach adopted will depend largely on the atmosphere surrounding the workpiece. Brazing processes are conducted in one of three types of atmosphere, defined according to the reaction that occurs between the atmosphere and the constituent materials, as:

- Oxidizing (e.g., air)
- Essentially inert (e.g., nitrogen, vacuum)
- Reducing (e.g., hydrogen, carbon monoxide, halogen containing)

The implications associated with using each of these atmospheres are considered next.

Oxidizing atmospheres. Air is the most common oxidizing atmosphere. The principal advantages of joining in air are that no special gas-handling measures are required and that there are no difficulties associated with access to the workpiece during the brazing operation. However, because most component surfaces and those of the filler metal are likely to form oxide scale when heated in air, normally fluxes must be applied to the joint region. A flux is capable of chemically and/or physically removing an oxide film. The flux may be applied either as a separate agent or may be an integral constituent of the joining alloy. The subject of fluxes is discussed in detail in Chapter 3, section 3.2.

Gold and some of the platinum-group metals do not oxidize when heated in air. These precious metals are therefore sometimes applied as metallizations to the surfaces of the components being joined in fluxless processes. The use of wettable metallizations is discussed in Chapter 4, section 4.1.2.1. Brazes that contain significant proportions of precious metals, including silver, are generally less susceptible to oxidation than other alloys, enabling mild fluxes to be used.

An oxidizing atmosphere is occasionally desirable during brazing. Not only do some fluxes require the presence of oxygen in order to function, but in some instances, it is a prerequisite for successful joining that oxygen be present. An example is provided by the copper-copper oxide eutectic brazing process in which copper is brazed to oxide ceramic materials, such as alumina, by a eutectic that is formed in situ between copper and Cu_2O just below the melting point of copper [Schwartz 1990]. This process is used to manufacture so-called direct-copper-bonded substrates of controlled coefficient of thermal expansion; see Chapter 4, section 4.2.1.3.

Inert atmospheres. From a practical point of view, an atmosphere is either oxidizing or reducing. This is because it is not possible to remove and then totally exclude oxygen from the workpiece, except perhaps under rigorous laboratory conditions. Thus, when defining an atmosphere as inert, it must be taken as meaning that the residual level of oxygen present is not sufficient to adversely affect the joining process under consideration. An atmosphere that might be suitable for brazing silver jewelery may be inadequate for joining nickel-base superalloys.

Because the "inertness" of an atmosphere is judged relative to the specific application, it is necessary to define a quantitative measure of the oxygen present. This parameter is the oxygen partial pressure. Partial pressure provides a measure of the concentration of one gas in an atmosphere containing several gases. The partial pressure of a gas in a mixture of gases is defined as the pressure it would exert if it alone occupied the available volume. Thus, dry air at atmospheric pressure (0.1 MPa, or 14.5 psi) contains approximately 20% oxygen by volume so that the oxygen partial pressure in air is 0.02 MPa (2.9 psi).

Typical inert atmospheres among the common gases include nitrogen and argon. The oxygen partial pressure in standard commercial-grade bottled gases is of the order of 10 mPa (1.5×10^{-6} psi). Higher quality grades are available, but their cost is usually too prohibitive to permit their use in most industrial applications.

Vacuum is frequently used as an inert, protective environment for filler metal joining processes. Vacuum offers several advantages compared with a gas atmosphere, particularly the ability to readily measure and control the oxygen partial pressure. In a substantially leak-free system, the oxygen partial pressure is one-fifth of the vacuum pressure, which is relatively easy to determine. Although a roughing vacuum of 100 mPa (1.5×10^{-5} psi) will provide an atmosphere with the same oxygen partial pressure as a standard inert gas, it is possible to improve on this value, by several orders of magnitude, using a high-vacuum pumping system. Alternatively, a low-oxygen partial pressure may be achieved by obtaining a roughing vacuum, back filling with an inert gas and then roughing out again. The effect of the second pumping cycle will be to reduce the oxygen partial pressure to less than typically one-thousandth of that in the inert gas, that is, approximately 100 μPa (1.5×10^{-8} psi). This estimate assumes that the furnace chamber is completely leak-tight and does not outgas from interior surfaces, nor does any oxygen or water vapor backstream through the pump.

The disadvantages of using a vacuum system for carrying out a joining process are, principally, restricted access to the workpiece and the inadvisability of using either fluxes or filler metals with volatile constituents, such as cadmium, because the vapors can corrode the vacuum chamber, degrade its seals, and contaminate the pumping oils. This problem is not limited to the well-known volatile elements. Many metals that have negligible vapor pressure at normal ambient temperatures will volatilize during high-temperature brazing processes (>1000 °C, or 1830 °F), particularly when these entail using reduced pressure atmospheres. Manganese-containing brazes and base materials fall into this category because the vapor pressure of this element is 1 Pa (1.5×10^{-4} psi) at 1000 °C (1830 °F).

A frequently overlooked consideration in reduced-pressure atmospheres is adsorbed water that exists naturally on surfaces that are exposed to ambient atmospheres. The continuous streaming of water vapor that desorbs from surfaces and flows past the workpiece as the pressure in a vacuum chamber is reduced is a source of oxidation. In a vacuum system operating at 10 mPa (1.5×10^{-6} psi), the desorbing water vapor constitutes the major proportion of the residual atmosphere. An adsorbed monolayer of water vapor of just 100 mm^2 (0.16 $in.^2$) in area desorbs to a gas pressure of 4 mPa (6×10^{-7} psi) per liter of chamber volume. The surfaces of the chamber should therefore be smooth to minimize the surface area, and also dry. In order to reduce this problem further, the walls of the vacuum chamber should be heated and the system should always be vented to a dry atmosphere. To effectively desorb water vapor, the bakeout temperature should be at least 250 °C (480 °F), which may be difficult to achieve in practice owing to design constraints and the employment of seals of rubber and other organic materials.

Another source of oxidizing contamination in a vacuum system is oil vapor mixed with air and water vapor, backstreaming from a rotary pump. This can occur whenever the pressure inside the vacuum chamber drops below 1 Pa (1.5×10^{-4} psi) but can be largely eliminated by employing a foreline trap, or by isolating the pump from the chamber once the required pressure reduction has been obtained. The widespread adoption of turbomolecular pumps on modern vacuum equipment has rendered this problem obsolete.

The practice of relying on an open gas shroud, as in various welding processes, to provide an inert atmosphere for brazing is often unsatisfactory because it is extremely difficult to control such an atmosphere reliably. For example, turbulence in the inert gas shroud can result in a supply of air actually being directed at the workpiece. Recent advances in furnace technology now permit open furnaces, which often take the form of belt furnaces intended for continuous brazing processes, such as jewelery chain making. These are capable of achieving very high specification atmospheres in the working zone, through careful design of the gas flow at the open portals.

Reducing atmospheres. A reducing atmosphere is one that is capable of chemically removing surface contamination from metals. Gases that provide reducing conditions are, principally, hydrogen and carbon monoxide, and generally, proprietary mixtures that liberate halogen radicals. Specific gas-handling systems are usually needed for these in order to satisfy health and safety legislation.

For a few metals, notably copper and silver, hydrogen is satisfactory as a reducing atmosphere in a brazing furnace. No less important for meeting its functional requirement than the oxygen partial pressure of the gas is its water content. Hydrogen is a relatively difficult gas to dry, and the water vapor present can present a serious problem. A frost point of −70 °C (−95 °F) is equivalent to a water content of 0.0002% by volume—that is, an oxygen partial pressure of about 10 mPa (1.5×10^{-6} psi). There is also the risk of explosion when dealing with hydrogen at high temperatures, and hydrogen can embrittle some materials. A more detailed treatment of reducing atmospheres and their use is given in Chapter 3, section 3.1.

1.3.2.6 Coatings Applied to Surfaces of Components

Occasionally, the desired joining alloy (chosen on the basis of melting temperature and physical properties) is metallurgically incompatible with the substrate in the sense that the filler either does not wet the substrate, will wet nonuniformly, or forms embrittling phases by reaction. A solution is to apply a surface coating that will promote wetting by the braze and reacts with it in a benign manner. For example, the native oxide on titanium prevents this metal's being wet by low-melting-point silver-base brazes in an inert atmosphere. However, a gold coating, applied by electroplating and diffused in by heat treatment, akin to that used in sheradizing (sherardizing), substitutes for the oxide and renders the titanium wettable. Stainless steels are often plated with nickel for similar reasons. As a general guide, plating of stainless steels and other heat-resistant alloys to facilitate brazing is generally recommended when the titanium content of the components exceeds 0.7%, the aluminum content exceeds 0.4%, or the combined aluminum and titanium contents exceed 0.7%. The nickel plating will need to be at least 10 μm (400 μin.) thick to prevent these species' diffusing through to the surface during the heating stage of a typical brazing cycle and also to accommodate the depth of alloying that usually takes place between a braze and a substrate. Coatings can be applied by a variety of techniques and to thicknesses that suit the particular application.

On substrates destined for brazing, it is normal to apply coatings that are required by wet-plating methods, which are quick, economical, and flexible with regard to the coating thickness. If the substrate is refractory in character, adhesion of metal coatings tends to be poor or impossible unless the braze contains a suitable activating ingredient; see Chapter 4, section 4.1.2.2. An alternative is to use a thick-film coating. A conventional thick film comprises metal particles embedded in a glass or ceramic matrix. On firing, segregation of the constituents occurs so as to leave a metal-rich surface that is wettable by molten brazes, while the nonmetallic part accumulates at the interface with the substrate and bonds to it in the manner of a glaze. These and other metallizations, and the principles on which they are designed, are described in Chapter 4, section 4.1.2.1.

1.3.2.7 Cleaning Treatments

The surfaces of the components to be joined and the filler metal preforms must be free from any nonmetallic films, such as organic residues and metal oxides, to enable the molten filler metal to wet and alloy with the underlying metal. Fluxes are often capable of removing surface oxides, provided they are reasonably thin.

Organic films can be removed with petroleum and chlorinated solvents, which obviously should not react with the underlying materials. Although organic films will burn to leave carbon at the brazing temperature, their presence can cause difficulties with flux application as these are usually water-base materials. Thick oxides and other nonmetallic surface layers can be removed mechanically and chemically. Chemical cleaners can be either alkaline or acidic and may employ electrolytic activation. Iron and stainless steel are best cleaned chemically and mechanically to ensure complete removal of organic matter and thin native oxides as far as is practicable. A wide variety of chemical agents, including sulphuric, nitric and hydrochloric acids, phosphate-based solutions, and salt baths, can be used. Removal of copper oxides can also be achieved mechanically and by pickling. The formulation of the pickle will vary with the alloy type, but most are based on sulfuric acid at a concentration of 5 to 15%. Nickel alloys are chemically cleaned prior to brazing using nitric acid, often containing sodium chloride and sulfuric acid in addition. Any residual etchant left after rinsing is neutralized by a dilute sodium hydroxide solution. Molybdenum and its alloys can be cleaned by inorganic etches as well as in molten salt baths such as 70% sodium hydroxide and 30% sodium nitrate at about 300 °C (570 °F). Light oxide films can be removed from tungsten by a hydrogen atmosphere at 1065 °C (1950 °F). For most metals, mechanical cleaning, followed by a light acid etch generally gives the best results. Carbides and ceramics do not require cleaning because they do not form "native" oxides, but they benefit from chemical treatments to remove residues from earlier processes, particularly embedded fine particles picked up from grinding, cutting, and polishing media that otherwise interfere with spreading of the braze.

1.3.2.8 Heat Treatments Prior to Joining

Prejoining heat treatments are occasionally useful in providing stress relief and thereby preventing unpredictable distortion during heating of the components to the bonding temperature. Brasses and bronzes have a tendency to form stress cracks when cold worked and then heated rapidly. Although any cracks in the vicinity of the joint will be filled by molten braze, the integrity of the assembly in such a case is likely to be unreliable. It is therefore normal to perform a stress relief by heat treatment at about 300 °C (570 °F) prior to brazing. In a controlled atmosphere process, this step can be integrated into the heating cycle of the brazing operation. Other situations where prejoining heat treatments can be beneficial include those involving components with metallic or nonmetallic surface films that are thermally unstable. In the case of silver, for example, the oxide will dissociate readily when heated above 190 °C (375 °F) in an ambient atmosphere. Likewise, silver sulfide dissociates on heating above 842 °C (1548 °F).

An additional note of caution needs to be sounded with regard to the brazing of stressed components. The joining operation can lead to brittle failure of stressed components through a mechanism known as *liquid metal embrittlement.* For example, stainless steels are exceptionally prone to embrittlement by copper and high-copper brazing alloys [Heiple, Bennette, and Rising 1982].

1.3.2.9 Heating Cycle of the Joining Operation

The prepared components and filler metal, possibly mounted in jigs, are joined by applying heat. The heating cycle involves four important processing parameters: the heating rate, the peak bonding temperature, the holding time above the melting point of the filler, and the cooling rate.

In general, it is desirable to use a fast heating rate to limit reactions that can occur below the prescribed bonding temperature. However, the maximum heating rate is normally constrained by adverse temperature gradients developing in the assembly. These gradients can produce distortions in the components and give rise to nonuniform reactions between the filler and the pair of joint surfaces. Also, temperatures are difficult to measure reliably during fast heating schedules. A better practice, especially when joining in a vacuum or controlled-atmosphere furnace, is to heat the assembly rapidly to a preset temperature that is just below the melting point of the filler metal and then hold it at this temperature for sufficient time (which can range from a

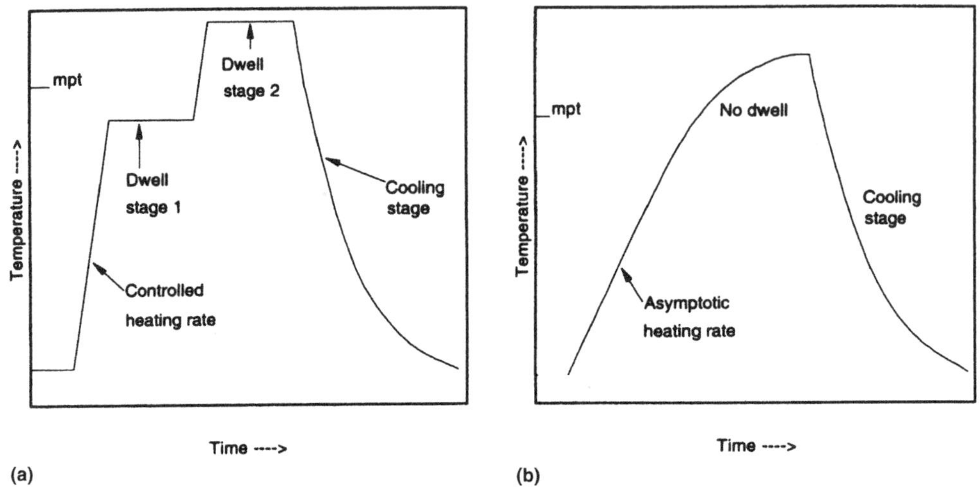

Fig. 1.30 Profiles of typical temperature cycles. (a) Heating cycle with a controlled profile incorporating dwell stages to reduce thermal gradients. (b) Heating cycle defined solely by attainment of a peak temperature

few seconds to over one hour, depending on the size of the assembly and the heating method) to allow the assembly to thermally equilibrate and for water vapor to flush out of the joint. Following this dwell, the assembly may then be rapidly heated to the bonding temperature. Profiles of typical temperature cycles are shown schematically in Fig. 1.30.

The joining temperature should be such that the filler is guaranteed to melt but at the same time should not be so high that the filler degrades through the loss of constituents or by reaction with the furnace atmosphere. The optimum temperature is normally determined by metallurgical criteria, most importantly, the nature and extent of the filler-substrate interaction. The peak process temperature is frequently set at about 50 to 100 °C (90–180 °F) above the melting point of the braze because accurate temperature measurement and control is not always possible, especially in reduced-pressure atmospheres, where bulky jigging is used or where conduction between the heat source and the workpiece is poor. Moreover, the reported melting temperatures of some filler metals are not always based on accurate measurements, and it is prudent to make some allowance for this uncertainty. The minimum time that the assembly is held above the melting point must be sufficient to ensure that the filler has melted over the entire area of the joint, and the maximum time is usually a compromise based on practical and metallurgical considerations. Extended dwell times tend to result in excessive spreading by the molten filler and reaction with the substrate, oxidation gradually taking place (depending on the nobility of metallic workpieces), and deterioration of the properties of the parent materials.

The variables of heating rate and peak process temperature combine to give a window for maximum spread by the filler metal that depends on the filler/substrate combination, volume of the filler metal, and process atmosphere or flux. Isospread contours for silver-copper braze on nickel are shown in Fig. 1.31. From this it can be seen that the conditions for maximum spreading are

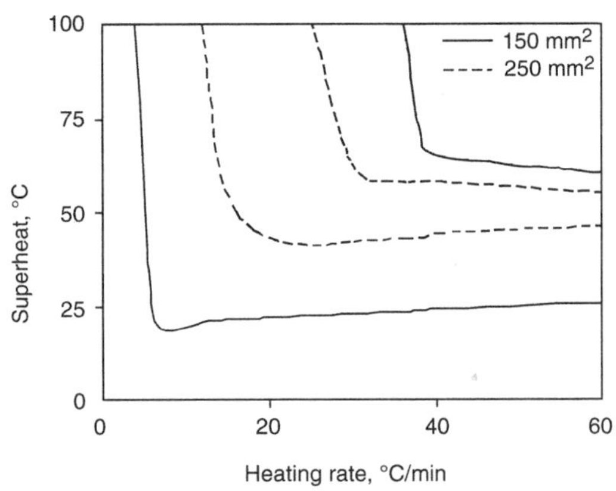

Fig. 1.31 Combined effects of superheat and heating rate on the spreading of 0.5 mg spheres of Ag-29Cu braze on nickel substrates in a nitrogen-10% hydrogen atmosphere. Area of isospread contours in mm^2 are shown. Adapted from Weirauch and Krafick [1996]

a peak temperature of 830 °C (1526 °F), reached at a rate of 25 °C/min (45 °F/min).

The cooling stage of the cycle is seldom controlled by the operator but tends to be governed by the thermal mass of the assembly and jig. Forced cooling can lead to problems such as exacerbating mismatch stresses. Occasionally, one or more dwell stages are required, either to provide stress relief to the bonded assembly or to induce some requisite microstructural change. An example of the latter is joining of tool steels with copper. The brazing operation is performed at 1120 °C (2050 °F) for a few minutes, but the assembly is cooled to only approximately 850 °C (1560 °F). This temperature is maintained for about an hour in order to convert the base material to the austenitic phase. The component is then quench-cooled in oil and finally tempered at about 500 °C (930 °F) to achieve the desired hardness. Another example is the solution treatment, quench and temper, steps necessary to endow copper-beryllium alloys with their "springlike" properties. In this process the components are solution treated at 760 to 790 °C (1400 to 1455 °F), water quenched, and then aged at 315 to 345 °C (600 to 655 °F).

A heat treatment temperature of about 75% of the melting point (solidus temperature) of the filler metal in Kelvin usually provides the optimal relief of residual stresses. For many steel components, controlled cooling is vital in order to achieve the desired microstructure and hence the correct temper. Stainless steels are susceptible to surface cracking if a martensitic transition is allowed to occur, owing to the associated volumetric strain. This phase change can be prevented by performing an isothermal anneal above the transition temperature; thereafter, the component should be cooled sufficiently slowly to prevent cracks from developing.

There are also situations in which a particular bonding cycle is required to dovetail with other thermal processes that are required in the fabrication of the assembly, for example spheroidization of the graphite particles in cast iron, to give it a steellike strength and ductility. In such instances, the economics of production of the complete assembly often need to be taken into account when specifying the parameters of the process cycle.

1.3.2.10 Post-Joining Treatments

Various types of post-joining treatments can be applied, by far the most common of which is some form of cleaning operation. A cleaning schedule is used to remove flux residues and tarnishing from the components. Flux residues tend to be corrosive, particularly those containing halide species as active ingredients. These become reactive in moist air and they can affect the long-term reliability of components in service. Both chemical and mechanical means of flux residue removal are employed. Most brazing flux residues are soluble in hot water to a greater or lesser extent, particularly if the solution is mildly acidic. If the assembly can withstand quench cooling, the thermal shock helps crack and craze the solidified glassy flux residue, facilitating its removal. Ultrasonic agitation can also be used. If the brazing operation is conducted in air, components are often disfigured with oxide bloom. This is usually removed by an acid pickle and/or mechanical polishing. Stop-off compounds mostly have to be removed by mechanical means.

There are two alternatives to cleaning: use a fluxless process, or simply not clean at all.

At face value, fluxless processes are highly attractive because no flux is present and therefore no cleaning of residues is required. However, fluxless processes are more difficult to implement and tend to be incompatible with low cost, volume manufacturing. In order to obtain satisfactory wetting and spreading, the process atmosphere needs to be depleted of oxygen and water vapor to levels that can be obtained only in closed vessels. Details on fluxless brazing are given in Chapter 3, section 3.4.

The ultimate alternative to cleaning is simply not to do it. Cleaning adds to the capital and consumable process cost and results in an assembly yield loss. Provided the flux residues do not adversely impact the reliability of the product, then the merits of cleaning are questionable. Aluminum heat exchangers are manufactured by the tens of thousand per annum using a process that uses a flux with residues that are stable under normal environmental conditions for the typical life of such products. However, the presence of halides in many brazing fluxes presents a significant corrosion risk because these compounds are aggressive toward most base metals, especially in the presence of moisture. It is possible to manage this risk, to a degree, by sealing the joint area with an impervious lacquer or paint. Obviously for life- or mission-critical products, cleaning will probably always be undertaken, but in many instances, accepting a compromise of a finite life may be considered a technically and economically sound approach.

1.3.2.11 Statistical Process Control

All processes are subject to variation, and achieving stability of processes is an important step in any quality improvement program. The application of statistics to monitor and control the variability is termed *statistical process control* (SPC) [Ledolter and Burrill 1999]. Many industrial brazing processes are subject to SPC. A modern jewelery chain assembly line can achieve joint defect rates of a few parts per million (ppm). This means that, quite literally, only one or two brazed joints in every million made would fail a quality inspection. This degree of manufacturing consistency is achieved only through SPC.

A fundamental tool in SPC is a graphical display, known as a control chart. This chart provides the basis for deciding whether the variation in the output of a process is due to common, randomly occurring variations or to unusual causes, which requires investigation and action. The control chart is a chronological plot of particular characteristics, such as joint strength or peak reflow temperature, sampled at periodic intervals. This information furnishes data on the process stability and provides an understanding of improvements, where made. Whenever a significant deviation from the norm is detected, a decision can be made to adjust a process variable in order to restore the output to the required quality level. Obviously, to accomplish this, there must be a proper understanding of the relationship between the process variables and the output.

There are different types of control charts, designed for different situations, which are classified by the type of data they contain. Control charts designed to monitor the proportion of defective items are referred to as p-charts, while charts that track the number of defects in the product are known as c-charts. Both are used to describe attribute data, that is, a record of the presence and absence of certain characteristics. Quantitative data are monitored using a mean or x-bar chart, while process variability is measured using range charts (r-chart) and standard deviation charts (s-chart).

The basis of SPC is, for each parameter tracked, to select upper- and lower-control limits (see Fig. 1.32). These limits are set at a multiple number of standard deviations from the mean such that there will be a high probability that the data will fall between these limits when the process is working as desired. Intervention is required only when the process metrics drift further from the acceptable average.

It is possible to apply SPC to virtually any process or machine output. In many instances this methodology provides a highly effective basis for controlling manufacturing processes. As with any tool, it is necessary to use some discretion and critical thought to ensure that SPC is appropriate and the cost of implementing and sustaining it is justified.

1.3.3 Health, Safety, and Environmental Aspects of Brazing

Brazing encompasses the use of a large number of different materials, covering metallic and nonmetallic elements for the fillers and the parent materials, and organic and inorganic chemicals used in fluxes, controlled atmospheres, and for removing flux residues. Several of these materials are hazardous in varying degrees to the operators, or to the environment [Sax and Lewis 1989]. Accordingly, they must be handled, used, and disposed of as appropriate, according to national codes of practices or regulations governing hazardous substances. Official listings produced by national health and safety authorities classify materials according to their toxicity level, e.g., the exposure limits for hazardous vapors and dusts.

The main problem with brazes and brazing fluxes arises when they are heated to make a joint in air. The fume contains a cocktail of gases that can cause eye and nose irritation, dermatitis, asthma, and respiratory problems. The fume contains fine particles, in the range 0.1 to 1 μm (4–40 μin.), which is the most dangerous size distribution for causing long-term lung damage. The recommended solution is to ensure that the workplace or work chamber is ventilated using an appropriately designed extraction system that is able to exhaust the gases and trap the particulates [Jakeway 1994]. Preventing exposure to the hazard by appropriate measures should always be given higher priority than protective measures.

All materials that are likely to be encountered in a joining context will have an assigned value of maximum exposure limit, usually in weight per unit volume (normally in mg/m^3). Examples

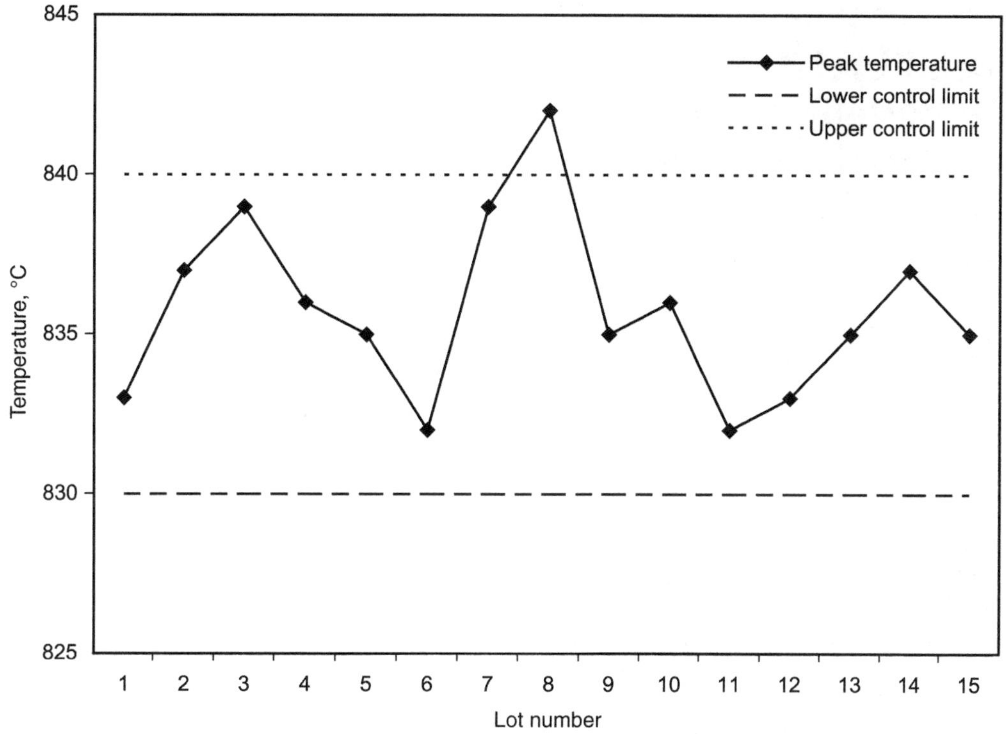

Fig. 1.32 Statistical process control (SPC) chart for peak process temperature measured at a test point on a component, showing the upper- and lower-control (i.e., intervention) points for this process

of occupational exposure limits for metals commonly used in brazing alloys, constituents of brazing fluxes, and fumes from brazing torches are given in Table 1.6. The difference between the long- and short-term exposure limits reflects the capability of the average constitution to excrete ingested material. The form of the material is also relevant. Powders and dusts are more hazardous than nonvolatile liquids and monolithic solids. Both are ranked according to the maximum inhalable quantity in mg/m^3, time weighted over a period of time, either short term, meaning minutes, or over a longer period of many hours. For correct interpretation of the rules, regulations and audits, reference should always be made to a qualified safety practitioner because there are often legal aspects to consider also.

Care must be taken also in the storage of materials both prior to use and in the procedures for the subsequent disposal of residues, exhaust emissions, and other associated effluent, such as solutions containing rinsed fluxes. These are usually subject to statutory controls. For many organic chemicals and gases, in this context binders used in flux pastes and halogenated gases, there may also be fire risks to consider. The flammability is rated according to flashpoint, which is the lowest temperature at which the substance can spontaneously ignite when it is in a saturated condition.

Table 1.6 Guidance occupational exposure limits for metals commonly used in brazing alloys, constituents of brazing fluxes, and fumes from brazing torches

Element or compound	8 h exposure limit, mg/m^3	10 min exposure limit, mg/m^3
Cadmium oxide fume	0.05	0.05
Copper fume	0.2	. . .
Manganese fume	1.0	3.0
Nickel	0.5	. . .
Silver	0.1	. . .
Zinc oxide fume	5.0	10.0
Hydrogen fluoride fume	. . .	2.5
Fluorine	2.5	. . .
Boron trifluoride	. . .	3.0
Nitric oxide	30	45
Nitrogen dioxide	5	9

The user should consult suppliers' materials safety data sheets (MSDS) and national health and safety standards. Ellipses indicate that there is no established limit. Values are time-weighted averages.

Appendix A1.1 Relationships Among Spread Ratio, Spread Factor, and Contact Angle of Droplets

Expressions describing the spread of a molten metal droplet are derived under the following set of idealized conditions:

- The original metal pellet is in the form of a spherical bead of radius a (and diam $D = 2a$).
- The droplet resolidifies after spreading on the substrate as a spherical cap of radius R and height h, its interface with the substrate having a diameter $2A$, as shown in Fig. 1.33.
- The volume of the original pellet is equal to the volume of the resolidified droplet. This means that any volatilization of the molten droplet and reaction with the substrate do not affect its volume measurably.

The volume of the spherical cap is

$$V = \frac{1}{6}\pi h(h^2 + 3A^2)$$

Spread Ratio and Contact Angle

The spread ratio, S_r is defined as:

$$S_r = \frac{\text{Plan area of spread on the substrate surface}}{\text{Plan area of the original spherical pellet}}$$

A and a are related by the conservation of the volume of the droplet, that is:

$$V = \frac{4}{3}\pi a^3 = \frac{1}{6}\pi h(h^2 + 3A^2)$$

Therefore:

$$a = \frac{1}{2}[h(h^2 + 3A^2)]^{1/3}$$

and

$$S_r = \frac{4A^2}{[h(h^2 + 3A^2)]^{2/3}}$$

From the geometry (Fig. 1.33), $A = R \cdot \sin\theta$ and $h = R \cdot (1 - \cos\theta)$:

$$S_r = \frac{4A^2/h^2}{(1 + 3A^2/h^2)^{2/3}} = \frac{4\cot^2/\theta/2}{(1 + 3\cot^2\theta/2)^{2/3}}$$

for $0° < \theta < 180°$

Spread Factor and Contact Angle

The spread factor is defined by the formula:

$$S_f = \frac{D - h}{D} = \frac{(h^3 + 3A^2h)^{1/3} - h}{(h^3 + 3A^2h)^{1/3}}$$
$$= 1 - \frac{1}{(1 + 3A^2/h^2)^{1/3}}$$
$$= 1 - \frac{1}{(1 + 3\cot^2\theta/2)^{1/3}}$$

for $0° < \theta < 180°$

Contact Angle and the Dimensions of the Solidified Pool of Filler

From Fig. 1.33 it can be seen that:

$$A^2 + (R - h)^2 = R^2$$

according to the Pythagorean Theorem.

Rearranging this equation:

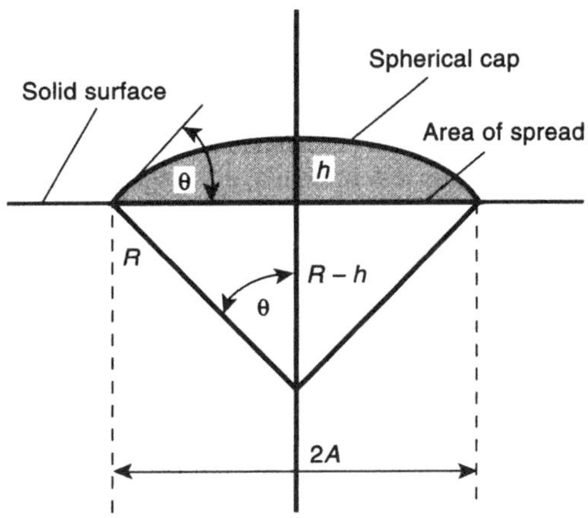

Fig. 1.33 Spherical cap geometry

$$R = \frac{A^2 + h^2}{2h}$$

Therefore:

$$\sin\theta = \frac{2}{(A/h) + (h/A)}$$

REFERENCES

- Ambrose, J.C., Nicholas, M.G., and Stoneham, A.M., 1992. Kinetics of Braze Spreading, *Proc. Conf. British Association for Brazing and Soldering,* 1992 Autumn Conference, Coventry, U.K.
- Andrade, E.N.C., 1952. The Viscosity of Liquids, *Proceedings of the Royal Society,* A, Vol 215, p 36–356
- Asthana, R. and Sobczak, N., 2000. Wettability, Spreading and Interfacial Phenomena in High Temperature Coatings, *J. Mater. Sci.,* Vol 52 (No. 1), http://www.tms.org/pubs/journals/JOM/0001/Asthana/Asthana-0001-text.html
- Bakulin, S.S., Shorshorov, M.Kh., and Shapiro, A.E., 1992. A Thermodynamic Approach to Optimising the Width of the Brazing Gap and the Amount of Brazing Alloy, *Weld. Int.,* Vol 6 (No. 6), p 473–475
- Bever, M.B., Ed., 1986. *Encyclopedia of Materials Science and Engineering,* Pergamon Press
- Brakke, K.A., 2003. "Surface Evolver," Version 2.20, Susquehanna University, PA, http://www.susqu.edu/brakke
- de Gennes, P.G., 1985. Wetting: Statistics and Dynamics, *Reviews of Modern Physics,* Vol 57 (No. 3), p 827–863
- Elmer, J.W. and Kautz, J.D., 1993. Fundamentals of Friction Welding, *ASM Handbook,* Vol 6, *Welding, Brazing, and Soldering,* ASM International
- Eustathopoulos, N. and Coudurier, L., 1979. Wettability and Thermodynamic Properties of Interfaces in Metallic Systems, *Proc. Conf. British Association for Brazing and Soldering,* Third International Conference (London, U.K.), Paper 5
- Feature, 1976. Diffusion Bonding—Tomorrow's Low-Cost Fabrication Tool, *Met. Mater.,* Feb, p 37–39
- Feduska, W., 1959. High-Temperature Brazing Alloy—Base Metal Wetting Reactions, *Weld. J. Res. Suppl.,* Vol 38 (No. 3), p 122s–130s
- Fleetwood, M.J. et al., 1988. Control of Thin Strip Casting, *Proceedings of the Second International Conference on Rapidly Solidified Materials,* March 7–9 (San Diego, CA)
- Hawk, J.A. et al., 1993. Self-Propagating, High Temperature Synthesis as a Technique to Join Metals, *Materials Research Society Symposium Proceedings,* Vol 314, p 183–196
- Heiple, C., Bennette, W., and Rising, I., 1982. Embrittlement of Several Stainless Steels by Liquid Copper and Liquid Braze Alloys, *Mater. Sci. Eng.,* Vol 52, p 277–289
- Hill, A. and Wallach, E.R., 1989. Modeling Solid State Diffusion Bonding, *Acta Mater. Metall.,* Vol 37 (No. 9), p 2425–2437
- Howe, J.M., 1993. Bonding, Structure and Properties of Metal/Ceramic Interfaces: Part 1 Chemical Bonding, Chemical Reaction and Interfacial Structure, *Int. Mater. Rev.,* Vol 38 (No. 5), p 233–256
- Humpston, G. et al., 1992. Recent Developments in Silicon/Heat-Sink Assemblies for High Power Device Applications, *GEC Review,* Vol 7 (No. 2), p 62–78
- Jakeway, P., 1994. The Healthy Solderer, *The Health and Safety Practitioner,* April, p 20–23
- Jarman, R.A., Linekar, G.A.B., and Booker, C.J.L., 1975. Interfacial Corrosion of Brazed Stainless Steel Joints in Domestic Tap Water, *Brit. Corros. J.,* Vol 8, p 33–37; Vol 10, p 150–154
- Jones, H., 1982. "Rapid Solidification of Metals and Alloys," Monograph No. 8, The Institution of Metallurgists
- Karavolis, M., 1993. Self-Brazing Metals Open up Design Options, *Weld. Des. Fabr.,* Vol 66 (No. 12), p 36–37
- Keller, D.L. et al., 1990. Wettability of Brazing Filler Metals, *Weld. J. Res. Suppl.,* Vol 69 (No. 10), p 31s–34s
- Kritsalis, P., Coudurier, L., and Eustathopoulos, N., 1991. Contribution to the Study of Reactive Wetting in the CuTi/Al_2O_3 System, *J. Mater. Sci.,* Vol 26, p 3400–3408
- Landry, K., Rado, C., and Eustathopoulos, N., 1996. Influence of Interfacial Reaction Rates on the Wetting and Driving Force in Metal/Ceramic Systems, *Metall. Mater. Trans. A,* Vol 27A, p 3181–3186
- Laurent, V., Chatain, D., and Eustathopoulos, N., 1991. Wettability of SiO_2 and Oxidized SiC by Aluminum, *Mater. Sci. Eng. A,* A135, p 89–94

- Ledolter, J. and Burrill, C.W., 1999. *Statistical Quality Control—Strategies and Tools for Continual Improvement,* John Wiley and Sons
- Lee, A.S. and Norris, P.M., 1997. A New Optical Method for Measuring Surface Temperature at Large Incident Probe Angles, *Rev. Sci. Instr.,* Vol 68, p 1307–1311
- Levi, G. and Kaplan, W.D., 2003a. Aluminium-Alumina Interface Morphology and Thermodynamics from Dewetting Experiments, *Acta Mater.,* Vol 51 p 2793–2802
- Levi, G. and Kaplan, W.D., 2003b. Adhesion and Microstructure at Metal-Ceramic Interfaces, *Proc. Conf. International Symposium on Microelectronics and Packaging,* June 18, Herzliya, Israel
- Li, J.G. and Hausner, H., 1991. Wettability of Silicon Carbide by Gold, Germanium and Silicon, *J. Mater. Sci. Let.,* Vol 10, p 1275–1276
- McDonald, M.M. et al., 1989. Wettability of Brazing Filler Metals on Molybdenum and TZM, *Weld. J. Res. Suppl.,* Vol 70 (No. 10), p 398s–395s
- Meier, A., Javernick, D.A., and Edwards, G.R., 1999. Ceramic-Metal Interfaces and the Spreading of Reactive Liquids, *JOM* (No. 1), p 44–47
- Messler, R.W., 1999. *Principles of Welding: Processes, Physics, Chemistry and Metallurgy,* John Wiley and Sons
- Milner, R.D., 1958. A Survey of the Scientific Principles Related to Wetting and Spreading, *Brit. Weld. J.,* Vol 5, p 90–105
- Nicholas, M.G., 1989. Metal Surface Energies and Capillarity, Pub. 466, *Physical and Electrical Characterisation of Metals,* M. McLean, Ed., The Institute of Metals, London, U.K., p 177–227
- Nicholas, M.G., 1998. *Joining Processes: Introduction to Brazing and Diffusion Bonding,* Kluwer Academic
- Nicholas, M.G. and Crispin, R.M., 1986. Some Effects of Anisotropic Roughening on the Wetting of Metal Surfaces, *J. Mater. Sci.,* Vol 21, p 522–528
- O'Brien, R.L., Ed., 1991. *Welding Handbook,* Vol 2, *Welding Processes* (8th ed.), American Welding Society
- Okamoto, I., Takemoto, T., and Den, K., 1976. Vacuum Brazing of Aluminium Using Al-12%Si System Filler Alloy, *Trans. Jap. Weld. Res. Inst.,* Vol 5 (No. 1), p 97–98
- Otooni, M.A., Ed., 1995. *Science and Technology of Rapid Solidification and Processing,* Dordrecht: Kluwer Academic
- Sax, N.I. and Lewis, Sr., R.J., 1989. *Dangerous Properties of Industrial Materials,* 7th ed., Van Nostrand Reinhold
- Schwartz, M.M., 1990. *Ceramic Joining,* ASM International
- Schwartz, M.M., 2003. *Brazing,* 2nd ed., ASM International
- Seah, K.H.W., Hemanth, J., and Sharma, S.C., 1998. Effect of the Cooling Rate on the Dendrite Arm Spacing and the Ultimate Tensile Strength of Cast Iron, *J. Mater. Sci.,* Vol 33, p 23–28
- Sloboda, M.H., 1961. Design and Strength of Brazed Joints, *Weld. Met. Fabr.,* Vol 7, p 291–296
- Srivatsan, T.S. and Sudarshan, S., 1993. *Rapid Solidification Technology: An Engineering Guide,* Technomic
- Tunca, N., Delamore, G.W., and Smith, R.W., 1990. Corrosion of Mo, Nb, Cr, and Y in Molten Aluminium, *Metall. Trans. A,* Vol 21A (No. 11), p 2919–2928
- Tylecote, R.F., 1967. Diffusion Bonding: Part 1, *Weld. Met. Fabr.,* Vol 35 (No. 12), p 483–489
- Tylecote, R.F., 1968. *The Solid Phase Welding of Metals,* Edward Arnold
- Wang, X.H. and Conrad, H., 1995. Kinetics of Wetting Ag and Cu Substrates by Molten 60Sn40Pb, *Metall. Mater. Trans.,* Vol 26A, p 459–469
- Weeks, J.R. and Gurinsky, D.H., 1958. *Liquid Metals and Solidification,* American Society for Metals
- Weirauch, Jr., D.A. and Krafick, W.J., 1996. The Spreading Kinetics of Ag-28Cu on Nickel: Part 1, Area of Spread Tests on Nickel Foil, *J. Mater. Res.,* Vol 11 (No. 8), p 1897–1916
- Xian, A., 2000. Thermodynamic Discussion on Young's Equation in Wetting, *Zeitschrift für Metall.,* Vol 91 (No. 4), p 316–322
- Yost, F.G. and Romig, A.D., 1988. Thermodynamics of Wetting by Liquid Metals, *Materials Research Society Symposium Proceedings,* Vol 108, p 385–390
- Yunchen, Y., Xiaoming, Q., and Hongying, W., 1992. Research on the Welding Behaviour of the Liquid Phase of Eutectic Reaction, *China Weld.,* Vol 1 (No. 1), p 75–82
- Zvolinskii, I.V. et al., 1995. Special Features of Brazewelding Metal Matrix Composites, *Weld. Int.,* Vol 9 (No. 1), p 41–43

CHAPTER 2

Brazes and Their Metallurgy

THIS CHAPTER presents an overview of families of brazing alloys that one is likely to encounter in a manufacturing environment. Necessarily, the survey must include consideration of the parent material with which the filler is used because the suitability of a braze for a particular joining process will depend largely on its compatibility with the parent (or base) materials.

Extensive reference is made to phase diagrams in order to highlight particular points. An introduction to alloy constitution and phase diagrams designed for those with little background in the subject is presented in section 2.3, which covers interpretation of phase diagrams and the associated terminology. Methods for their determination are summarized in the literature [Humpston and Jacobson 1993, Chapter 3].

For a braze to be compatible with a particular parent material, it must exhibit these characteristics:

- It must have a liquidus temperature below the melting point (solidus temperature) of the parent materials. Usually, a margin is required between these two temperatures in order to achieve adequate fluidity of the molten filler. In many cases, the fluidity of brazes is not a strong function of temperature, but the overall flow behavior, which is of immediate practical interest, often does depend substantially on the joining temperature and the rate of heating to this temperature.
- It must be capable of producing joints at temperatures at which the properties of the base materials are not degraded. For example, many work-hardened and precipitation-hardened alloys cannot withstand elevated temperatures without loss of their beneficial mechanical properties. The first type of hardening involves subjecting the alloy to mechanical deformation such as rolling or hammering when reasonably cold. As the temperature is raised, the deformation damage is removed by atomic rearrangement in the metal and the associated removal of dislocations, which are the normal effects of the annealing process. Precipitation hardening is accomplished by creating a finely divided phase within the material, which can be thought of as akin to a composite material. The dispersed phase is precipitated by means of an appropriate heating schedule and its strengthening effect is likewise degraded by high temperatures.
- It must wet the parent materials or a metallization applied to the parent materials in order to ensure good adhesion through the formation of metallic bonds. As explained in the preceding chapter, it is not possible to obtain spreading without wetting, but an absence of spreading does not automatically imply a lack of wetting. For example, silver brazes, such as the 44Ag-30Cu-26Zn alloy, exhibit greatly superior spreading on mild steels than do silver-free brazes, such as 54Cu-35Zn-6Ni-4Mn-1Si, but the wetting and adhesion to the steel parts are comparable, as indicated by the closely similar strengths of the brazed joints [Jacobson et al. 2002].
- It must not excessively erode the parent metals at the joint interface. The associated alloying reactions, which must occur to form a metallic bond, should not result in the formation of either a large proportion of brittle phases within the joint or of significant concentrations of brittle phases along interfaces or other critical regions of the joint. Even ductile phases can have weak interfaces with solidified filler alloys. Similarly, the products of interalloying must not generate other forms of weakness such as voids in the joint or a susceptibility to corrosion.

- It must not contain constituents or impurities that might embrittle or otherwise weaken the resulting joint. Likewise, the parent material must not contribute constituents or impurities to the braze that will have a similar effect.

Besides being compatible with the parent material, the brazing alloy and the joining process used must be mutually suited. For example, brazes containing zinc and other volatile constituents are not usually appropriate for furnace brazing operations, especially when these involve reduced pressures. One of the few applications for zinc, as a filler metal, is for joining beryllium and its alloys by dip brazing. Zinc has a sufficiently low melting point to make it possible to circumvent problems of its volatility. Zinc wets beryllium well and does not undergo any unfavorable metallurgical reactions with it. Beryllium forms intermetallic compounds with the higher atomic weight metals that are major constituents of most brazing alloys, silver being an exception. But, the high temperatures required for silver-base brazes cause recrystallization of the beryllium and degradation of mechanical properties.

The degree of temperature uniformity that can be achieved over the joint area will have an influence in determining the minimum temperature difference that can be tolerated between the melting temperature of the filler and the melting (or degradation) temperature of the parent material. This consideration is particularly relevant to the joining of aluminum alloy components at about 600 °C (1100 °F), with the well-known Al-12Si eutectic braze that melts at hardly more than 20 °C (35 °F) below this temperature. Moreover, aluminum possesses a low thermal heat capacity, and stainless steels have a relatively low thermal conductivity, which make these materials fairly difficult to join using local heat sources that do not envelop the workpiece.

The condition of the surface of the parent material may affect its compatibility with brazes, especially when fluxes are not used. As an obvious example, filler metals will less readily wet an oxidized surface than a freshly cleaned metal surface. This consideration often determines the acceptable shelf life of components prior to joining. The term "atomically clean" should not generally be used to describe cleaned surfaces. It is an abstract ideal and is not normally relevant in practical situations.

In order to establish whether a particular parent metal (or nonmetal) is compatible with a given braze, it is necessary for the appraisal to be carried out under conditions that are likely to be representative of those used in any practical implementation of the joining process. Parameters such as process time, temperature, heating rate, and atmosphere can be critical in this regard. Storage shelf life of the components (and the filler, especially when in the form of a braze paste) is another relevant factor that needs to be taken into account but which is often neglected during transfer of a process from the laboratory to the factory.

The properties of the braze and the joint that it is used to make must also be compatible with the service requirements. These are likely to involve a combination of at least some of the following considerations:

- The strength and ductility of the joint should meet certain minimum requirements over the range of service temperatures. Brazed joints are primarily used in applications where mechanical strength is an important criterion.
- The design of the joint should not introduce stress concentrations in the assembly that might arise through solidification shrinkage or formation of intermetallic phases. Likewise, the design should not result in undue distortion of the assembly through excessive thermal expansion mismatch strains or the filler metal being too unyielding.
- The joint is normally required to be resilient to the service environment, in terms of corrosion and oxidation resistance and compatibility with vacuum, in accordance with functional requirements. Zinc-containing brazes are particularly restrictive in this regard.
- The filler must be compliant with statutory needs. These include hallmarking regulations for precious metals, health restrictions on lead and cadmium for certain culinary and medical applications, and nickel in applications that involve prolonged skin contact, such as spectacle frames.
- Aesthetic requirements are often important, for example, color, and color matching in jewelery and utensils. The ability of joints to accept surface finishes such as paints and electroplatings may also be critical. Good fillet formation is often demanded for aesthetic reasons and also as a criterion of acceptable joint quality. The reader is cautioned that the inferences drawn about joint quality from external appearance can be misleading, particularly in connection with large area joints.

- Joints are sometimes required to possess certain thermal and electrical properties. In electrical and optical assemblies, conformance of these characteristics normally determines their fitness for purpose.

The simultaneous attainment of several of these desired characteristics is frequently achievable with common filler metals, provided that basic design guidelines and process conditions are satisfied. Silver- and nickel-base brazes account for a significant proportion of brazed joints in industrial production. It is the more exceptional and demanding service requirements for specific applications that have given rise to the development of the hundreds of additional filler metal compositions. However, the commercially available fillers and matching fluxes have been designed so that, when used in conjunction with the common engineering parent materials, they meet many, if not all, of the listed requirements.

2.1 Survey of Brazing Alloy Systems

There are literally hundreds of different brazing alloy compositions available on the market. After all, in its broadest definition, a braze is any combination of metals that produces a liquid phase suitable for a joining operation below the melting temperature of a different combination of elements that constitute the parent material.

The survey will begin with consideration of silver-base brazing alloys. These refer mostly to those based on the silver-copper-zinc alloy system, and they provide a particularly instructive example of how fillers can be progressively tailored to suit particular requirements through modifications to the composition. Silver and copper generally constitute 60% by weight or more of these brazing alloys and both elements are occasionally used on their own as filler metals, or together in the eutectic alloy. Our survey of this alloy system, therefore, starts with these two elements.

The remaining brazes fall into a relatively small number of alloy families, defined according to the major metal constituent (see Chapter 1, Fig. 1.3). Only a few brazes are based on eutectic systems. More commonly, the constituent elements form solid solutions, enabling a large number of brazes with different compositions to be developed from each alloy family. Within each alloy family, the differences in composition usually reflect only slight variations in user requirements, as is shown later. The brazing alloys are treated on the basis of this classification, with attention placed on delineating the principal features of each alloy system. For a more detailed coverage of the available alloys, the reader should consult reference publications [e.g., Schwartz 2003] and National Standards documents, as well as data sheets and manuals supplied by the manufacturers and suppliers of brazing alloys, which are readily accessible on the Internet.

2.1.1 *Pure Silver*

Pure silver (melting point = 962 °C, or 1764 °F) is seldom used as a braze, owing to its relatively high cost, although this metal is perfectly satisfactory in most other respects. It possesses reasonable mechanical properties; is metallurgically compatible with most parent metals; has a low vapor pressure; does not oxidize; and exhibits excellent fluidity when heated above its melting point, enabling very narrow and strong joints to be formed with steel components (see Chapter 1, section 1.2.7 and Chapter 4, section 4.3.3.1). Silver does not alloy significantly with iron, but where alloying with parent materials does occur, the mechanical properties of joints made with this noble metal will generally benefit from the alloying.

Pure silver is used mostly as a braze for bonding reactive (or active) metals such as beryllium and titanium. Silver is chosen for this application because its oxide is not stable above 190 °C (375 °F), even in air. Hence, molten silver will usually form a metallic bond to these chemically active metals. Where less noble metals, including copper, are used as the filler metal, there is a tendency for a layer of oxide to remain at the joint interface with beryllium and other refractory, to the detriment of mechanical properties, unless an active metal, such as titanium, or a rare earth, is also present in the braze (see Chapter 7.2).

2.1.2 *Pure Copper*

Copper (melting point = 1085 °C, or 1985 °F) is appreciably cheaper than silver, but this advantage is offset by the higher process temperatures required and, more particularly, by the need to pay closer attention to quality of the atmosphere in order to prevent excessive oxidation of the braze and the associated components.

The largest use of pure copper as a braze is for fluxless joining of mild steel in vacuum and reducing atmospheres. This combination of materials is almost ideal from a metallurgical point of view. As with silver (see previous section), molten copper wets steel extremely well, is very fluid, and produces negligible substrate erosion. Therefore, strong joints can be made between components that have an interference fit even when the copper filler is required to flow relatively long distances (several cm, or an inch or so).

However, the benefits accruing from the lack of extensive alloying with mild steel mean that the joints effectively comprise pure copper, which is not a particularly strong, hard, or fatigue-resistant metal. These characteristics can be beneficial from the point of view of affording stress relief for joined components possessing different coefficients of thermal expansion (CTE). However, where the joint is a loose fit, and the gaps are relatively large, the strength of the joint is substantially determined by the properties of the filler, and minor additions of elements such as nickel, chromium or cobalt (in the range 0.5 to 5%) are often made to improve the mechanical properties of the braze, through solid solution and precipitation strengthening. This is not necessary if the parent metal is stainless steel, which contains one or more of the elements, chromium, nickel or cobalt, because the wetting by molten copper will tend to incorporate these elements into the braze. Note that not all types of stainless steels are compatible with copper brazes because a concentration of certain alloying additions at grain boundaries in the steel can lead to excessive attack by the braze in these regions.

Another modification that can be made to improve the characteristics of pure copper as a braze is the addition of either boron with nickel to produce an alloy (Cu-2B-1Ni) melting in the range 1080 to 1100 °C (1975–2010 °F), or phosphorus (Cu-8P, melting at 714 °C, or 1317 °F). The metalloid elements boron and phosphorus depress the melting point of copper, through the formation of copper-rich eutectics, as can be seen from the copper-phosphorus phase diagram (Fig. 2.1). They also tend to flux surface oxides, as described in Chapter 3, section 3.3, and are often referred to as *self-fluxing brazes.*

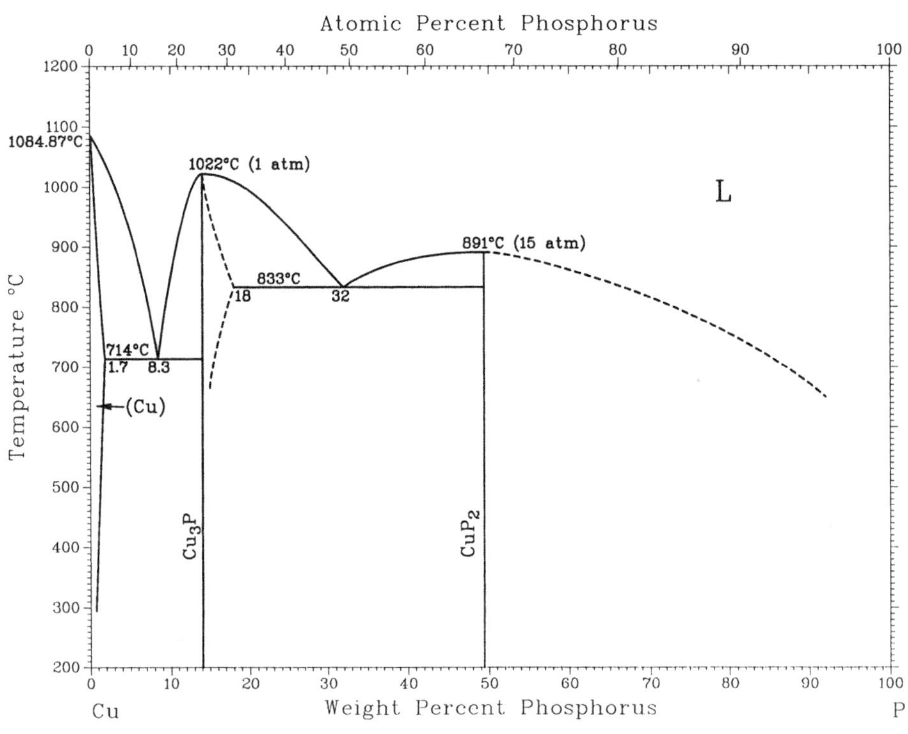

Fig. 2.1 Copper-phosphorus phase diagram

A number of joining processes use thin copper metallizations and rely on the generation of a molten filler, at temperatures below the melting point of copper, through alloying with the parent metal. Electroplatings about 5 μm (200 μin.) thick are used as the "braze" for the fluxless joining of silver alloys that are free of zinc and other volatile metals, at about 800 °C (1470 °F) [Tuah-Poku, Dollar, and Massalski 1988]. The formation of liquid at these temperatures relies on solid-state diffusion to establish a thin layer of the silver-copper eutectic composition alloy at the joint interface, which is molten at the process temperature. Aluminum alloy components can be joined similarly using this approach at temperatures close to 575 °C (1065 °F) [Niemann and Wille 1978]. Such joints tend to be strong, being both thin and usually extremely well filled. Because of the exceptional thinness of the layer of braze, extended bonding times can result in the copper diffusing completely into the parent material. This can forestall the problem of galvanic corrosion that might otherwise arise when there is an abrupt interface between two dissimilar metals. Joint remelting during service is also prevented because the zone of low-melting-point filler metal is dissipated completely in this isothermal bonding process, through interdiffusion. Further details on this type of process, known as *diffusion brazing,* are given in Chapter 6.

2.1.3 Silver-Copper

Silver and copper enter into eutectic equilibrium at 779 °C (1435 °F) when 28% of copper is added to pure silver. The silver-copper phase diagram is shown in Fig. 2.2. The eutectic microstructure is lamellar, with a silver-rich phase interspersed with a copper-rich phase. This alloy system forms the basis of the widely used ternary, quaternary, and quinary silver-base brazes, although it is sometimes used on its own.

Silver-copper eutectic is a malleable alloy and can be readily worked into a wide variety of preform geometries. It is employed mostly in the fluxless joining of copper-base alloys in vacuum; there is insufficient thermal activation in such an environment to reduce the iron, chromium, and cobalt oxides that are present on the surface of ferrous alloys at the typical temperatures at which the silver-copper braze is used (800 to 850 °C, or 1470 to 1560 °F). For this reason, reducing atmospheres or fluxes are needed for brazing other materials.

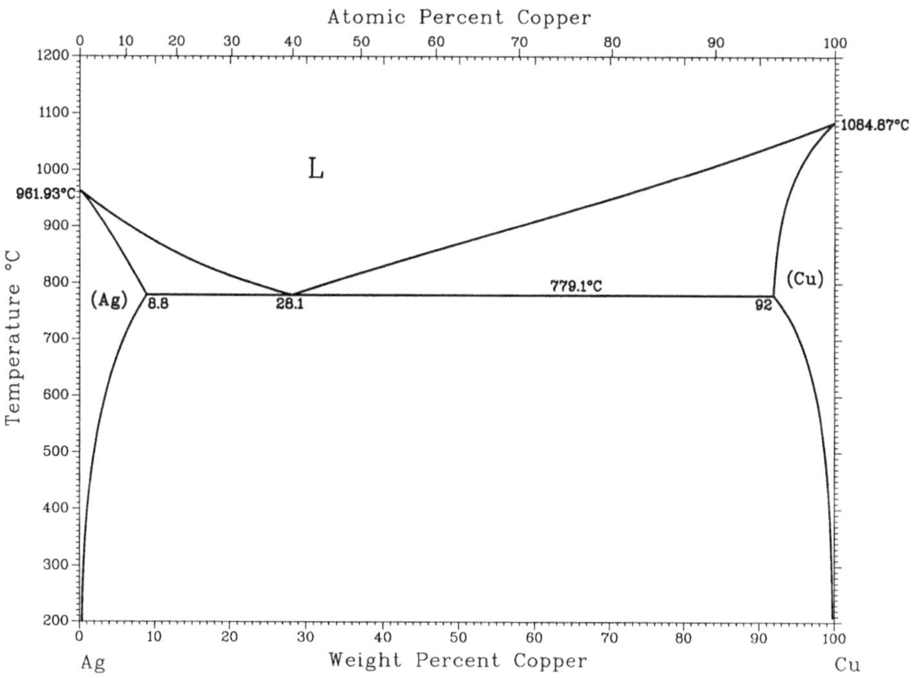

Fig. 2.2 Silver-copper phase diagram

It is appropriate to mention briefly a group of silver-copper alloys that contain additions of phosphorus. These alloys are known as *self-fluxing filler metals* in that they can be used in air without the addition of flux to the joint region, provided the parent metal is not too refractory. The silver content confers both enhanced strength and elongation to failure to brazes of this group because the Cu_3P phase is brittle, but this must be balanced against the cost premium of silver. They are discussed in further detail in Chapter 3, section 3.3, where a description is also given of the self-fluxing mechanism. One composition, 74.75Cu-18Ag-7.25P, is a eutectic, with a melting point of 644 °C (1071 °F) so that in addition to being self-fluxing, it has two other possible advantages over the silver-copper eutectic braze, namely, a melting point that is lower by almost 140 °C (250 °F), and a much lower silver content, which considerably reduces its intrinsic cost. However, the range of application of these phosphorus-containing alloys is considerably more limited; in particular, they are unsuitable for use with steels where there is a load-bearing requirement placed on the brazed assembly, due to the formation of brittle phosphides. The range of alloy compositions that are used most widely as brazes, which encompasses all the AWS-designated compositions listed in Table 3.5 (in Chapter 3), is mapped on the liquidus projection of the Ag-Cu-Cu_3P partial ternary system in Fig. 2.3. Silver-copper-lithium brazes are also self-fluxing and therefore can be used with certain steels to avoid problems arising from the formation of brittle phosphides were the phosphorus-containing brazes to be employed (see Chapter 3, section 3.3).

An "active" braze based on the silver-copper eutectic has an addition of up to 5% titanium [Mizuhara and Mally 1985]. This braze is prepared as either a tri-foil, consisting of a titanium sheet that is roll-clad with the silver-copper eutectic braze, a silver-copper alloy wire with a titanium core or, more recently, in the form of a foil of the ternary alloy. The ternary alloy foil is prepared most readily by rapid solidification directly from the melt because cast ingots of the alloy have poor mechanical properties. Rapidly solidified alloys of this type, some containing additional elements, have been developed commercially and been assigned registered names.

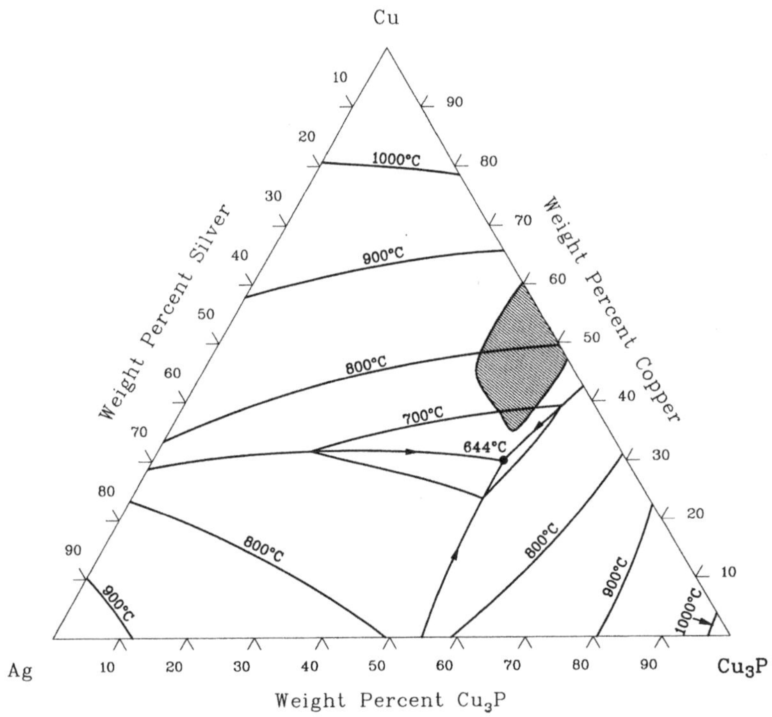

Fig. 2.3 Liquidus surface of the Ag-Cu-Cu_3P partial ternary system. The range of alloy compositions that are widely used as self-fluxing brazes is indicated by the shaded region.

For example, the alloy of composition Ag-26.7Cu-4.5Ti is marketed under the name Ticusil. This alloy and a selection of other commercially available active brazing alloys, produced by rapid solidification, are listed in Table 2.1.

The purpose of adding titanium is to introduce a highly reactive metal that is capable of directly wetting and bonding to nonmetallic materials, particularly various ceramics and graphite. At elevated temperatures, typically 900 °C (1650 °F) and above, the titanium reacts with the nonmetallic parent material to produce a complex interfacial layer that is wettable by the silver-copper braze. So-called active filler metals, containing titanium and other reactive metals, are discussed in further detail in Chapter 7, section 7.2.

The relatively high melting point of brazes based on the silver-copper eutectic has several disadvantages. These are, principally, the energy costs associated with heating the components to the elevated process temperature, the tendency to thermally degrade the mechanical properties of many parent metals (particularly if they are work- or precipitation-hardened), and the costs relating to the removal of the silaceous fluxes that are required for torch brazing in air and that are not soluble in water (see Chapter 3, section 3.2).

The Cu-36Ag-3Si-0.15Sn braze was developed as a form-fit-function replacement for silver-copper eutectic, but with half the silver content, and it is therefore considerably cheaper to produce. This brazing alloy has a similar application temperature range as the silver-copper eutectic and, moreover, it can be just as readily mechanically reduced to foil and wire. Also, the mechanical and physical properties of joints made with this 36% silver alloy are largely indistinguishable from those obtained with the 72% binary alloy [Dev et al. 1992].

2.1.4 Copper-Zinc and Silver-Zinc Brazes

Binary copper-zinc alloys (brasses) have been used for centuries as brazes, but those compositions that have a narrow melting range contain approximately 40 to 50% zinc, as can be seen from the copper-zinc phase diagram given in Fig. 2.4. Their melting points are all above 800 °C (1470 °F), and they possess reasonable ductility. Usually, copper-zinc brazes contain small additions of other elements, principally iron, manganese, nickel, tin, and silicon, in order to improve the mechanical properties. These copper-zinc alloys are used mostly for joining steel and cast iron. The high percentage of volatile zinc is undesirable for furnace brazing operations and means that, particularly in atmospheric brazing, the process temperature must be tightly controlled. Copper-zinc brazes tend to be used at low superheats, often 20 °C (36 °F) or less.

The silver-zinc binary phase diagram is shown in Fig. 2.5. Silver-zinc binary alloys exhibit similar features but have the added disadvantage of higher cost of silver in the alloys that have a narrow melting range. These typically contain 60 to 70% silver. However, the combination of high silver content; matching color to hallmarked silver; and resistance to silver cleaning fluids, most of which contain ammonia, do

Table 2.1 Examples of some commercially available rapidly solidified brazes

Composition	Melting range, °C	Structure	Typical applications
Ag-28Cu	779	Microcrystalline	Most engineering materials
Ag-38Cu-5Ti	775–790	Microcrystalline	Engineering ceramics
Al-12Si	577	Microcrystalline	Aluminum alloys
Cu-30Sn-10M	640–700	Amorphous	Copper alloys and mild steel
Cu-10Ni-8P-4Sn	610–645	Microcrystalline	Copper alloys and mild steel
Cu-15P	714	Amorphous	Copper alloys and mild steel
Cu-20Sn	770–925	Microcrystalline	Copper alloys and mild steel
Co-19Cr-19Ni-8Si-1B	1120–1150	Amorphous	Cobalt-base alloys and superalloys
Ni-10P	880	Amorphous	Steels, stainless steels, and superalloys
Ni-32Pd-8Cr-3B-1Fe	940–990	Amorphous	Steels, stainless steels, and superalloys
Ni-14Cr-5Si-5Fe-3B	970–1075	Amorphous	Steels, stainless steels, and superalloys
Ni-15Cr-3B	1020–1065	Amorphous	Steels, stainless steels, and superalloys
Ni-41Pd-9Si	712–745	Amorphous	Stainless steel and super-cemented carbides
Pd-38Ni-8Si	830–875	Amorphous	Stainless steel, superalloys, and cemented carbides
Ti-15Cu-15Ni	902–932	Amorphous	Superalloys and engineering ceramics
Ti-20Zr-20Ni	848–856	Amorphous	Superalloys and engineering ceramics
Zr-17Ni	961	Amorphous	Titanium-base alloys
Zr-28V-16Ti	1193–1250	Amorphous	Titanium-base alloys

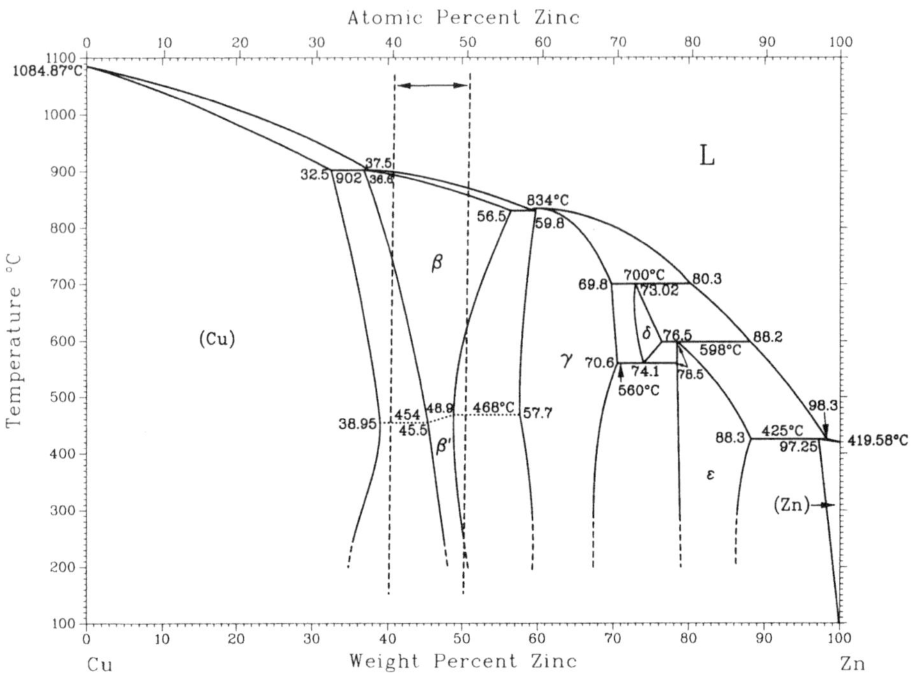

Fig. 2.4 Copper-zinc phase diagram. The range of alloy compositions used as brazes is indicated. These are characterized by having narrow melting ranges.

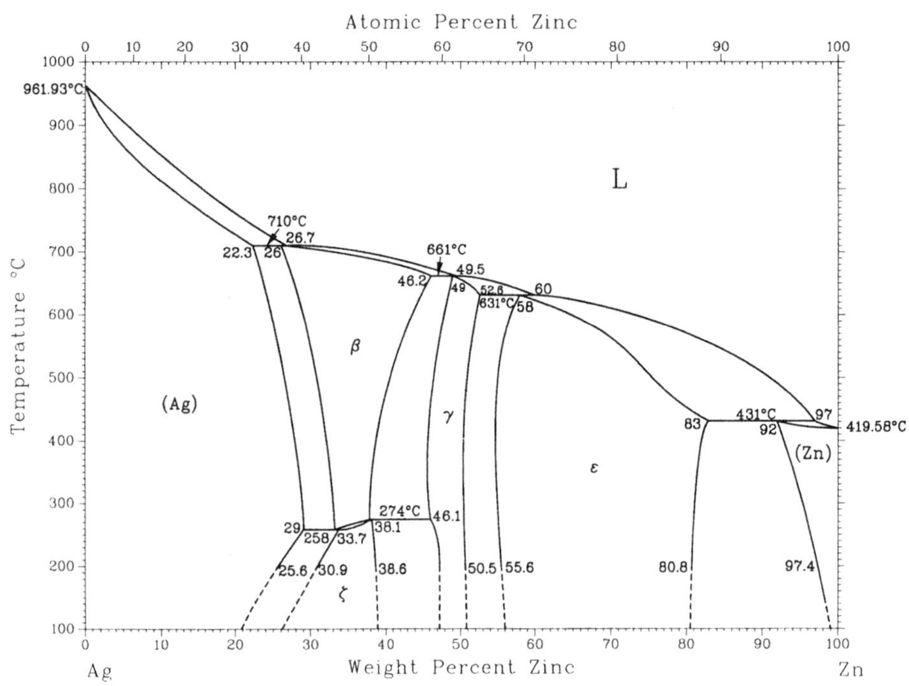

Fig. 2.5 Silver-zinc phase diagram

make them attractive for joining certain types of silverware. By comparison, ammonia can induce stress-corrosion cracking in copper-rich alloys.

2.1.5 Silver-Copper-Zinc

Silver, copper, and zinc form a eutectic alloy in the proportions 56Ag-24Zn-20Cu at 665 °C (1229 °F). The lowest melting-point alloys in this system extend from the ternary eutectic point toward the zinc corner of the phase diagram. However, the alloys that contain more than about 40% zinc suffer a drastic drop in their mechanical strength and ductility. The fabrication of these alloys as wrought products is difficult and the joints formed using them tend to be weak. Hence, they are seldom used industrially. This weakness is accounted for by the phase relationships in this ternary alloy system. For alloys of silver and copper containing up to about 25% zinc, the major phases in the solid state are face-centered cubic (fcc) silver and copper solid solutions, which are inherently ductile. Above this limit, body-centered cubic (bcc) silver-zinc and copper-zinc phases appear in the alloy microstructure, which are less responsive to mechanical working.

The range of compositions that lend themselves to brazing applications is restricted further with respect to the copper and silver contents. Alloys that contain more than about 60% copper have a liquidus temperature above 900 °C (1650 °F) and reduced strength, while silver concentrations above 85% are undesirable for similar reasons: high silver-content alloys are also expensive. The liquidus surface of the silver-copper-zinc ternary phase diagram is represented in Fig. 2.6. The range of alloy compositions that are widely used as brazes has been mapped on this projection.

Silver-rich alloys in this ternary system are used as brazes in the jewelery industry. Although hallmarked sterling silver contains a minimum of 92.5% silver, until recently, it was not possible to match this specification with color-matched silver brazing alloys of sufficiently low melting point. For this reason, the hallmarking limits on the silver content for brazes used with silverware are statutorily relaxed to 67.5% silver. Silver-copper-zinc alloys are then capable of meeting the functional requirements with good color matching. However, with the advent of rapid solidification technology, an active silver braze without zinc, which

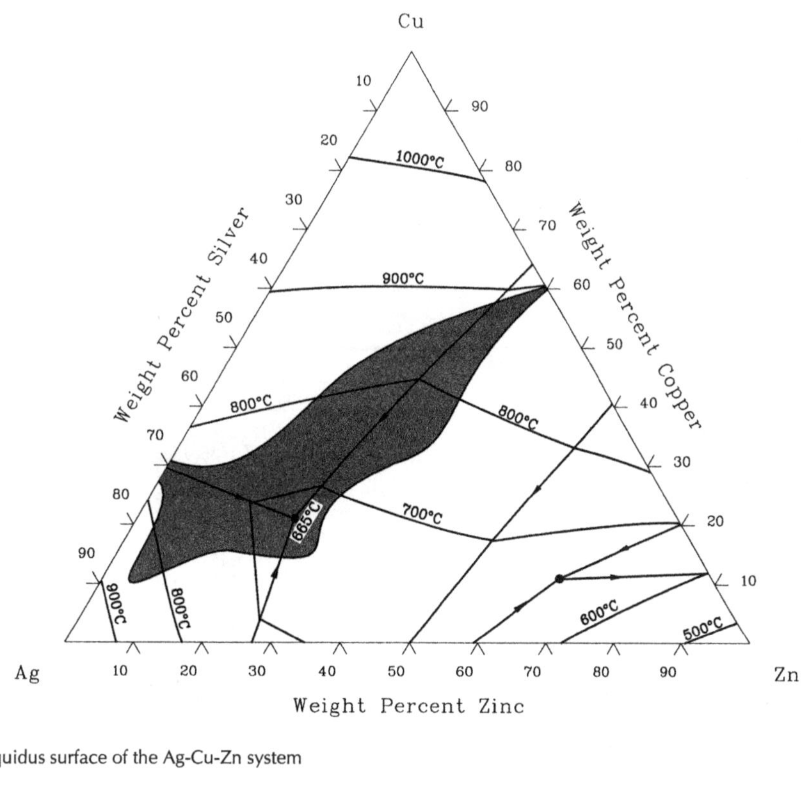

Fig. 2.6 Liquidus surface of the Ag-Cu-Zn system

is specifically tailored to meet the sterling silver standard, has become available. This is the silver-ABA alloy, of composition Ag-5Cu-1.25Ti-1Al (melting range 860–912 °C, or 1580–1673 °F). Active brazing alloys are treated in detail in Chapter 7, section 7.2.

For the joining of brass components, ternary alloys with a copper-to-zinc ratio close to 60:40 are used. These alloys possess the color of brass because they have the same majority phases in their microstructure. The silver content depresses the melting range, improves the ductility, fatigue, and corrosion resistance as well as the fluidity of the alloys.

The mechanical strength of the silver-copper-zinc brazes and of joints made with them can be enhanced, particularly for elevated-temperature service, by adding a small proportion of nickel, typically 2 to 4%. Nickel additions also promote wetting of carbides and improve the resistance of joints in steel components to crevice corrosion. There is evidence that the nickel forms interfacial phases that provide a reactive bond to refractory surfaces [Miller and Schwaneke 1978]. The surfaces of the fillets that form at the edge of the joints are nickel-rich, and therefore possess superior resistance to corrosion. The nickel-bearing alloys find use in the brazing of tungsten carbide cutting tool tips to steel shanks and for other small components where the reduced fluidity of the braze can be tolerated.

A further refinement is achieved by adding a fifth element, manganese, to form quinary alloys. Manganese has the ability to react with and dissolve carbon, which improves wetting of cemented carbides by the braze. For the same reason, the manganese-containing alloys are capable of wetting many grades of cast iron. The addition of the manganese also improves further the resistance to crevice corrosion of joints made to stainless steel. Manganese possesses unusual mechanical properties for a metal, among which is the ability to absorb high strain rate loading (impact). Its incorporation within brazes, even as a minor constituent, tends to increase the fracture toughness of joints. Both nickel and manganese are cheap in relation to silver and thus help to reduce the intrinsic materials cost of the braze.

A number of silver-copper-zinc brazing alloys used with silverware, brassware, and engineering alloys are listed in Table 2.2. The composition range and selected properties of these useful alloys are mapped out in Fig. 2.7 to 2.9.

2.1.6 Silver-Copper-Zinc-Cadmium

The addition of cadmium to silver-copper-zinc alloys confers three major benefits. One benefit is a reduction in the solidus temperature and the melting range; a second is a reduction in the silver content, as compared with the nearest cadmium-free equivalent alloy (Fig. 2.10). The liquidus projection of a section through the silver-copper-zinc-cadmium quaternary system at 40% silver is given in Fig. 2.11, which shows that the effect of cadmium additions to the silver-copper-zinc ternary eutectic is to reduce the liquidus and solidus temperatures until a transition reaction occurs at 615 °C (1140 °F). The quaternary alloy that lies at the composition Ag-25Cd-17Cu-16Zn (melting range, 610–620 °C, or 1130–1150 °F) is commonly referred to as the "pseudoeutectic" composition. Further cadmium additions widen the melting range of the alloys

Table 2.2 Silver-copper-zinc brazing alloys

Composition, wt%					Melting range		AWS	
Ag	Cu	Zn	Ni	Mn	°C	°F	designation	Notes
16.0	50.4	33.6	...	...	780–830	1435–1525	...	(a)
25.0	38.0	33.0	2.0	...	707–801	1305–1474	BAg-26	(b)
40.0	30.0	28.0	2.0	...	671–779	1246–1434	...	(b)
45.0	30.0	25.0	...	...	688–774	1270–1425	BAg-5	
49.0	16.0	23.0	4.5	7.5	690–801	1274–1474	BAg-22	(b)
50.0	34.0	16.0	...	...	677–743	1251–1369	BAg-6	
50.0	20.0	28.0	2.0	...	677–727	1251–1341	BAg-24	(b)
56.0	20.0	24.0	...	...	665	1229	...	(c)
65.0	20.0	15.0	...	...	671–713	1240–1315	BAg-9	
66.7	23.3	10.0	...	...	705–723	1300–1333	...	
67.5	23.5	9.0	...	...	700–730	1290–1345	...	(d)
74.0	19.2	6.8	...	...	720–765	1330–1410	...	(d)
81.0	14.0	5.0	...	...	730–800	1345–1470	...	(d)

(a) Alloy representative of the brazes used in brassworking. (b) Alloys used for brazing carbide tool tips. (c) Eutectic composition. (d) Alloys chiefly used by silversmiths

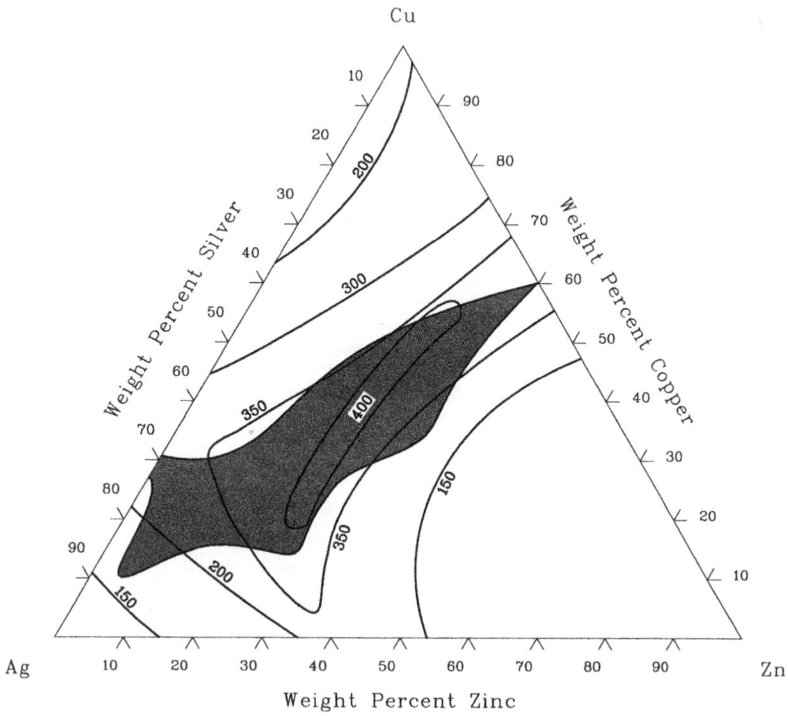

Fig. 2.7 Isostrength (MPa) contours for Ag-Cu-Zn alloys in the cast condition

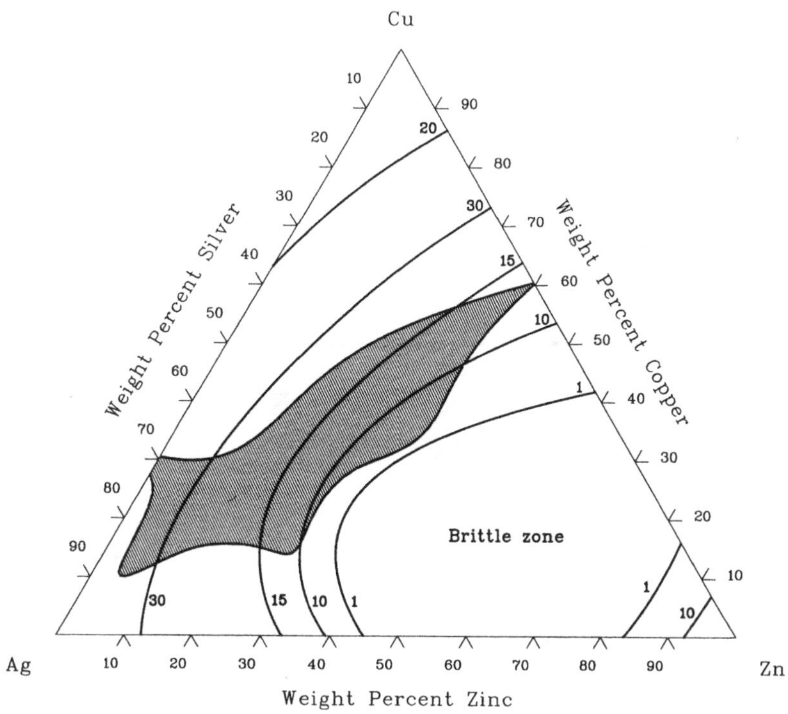

Fig. 2.8 Isoelongation (%) contours of as-cast alloys in the Ag-Cu-Zn system

substantially. The compositions of the principal brazes in this alloy family are listed in Table 2.3.

The third benefit brought about by the introduction of cadmium is an improvement in the fluidity of the alloys. A consequence of the enhanced fluidity is that it enables the production of narrower and better-filled joints that consequently exhibit improved mechanical properties. Braze compositions that are richer in silver, namely, those containing 67.5% of this element, are needed to satisfy the requirements on silver content of jewelery alloys. The alloys with a lower silver concentration are used primarily in engineering applications, where the silver con-

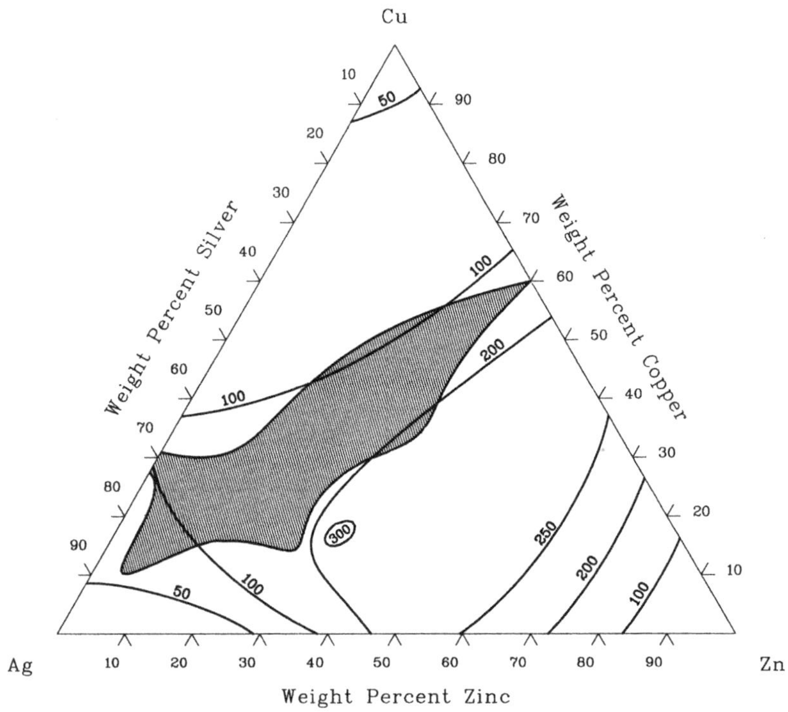

Fig. 2.9 Isohardness (Brinell) contours of as-cast alloys in the Ag-Cu-Zn system

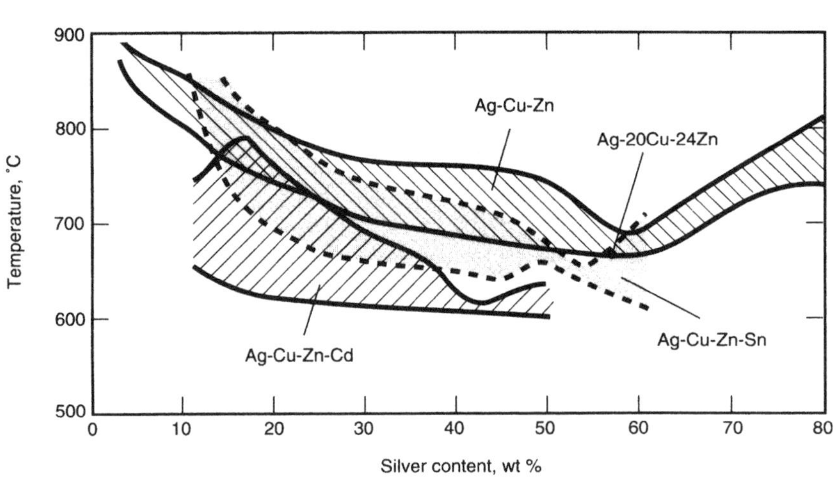

Fig. 2.10 Melting ranges of ternary Ag-Cu-Zn and quaternary Ag-Cu-Zn-Cd and Ag-Cu-Zn-Sn brazing alloys as a function of their silver content. Note that the liquidus and solidus lines do not correspond to fixed ratios of the other constituents.

tent can represent an important cost item. However, a reduction in the silver content is made at the detriment of the mechanical properties of the joints. Departure from the pseudoeutectic composition also widens the melting range of the braze, which then necessitates increasing the joining temperature.

The silver-copper-zinc-cadmium alloys are available also with nickel and manganese additions. These minor alloying constituents modify the characteristics of these brazes in the same way they do in the silver-copper-zinc family, described previously.

2.1.7 *Silver-Copper-Zinc-Tin*

The cadmium-bearing brazes were originally developed in the 1930s, before the health hazards of cadmium oxide vapor were widely appreciated. The safety threshold limit value (TLV) for the proportion of cadmium fume in the air is 0.05 mg/m^3. This level of operator pro-

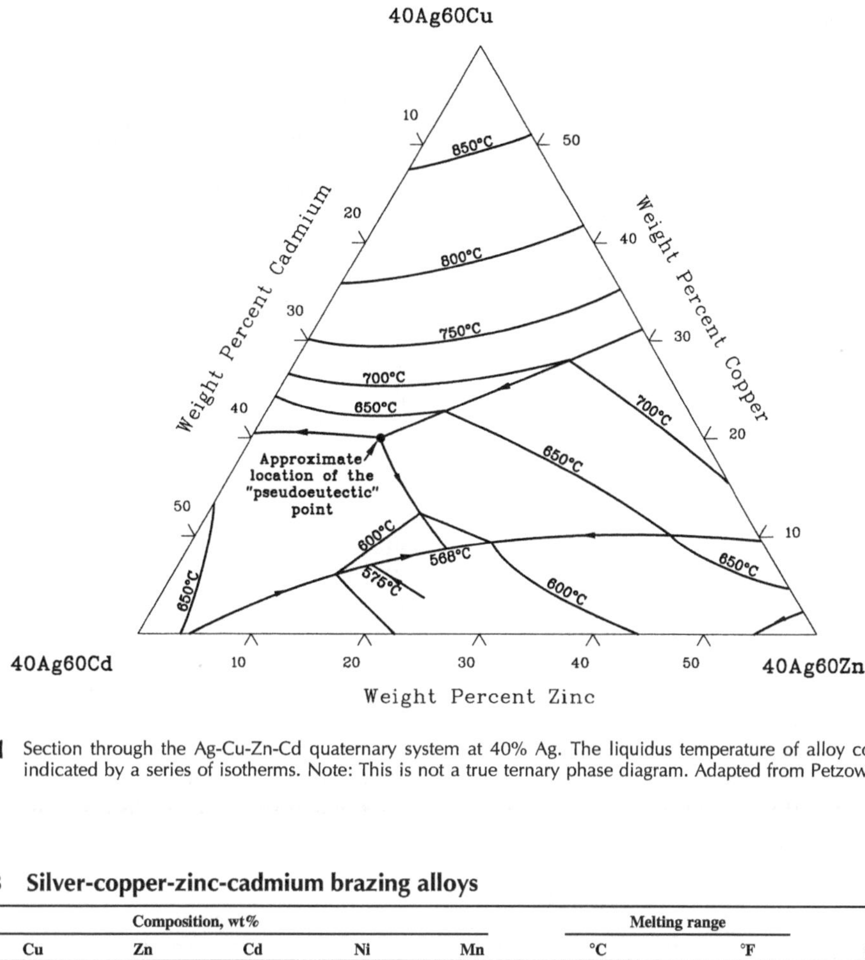

Fig. 2.11 Section through the Ag-Cu-Zn-Cd quaternary system at 40% Ag. The liquidus temperature of alloy compositions is indicated by a series of isotherms. Note: This is not a true ternary phase diagram. Adapted from Petzow and Aldinger [1968]

Table 2.3 Silver-copper-zinc-cadmium brazing alloys

Composition, wt%						Melting range		AWS designation/note
Ag	Cu	Zn	Cd	Ni	Mn	°C	°F	
25.0	40.0	15.0	17.0	0.5	2.5	625–770	1155–1420	. . .
25.0	35.0	26.5	13.5	. . .	. . .	607–746	1125–1375	BAg-27
30.0	27.0	23.0	20.0	. . .	. . .	607–710	1125–1310	BAg-2a
35.0	26.0	21.0	18.0	. . .	. . .	607–701	1125–1295	BAg-2
40.0	19.0	21.0	10.0	. . .	. . .	595–630	1105–1165	. . .
42.0	17.0	16.0	25.0	. . .	. . .	610–620	1130–1150	(a)
45.0	15.0	16.0	24.0	. . .	. . .	607–618	1125–1145	BAg-1
50.0	15.5	16.5	18.0	. . .	. . .	627–635	1160–1175	BAg-1a
50.0	15.5	15.5	16.0	3.0	. . .	632–688	1170–1270	BAg-3
67.5	11.0	11.5	10.0	. . .	. . .	635–720	1175–1330	(b)

(a) Pseudoeutectic composition alloy with the narrowest melting range. (b) Used for joining hallmarked silverware

tection can be achieved only in either an enclosed or a laminar flow space, both of which are expensive to maintain and clean. Over the past few decades, restrictions have been placed on the use of cadmium-containing brazes in most Western nations, which triggered a search for substitutes.

The outcome of this endeavor has been the development of a series of alloys in which tin replaces the cadmium [Timmins 1994]. The principal compositions are listed in Table 2.4. The tin largely preserves the good flowing characteristics conferred by the cadmium and also acts as a melting point depressant, although not to the extent that cadmium does. Moreover, the substitution broadens the melting range of the alloys, as can be seen in Fig. 2.10. Thus, for example, the melting range of the 55Ag-21Cu-22Zn-2Sn alloy is 30 °C (54 °F) (liquidus temperature = 660 °C, or 1220 °F; solidus temperature = 630 °C, or 1165 °F). This compares with just 10 °C (18 °F) for the cadmium-bearing alloy 42Ag-17Cu-16Zn-25Cd (liquidus temperature = 620 °C or 1150 °F; solidus temperature = 610 °C, or 1130 °F), which also contains significantly less silver. The low solid solubility of tin in zinc restricts the percentage of tin that can be added. Larger additions lead to the formation of embrittling intermetallic phases.

Nickel- and manganese-bearing versions of these alloys are also commercially available, as is a silver-copper-zinc-indium alloy. The substitution of tin by indium does not offer any particular improvement in the properties of the filler, and moreover, it increases the cost of the braze. However, all members of this brazing alloy family contain less than 25% zinc and hence are extremely ductile so that they can be rolled, drawn, and extruded with reduction ratios of 400:1, without intermediate annealing. Production of wire and foil preforms is, therefore, particularly easy. Other cadmium-free alternatives to silver-copper-zinc-cadmium brazes are mentioned in section 2.1.12.

2.1.8 Gold-Base Brazes

For many years, gold-bearing brazing alloys were used almost exclusively in the jewelery industry. These alloys, known as carat gold "solders" by the trade, were developed to provide goldsmiths with a wide range of relatively low melting point brazes that match the gold content and, especially, the color of carat jewelery alloys. Further details of brazes for gold-jewelery are given in Chapter 5.

An additional range of gold-bearing brazing alloys has been developed in response to technological demand, particularly from the electronics, nuclear power, and aerospace industries. Some examples are listed in Table 2.5. These brazing alloys are particularly suited for use in corrosion-resistant assemblies endowed with enhanced mechanical properties and that require joining alloys with matching properties. It is possible to enhance the oxidation resistance of the silver-bearing and nickel-bearing brazing alloy families by adding small percentages of certain elements such as aluminum, nickel, chromium, and manganese. The improved oxidation resistance stems from the ability of these elements to form relatively stable and inert oxide films. However, in chemically aggressive environments these modified alloys are susceptible to corrosion. By contrast, noble metals—especially gold and platinum—are, by their very nature, largely inert chemically and therefore capable of surviving in harsh environments.

The gold-base brazing alloys can be divided into three principal families, namely, gold-copper, gold-nickel and, gold-palladium. The gold content of some of these alloys is less than

Table 2.4 Silver-copper-zinc-tin brazing alloys

Composition, wt%						Melting range		AWS designation
Ag	Cu	Zn	Sn	Ni	Mn	°C	°F	
25.0	40.0	33.0	2.0	0.5	2.5	690–780	1275–1435	BAg-37
30.0	36.0	32.0	2.0	...	...	655–750	1210–1380	...
38.0	32.0	28.0	2.0	...	...	650–720	1200–1330	BAg-34
40.0	30.0	28.0	2.0	...	...	650–710	1200–1310	BAg-28
42.0	21.0	27.0	2.0	4.0	4.0	665–750	1230–1380	...
45.0	27.0	25.0	3.0	...	...	640–680	1185–1255	BAg-36
50.0	21.5	27.0	1.0	0.5	...	660–680	1220–1255	...
55.0	21.0	22.0	2.0	...	...	630–660	1165–1220	...
56.0	22.0	17.0	5.0	...	...	620–650	1150–1200	BAg-7
60.0	23.0	14.5	2.5	...	...	620–685	1150–1265	...
63.0	28.5	0	6.0	2.5	...	690–800	1275–1470	BAg-21

50%. Nevertheless, by convention they are classed as gold brazes because the presence of significant proportions of the precious metal constituent. The same applies to brazes containing the platinum-group metals.

The gold-bearing alloys designed for industrial use are superior to base-alloy brazes in the following respects:

- Improved resistance to oxidation
- Enhanced mechanical properties of joints at elevated temperatures
- Relatively low degree of erosion of component metal surfaces during the joining operation
- Good corrosion resistance in most chemical environments
- Require only mild fluxes, on account of the nobility of these alloys.

The three principal families of gold-bearing alloy brazes are considered briefly in turn.

2.1.8.1 Gold-Copper

The gold-copper binary alloy system, which also provides the basis for many of the brazes used in jewelery, is characterized by a continuous solid solution between the two constituent metals with a liquidus minimum at the Au-20Cu composition, where there is a single melting point (910 °C, or 1670 °F). The gold-copper phase diagram is shown in Fig. 2.12. All other alloy compositions in this system have narrow melting ranges, typically less than 20 °C (36 °F). This means that gold-copper brazes possess excellent fluidity and readily form good fillets. Alloys within the range of approximately 40 to 90% gold undergo ordering transformations at low temperature that produce a hardening effect. However, they are sufficiently ductile to be mechanically worked to foil and wire, suitable for making preforms. Gold-copper alloys will readily wet a range of base metals, including many refractory metals.

The addition of nickel to gold-copper brazes helps to improve their ductility: a typical composition is Au-16.5Cu-2Ni, which has a melting point of 899 °C (1650 °F). The nickel in the filler metal considerably improves its resistance to creep at elevated temperature (250–750 °C, or 480–1380 °F) [Stephens and Greulich 1995]. Adding silver to copper-gold binary alloys effects a significant reduction in the melting point, to 767 °C (1413 °F) at the composition 45Ag-29Au-26Cu. The liquidus surface of the phase diagram for this alloy system is shown in Fig. 2.13. However, for applications where superior corrosion resistance is required, the gold concentration needs to be kept above 60%, which increases the melting point to about 850 °C (1560 °F).

The elevated temperature properties (in particular, strength and corrosion resistance) can be enhanced further by introducing alloying elements such as chromium, manganese, molybdenum, and palladium, exemplified by the alloy Au-34Ni-4Cr-1.5Fe-1.5Mo. The improved mechanical properties at elevated temperature relate to the duplex microstructure, which consists of gold-rich and nickel-rich phases [Kang and Kim 1995]. Gold-base filler metals containing vanadium, such as the Au-15.5Ni-3V-0.7Mo alloy will wet alumina ceramics, particularly commercial purity grades that contain a glassy phase [Hosking 2000]. Molybdenum is known to increase the ductility of gold-base filler metals and is one reason for its inclusion in both of these formulations. These additions also promote wet-

Table 2.5 Industrial gold brazing alloys

Composition, wt%					Melting range		AWS designation
Au	Ag	Cu	Ni	Pd	°C	°F	
20.0	...	80.0	...	...	1020–1040	1865–1905	...
29.0	45.0	26.0	32.0	2.0	767	1413	...
30.0	...	...	36.0	34.0	1135–1165	2075–2130	BAu-5
35.0	...	62.0	3.0	...	975–1030	1785–1885	BAu-3
37.5	...	62.5	...	...	990–1015	1815–1860	BAu-1
50.0	...	...	25.0	25.0	1120–1125	2050–2055	...
60.0	20.0	20.0	...	...	845–855	1555–1570	...
62.5	...	37.5	...	...	930–940	1705–1725	...
70.0	...	...	22.0	8.0	1005–1045	1845–1915	BAu-6
75.0	...	...	...	25.0	1375–1400	2505–2550	...
80.0	...	20.0	...	...	910	1670	BAu-2
82.0	...	...	18.0	...	955	1750	BAu-4
82.5	...	16.5	2.0	2.5	899	1650	...
92.0	...	...	...	8.0	1180–1230	2155–2245	...

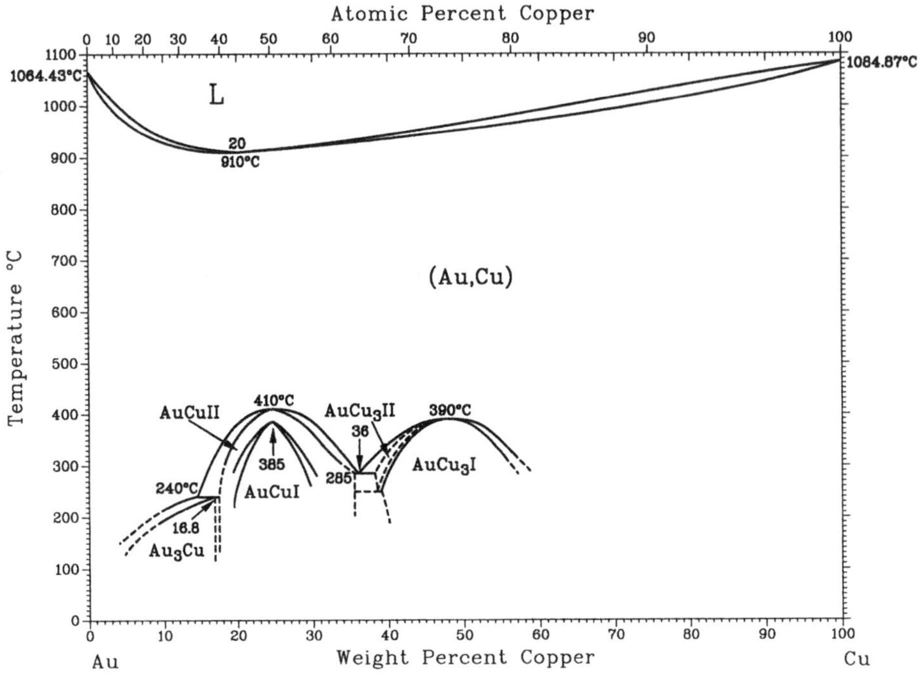

Fig. 2.12 Gold-copper phase diagram

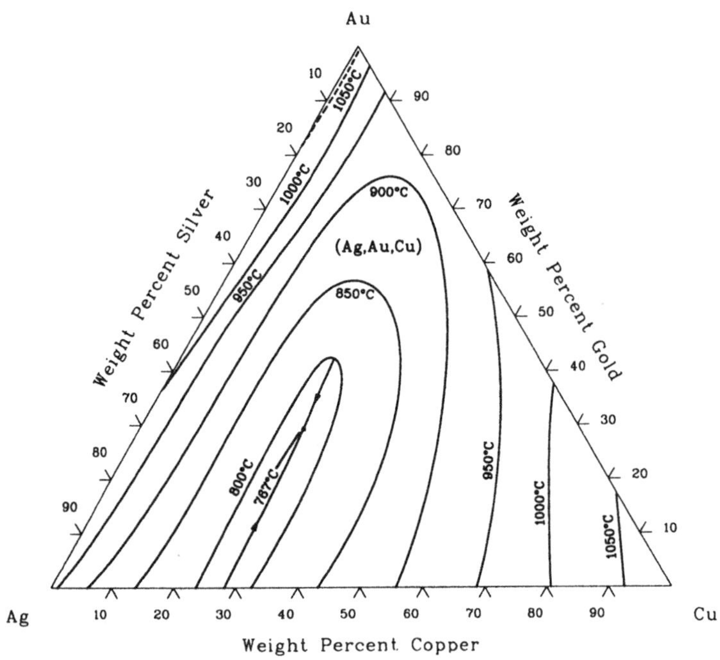

Fig. 2.13 Liquidus surface of the Ag-Au-Cu ternary system

ting of refractory materials, notably graphite and carbides. Another example is the Au-34Cu-16Mn-10Ni-10Pd alloy, which was developed for brazing components in the space shuttle main engine at about 1000 °C (1830 °F), in response to the need for joints with superior oxidation resistance.

2.1.8.2 Gold-Nickel

Gold forms a continuous series of solid solutions with nickel in a similar manner to the gold-copper alloys, except that the melting range tends to be wider. The liquidus/solidus minimum occurs at the Au-18Ni composition and 955 °C (1750 °F). The gold-nickel phase diagram is shown in Fig. 2.14.

A nickel content of 35% is generally the maximum used for brazing alloys, owing to the considerable widening of the melting range as the proportion of nickel is increased. The gold-nickel brazes possess many of the advantages of the gold-copper alloys, but provide the additional benefit of superior resistance to oxidation and improved wetting.

Extra constituents enable the proportion of costly gold in the alloy to be reduced significantly without sacrificing essential properties. Thus, the Au-22Cu-8.9Ni-1Cr-0.1B alloy, containing 68% gold (melting range 960–980 °C, or 1760–1795 °F) was developed as a cheaper equivalent to the minimum melting point Au-18Ni composition [Sloboda 1971]. This substitution provides an instructive example of the considerations that frequently apply in the development of new filler compositions. First, target properties are defined. In this case, the new alloy was required to:

- Readily wet heat-resistant steels, in particular martensitic stainless steels such as Jethete M-152 (Fe-12Cr-2.5Ni-1.7Mo-0.3V-0.12C), and produce sound joints to components of these materials
- Completely melt below 1000 °C (1830 °F) in order to be compatible with the special steels, which are heat treated (the beneficial effects of the heat treatment are lost above this temperature)
- Be free of volatile constituents to enable the braze to be used in (fluxless) vacuum joining operations
- Confer high strength and oxidation resistance to the joints
- Be sufficiently ductile so as to be mechanically workable to foil and wire for preforms

The alloy was designed such that copper substituted for some of the gold and the ensuing loss

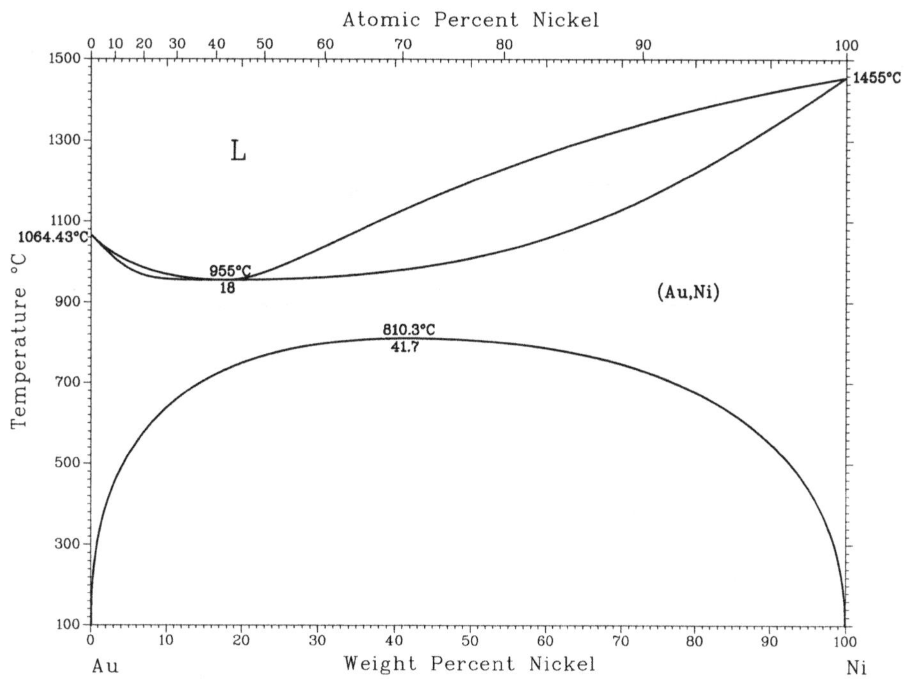

Fig. 2.14 Gold-nickel phase diagram

of oxidation resistance was compensated for by the chromium addition. The deterioration of the wetting characteristics due to the presence of the chromium was made good by the introduction of a small fraction of boron. Finally, the relative proportions of the constituents were then adjusted to minimize the melting range, while optimizing the mechanical properties.

2.1.8.3 Gold-Palladium

The addition of palladium to the gold-copper and gold-nickel alloys improves their resistance to oxidation at elevated temperatures. These alloys are therefore used mostly for joining superalloy and refractory metal components that need to serve in relatively aggressive environments such as exist in modern jet engines. Commercially available brazes of this type have melting temperatures that reach approximately 1200 °C (2190 °F). All these alloys are classified simply as gold-palladium alloys, drawing attention to the two precious metals in this series. Some examples are included in Table 2.5. Cobalt-base brazes are often substituted for high-temperature gold-base brazes (e.g., Co-20Cr-10Si, melting point 1180 °C, or 2156 °F) because they have similar application temperatures, coupled with good corrosion resistance and, of course, substantially lower cost. Several of these alloys contain small percentages of palladium, as can be seen from Table 2.6. Because cobalt-base brazes have inferior workability in bulk form, an economic means of fabricating preforms is by rapid solidification, as is the case for all of the compositions listed in Table 2.6.

2.1.9 Palladium-Base Brazes

Palladium is the major constituent of a series of brazing alloys that contain copper, nickel, and silver. These alloys possess many of the beneficial properties of the gold-bearing brazes but are cheaper. They confer on joints:

- Good mechanical integrity and freedom from brittle intermetallic compounds. This is a consequence of the fact that palladium forms solid solutions with most common engineering metals.
- Enhanced mechanical strength at elevated temperatures. In this respect they tend to be superior to the family of gold alloys that do not contain palladium or other platinum-group metals. As a general rule, strength at a particular temperature is related to the melting points of the major constituents, with higher melting point elements conferring superior elevated temperature service.
- High oxidation resistance at elevated temperatures, especially in the case of the palladium-nickel alloys
- Good corrosion resistance, although not as good as the gold-bearing brazing alloys
- Low vapor pressure at typical brazing temperatures, comparable to that of the gold alloys

Table 2.6 Cobalt-base brazing alloys produced as foils by rapid solidification

Composition, wt%							Melting range			
							Solidus		Liquidus	
Co	Cr	Ni	W	B	Si	Pd	°C	°F	°C	°F
Bal	21.0	. . .	4.5	2.15	1.6	. . .	1136	2077	1163	2125
Bal	21.0	15.0	4.5	4.4	4.2	. . .	1078	1972	1139	2082
Bal	21.0	15.0	4.5	4.4	4.4	3.0	1068	1954	1156	2113
Bal	21.0	15.0	4.5	4.4	4.4	5.0	1018	1864	1152	2106

Table 2.7 Palladium-bearing brazing alloys

Composition, wt%					Melting range	
Pd	Ag	Cu	Ni	Other	°C	°F
65	. . .	. . .	. . .	35Co	1230–1235	2245–2255
60	. . .	. . .	40	. . .	1237	2259
54	. . .	. . .	36	10Cr	1232–1260	2250–2300
25	54	21	. . .	. . .	900–950	1650–1740
21	. . .	. . .	48	31Mn	1120	2050
15	65	20	. . .	. . .	850–900	1560–1650
5	68.5	26.5	. . .	. . .	805–810	1480–1490

- A consistently narrow melting range, in most cases no more than 25 to 50 °C (45 to 90 °F).

Representative ranges of commercial palladium-bearing alloys are listed in Table 2.7. These brazing alloys find application in refractory metal structures where cost is more critical than with certain aerospace uses, which can justify the gold-bearing alloys. Platinum is superior to palladium in terms of its chemical inertness and on that account finds occasional use in brazing alloys, but the applications are restricted by the high cost of this metal that is compounded by its relatively poor mechanical workability.

2.1.10 Nickel-Bearing Filler Metals

Nickel-base brazing filler metals are more extensive in composition range and properties than even the silver-base alloys described previously. Their primary merit is the ability to endure high-temperature service (above 1100 °C, or 2010 °F), including in moderately aggressive environments, although their relatively low cost in comparison with silver-base alloys is also a considerable attraction. Their resilience to elevated temperatures can be attributed largely to the corrosion resistance of elemental nickel and its relatively high melting point (1455 °C, or 2650 °F).

Pure nickel is not widely used as a braze due to its high melting point, except for certain specialist applications, such as the joining of molybdenum and tungsten components intended for subsequent operation at elevated temperatures. A listing of the AWS designated nickel-base brazing alloys is given in Schwartz 2003, Table 4.7.

The melting point of nickel is depressed by alloying additions of an appreciable number of elements (e.g., boron, carbon, chromium, copper, manganese, phosphorus, silicon, and zinc), many of which are introduced in combination to give the variety of nickel-bearing filler metals that are commercially available. Of these, phosphorus and boron are particularly effective in low concentrations at promoting the wetting characteristics of nickel-base brazes. The nickel-boron and nickel-phosphorous phase diagrams are shown in Fig. 2.15 and 2.16, respectively. From these diagrams it can been seen that there is a "low"-melting-point eutectic between nickel and the metalloid (Ni_3B and Ni_3P) as there is between copper and Cu_3P. All of these elements are cheap in comparison with the constituents of noble brazing alloys. One of the most widely used nickel-base brazes, BNi-1, has the composition Ni-14Cr-5Fe-5Si-3B-0.7C, and melts in the range 977 to 1038 °C (1790–1900 °F).

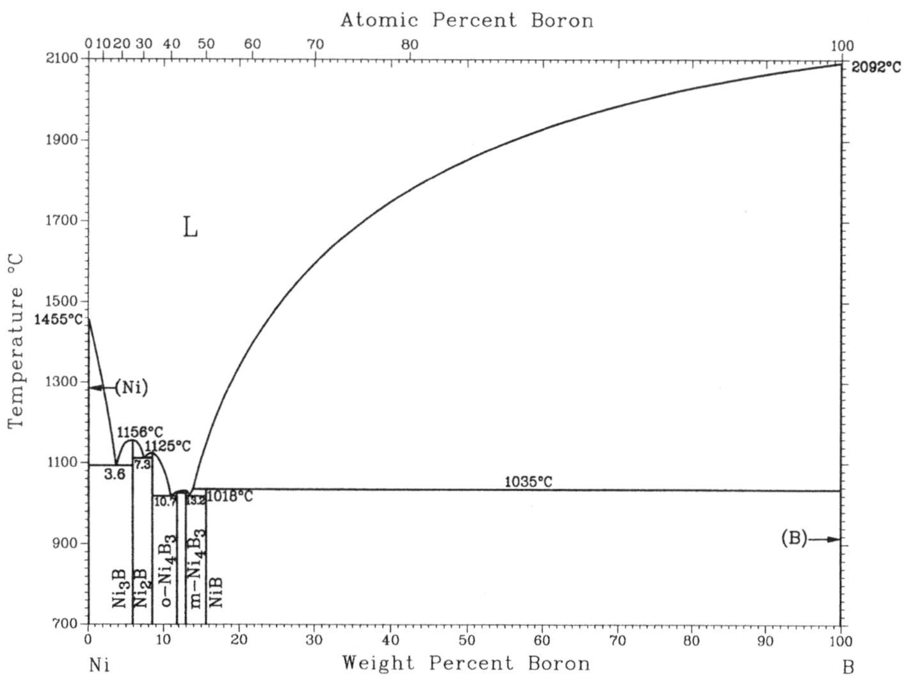

Fig. 2.15 Nickel-boron phase diagram

A feature of the filler metals containing boron, carbon, silicon, and phosphorus is that these elements will diffuse rapidly into many parent metals, thereby increasing the remelt temperature of the solidified braze. Therefore, the process is commonly referred to as *diffusion brazing* [Nicholas 1998]. A full description of this process is given in Chapter 6. This effect is exploited in the manufacture of engines with superalloy turbine blades because it enables the brazed component to be used in service at temperatures that exceed the peak brazing process temperature. However, care must be taken to select an appropriate braze/parent metal combination because the diffusion of certain elements into structural engineering alloys can result in intergranular embrittlement. Conversely, the formation of borides, carbides, phosphides, and similar compounds in the joint gap, through reaction with the constituents of the parent metals, can cause a reduction in joint ductility and impact strength. Filler metals containing boron may not be compatible with joining thin members because diffusion of boron from the braze into the components may depress the melting point of the latter to the extent that these thin components melt and consequently disintegrate.

One of the main applications of nickel brazes is joining of stainless steels and other heat-resistant alloys. Nickel and high-nickel alloys are embrittled by sulfur and many low-melting-point metals, including zinc. The braze alloy, therefore, must not contain such elements. They are also susceptible to stress cracking in the presence of molten brazes so that considerations of stress relief are important during both the heating and cooling stages of the brazing process. Many nickel alloys contain aluminum and titanium that will oxidize at the surface and interfere with wetting and spreading of the braze. Thus, common practice is to preplate the joint region with electroless nickel. This plating needs to be at least 10 μm (400 μin.) thick, depending on the concentration of the secondary elements in the parent materials, in order to avoid their diffusion through to the surface of the nickel plate during the heating cycle.

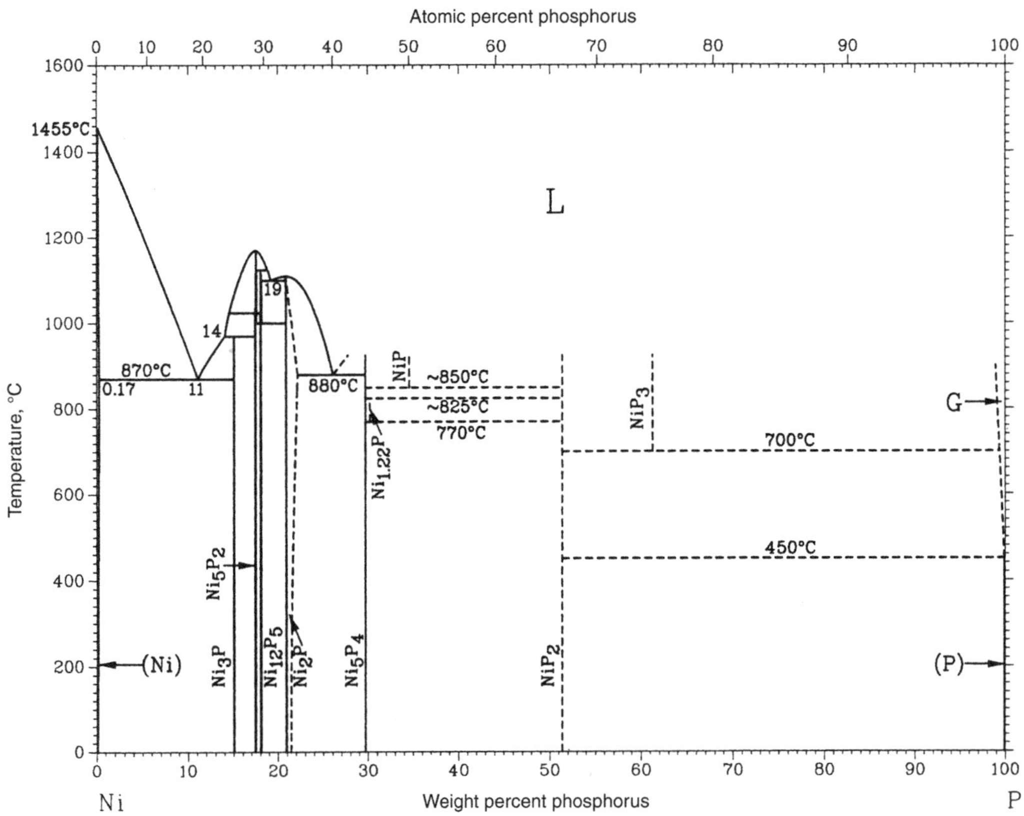

Fig. 2.16 Nickel-phosphorus phase diagram. Source: Lee and Nash [1991]

Nickel exhibits extensive intersolubility with many engineering base metals, notably iron, chromium, and manganese. Erosion of the parent metal by the braze during the joining process can therefore be severe unless close attention is paid to controlling the parameters of the brazing cycle, particularly the excess temperature above the melting point of the braze (i.e., the superheat) and also the duration of the temperature excursion, which, necessarily, includes the heating and cooling rate.

Nickel-base brazes are sometimes used to repair the face of components. Surface porosity in stainless steel castings can be filled with braze and out-of-tolerance or worn surfaces can be increased in thickness before machining to specification. This is possible because many nickel-base brazes possess mechanical properties and resistance to corrosion and oxidation compatible to that of many stainless steels and inconels.

Many nickel-base brazing alloys themselves contain significant proportions of phosphide and boride phases and therefore tend to be brittle. Until recently, this has meant that the compositions of nickel-base brazing alloys that could be fabricated as foils and wire were extremely limited. Instead, the compositions that are brittle were applied in the form of pastes, that is, alloy powder mixed in a fluid binder. Organic binders burn during the brazing cycle, which tend to introduce voids and carbonaceous residues that weaken joints. This restriction has been overcome by the development of rapid solidification casting technology (see Chapter 1, section 1.3.2.2). Because the alloys have low melting points in relation to those of the constituent elements, particularly when the brazing alloy contains boron and/or phosphorus, and because the cooling rate during rapid solidification typically exceeds 10^5 °C/s (2×10^5 °F/s), these nickel-base alloys solidify mostly with an amorphous structure [Rabinkin and Liebermann 1993; DeCristofaro and Henschel 1978]. Examples of rapid solidification casting and foils prepared by chill-bloc melt-spinning are shown in Fig. 1.24 and 1.25. Some nickel-base alloys that are available commercially as amorphous strip are included in Table 2.1. This family of alloys covers melting temperatures that range from about 700 °C (1290 °F) to over 1100 °C (2010 °F).

The use of this casting technology for producing foil and wire preforms not only enables the manufacture of brazes with compositions that are impossible to prepare by normal solidification, but it also confers a number of associated benefits:

- The brazing alloy is comparatively ductile in the amorphous state because there are no discrete phases or grain boundaries in the microstructure that might be sources of embrittlement. One might expect that after melting and resolidification, an amorphous alloy would revert to a crystalline one with a conventional dendritic microstructure. However, there is strong evidence in the published literature that the microstructure of joints made with both brazes and solders prepared by rapid solidification are finer and more uniform in composition and microstructure than joints made with conventionally prepared filler metals [DeCristofaro and Bose 1986, Bowen and Peterson 1987; Rabinkin and Lieberman 1993]. The reasons for this are not fully understood but are likely to be associated with a degree of atomic ordering in the liquid phase [Johnson 1990]. Evidence for this explanation comes from the gray/white allotropic transformation of tin. Gray tin does not resolidify as purely white tin after a cursory melting excursion, as one might expect if the liquid retained no "memory" of its previous state as a solid.
- The absence of a dendritic microstructure in both amorphous and microcrystalline metals means that a rapidly solidified alloy is metallurgically homogeneous, as cast, akin to a solid solution. The alloy therefore melts in a highly uniform manner, which helps to minimize local fluctuations in the erosion of the parent metals due to an absence of segregated phases in the braze.
- The refined microstructure of rapidly solidified filler metals results in their superior wetting and spreading on melting. This is a consequence of the discrete phases being finely divided in the microstructure so that the material tends toward instant homogeneity in the liquid state.
- Filler metals prepared by rapid solidification tend to be cleaner, both on the surface and within the bulk, because there is only a single processing step and that is usually carried out in a protective atmosphere. Metallic contamination from rolling and drawing machinery is eliminated, as is carbonaceous contamination from lubricating fluids emanating from the machinery.
- The homogeneous filler metal preforms produced by rapid solidification melt uniformly

and have a much narrower melting range, compared with the equivalent composition alloy prepared as a partitioned preform, comprising a core of titanium with a cladding of the other constituents. Thus, the 70Ti-15Cu-15Ni composition braze has a melting range of 902 to 932 °C (1656–1710 °F) as a homogeneous alloy and 912 to 1007 °C (1674–1845 °F) when prepared as a trifoil in the form of a titanium layer sandwiched between foils of Cu-50Ni and heated at 10 °C/min (18 °F/min). On account of their glassy or microcrystalline microstructure, rapidly solidified brazes are resistant to oxidation, as prepared, and therefore have an almost unlimited shelf life.
- Foils produced by rapid solidification can be made much thinner than those prepared in ingot form by normal solidification and then rolled down, and are typically 25 μm (1 mil) thick. The use of such thin foils also helps to minimize erosion of the parent metal by the braze.

When using nickel-base fillers containing metalloids, there is a tendency for embrittling nickel-metalloid compounds to precipitate in the joint if the thickness of the braze is more than about 50 μm (2 mils), which provides another incentive for using thin foils produced by rapid solidification. Where wide joints are made, this problem may be avoided by fitting into the joint gap a porous shim of a metal that will rapidly soak up the filler and dilute the metalloid [Lugscheider and Schittny 1988]. This technique is particularly useful in repair work. Further details about wide-joint-gap brazing are given in Chapter 4, section 4.3.4.2.

A family of brazes has been developed for joining 316L stainless steel offering strong resistance to corrosion by seawater and acids. The principal ingredients are nickel and chromium, which are good metals for resisting corrosion, where the chromium content is 10 to 16% and the melting range is roughly 1000 °C to 1100 °C (1830 °F to 2010 °F). The melting range and braze properties are tailored by the addition of boron, molybdenum, and silicon. The corrosion resistance of these brazes is imparted by the target microstructure, which is a nickel-chromium solid solution. To achieve this solid solution, the brazing cycle time needs to be extended in order to diffuse the other elements into the parent materials until they are sufficiently low in concentration that chromium borides and other deleterious phases cannot form [Rabinkin, Wenski and Ribaudo 1998].

Another group of nickel-bearing alloys also merits mention here. This family is based on nickel-titanium, which has been developed for joining titanium and its alloys. Titanium is one of the fastest growing industrial metals in terms of usage, due largely to its exceptionally high strength-to-weight ratio, low density, high toughness, and excellent corrosion resistance (because its tenacious refractory oxide, which affords effective protection). As mentioned in Chapter 1, section 1.1.7.3, titanium has excellent shaping and forming characteristics at elevated temperatures, arising out of its enhanced creep and superplasticity properties above 925 °C (1700 °F). There is a low melting point eutectic in the binary titanium-nickel system, at 27% Ni, which melts at 942 °C (1730 °F). Its two constituent phases are titanium and Ti_2Ni intermetallic compound, the latter imparting significant brittleness to the alloy. Rapid solidification is used to produce commercial foil of titanium-nickel alloys, and of lower melting point copper-nickel-titanium alloys, which have solidus temperatures as low as 890 °C (1630 °F). Because they contain over 60% titanium, strictly speaking, these are titanium rather than nickel-base alloys. Other brazes have been used successfully for joining components of titanium and its alloys, in particular silver-palladium, but this has the disadvantage of a significant cost premium.

2.1.11 High-Melting-Point Brazes

While low melting point is often an advantage for brazes in many applications, a high melting point is sometimes required for functional reasons. Also, some substrate materials are not degraded by high-temperature process cycles so that there is no necessity to compromise other parameters for the sake of a low brazing temperature. Examples of thermally robust materials include ceramics, graphite, and some superalloys. By contrast, the high-melting-point refractory metals are usually not brazed at very high temperatures. This is because heating above the recrystallization temperature initiates grain growth and causes a loss of mechanical properties. At the other end of the temperature scale, refractory metals undergo a ductile-to-brittle transition and must be handled carefully below this temperature. Ductile-to-brittle transition and recrystallization temperatures for some refractory metals are listed in Table 2.8.

Conventional, low-melting-point brazes will not wet and spread on nonmetals because there is no chemical affinity between the constituents of the braze and a nonmetal. It is therefore generally necessary to use either an "active braze" (see Chapter 7, section 7.2) or apply a wettable metallization to the faying surface, as described in Chapter 4, section 4.1.2. The ability to use a high-melting-point braze affords the possibility of making active elements major components of the filler metal, these mostly having high melting points.

Chromium, titanium, and zirconium are often principal constituents of high-melting-point brazes for nonmetals because the Gibbs free energy of their respective oxide and carbide formation implies strong and ready interaction between these metals and both oxide and carbide ceramics, as well as graphite. Graphite can be brazed by any filler alloy that contains metals with strong carbide forming tendencies. Table 2.9 lists some of the ternary alloys in this family and their recommended areas of application [Canonico, Cole, and Slaughter 1977].

The liquidus projection of the chromium-titanium-vanadium system is shown in Fig. 2.17. This alloy system exhibits many metallurgical features that make it suitable for application in brazing, namely, complete miscibility of the constituents in the liquid and solid state and a narrow melting range over the whole of the ternary composition triangle. A preferred composition was derived by selecting the titanium/vanadium ratio with minimum melting range and adding chromium as appropriate. It was found that a 25% chromium content resulted in ready wetting and spreading on nonmetals, but substantially more of this constituent compromised the ability to produce preforms by mechanical working. The brazing alloy composition was therefore fixed at Ti-25Cr-21V, giving a liquidus temperature of about 1500 °C (2730 °F). The excellent wettability of this braze is attested by its ability to spread over the surface of nonmetals and penetrate and fill surface opening asperities, pores, and cracks. Not only does this characteristic improve the mechanical properties of the assembly because there is no abrupt metal/nonmetal interface, but closure of surface defects actually improves the fracture toughness of the nonmetallic component.

Other high-temperature brazes include Ti-8.5Si, Ti-30V, and V-35Nb, which have brazing temperatures of 1400 °C (2550 °F), 1650 °C (3000 °F), and 1870 °C (3400 °F), respectively. All three of these brazes can be used to join molybdenum and its alloys and can be used for high-temperature service. Nickel brazes are not recommended for use with molybdenum because nickel can embrittle the parent material. Tungsten can be brazed using nickel-base brazes, but for particularly high temperature service, more exotic filler metals such as Nb-2.2B, Nb-20Ti, and Pt-11W-4B are used. The latter braze provides a joint remelt temperature of around 2200 °C (3950 °F). For even higher-temperature applications, alloys such as W-25Os, W-50M0-3Re, and Mo-5Os can be used as brazes. Because the components need to be heated above 2500 °C (4530 °F) in order to effect wetting by the braze, the heating must be very intense, and an electric arc is often employed as the source. Clearly, this joining process then shares many of the characteristics of welding.

2.1.12 Low-Melting-Point Brazes with Little or No Silver (Excluding Aluminum)

Notwithstanding the advantageous properties of silver-base alloys as brazes, in particular, their relatively low melting points (600–800 °C, or 1110–1470 °F), excellent flow characteristics, and wide metallurgical compatibility with base alloys, they are expensive. These brazes generally contain more than 40 wt% of silver and, with this metal being more than 75 times the price of copper and approximately 150 times that of zinc, on a weight basis, their value can represent a significant fraction of the overall manufacturing cost, especially for consumer products. Hence, there is a demand for cheaper filler metals to replace silver-base brazes.

One step in meeting this objective has been the development of low-silver braze alloys [Eagles, Mitchell and Rebbeck 1995]. The primary constituents of the proposed brazes are copper

Table 2.8 Ductile-to-brittle transition and recrystallization temperatures for some refractory metals

	Ductile-to-brittle transition temperature		Recrystallization temperature	
Metal	°C	°F	°C	°F
Molybdenum	150	300	1150	2100
Niobium	−200	−330	985	1800
Tantalum	−195	−320	1100	2010
Tungsten	260	500	1200	2190

and manganese with a few percent of silver and, optionally, tin to adjust the alloy properties in the particular context of brazing (Table 2.10). The solidified brazes have a microstructure composed mostly of cored-copper solid solution.

These filler alloys can be prepared in the form of either paste or rod and used in conjunction with standard commercial brazing fluxes. They function equally well in vacuum and hydrogen atmospheres. These brazes possess relatively high viscosity when molten, and therefore are most effective for wide gap brazing. With 0.2 mm (8 mils) joint clearance, ring and plug test samples in 316L stainless steel can exhibit tensile strengths in excess of 200 MPa (29 ksi) (Fig. 2.18). Results with joints with wider gaps have not been reported.

In the increasingly competitive consumer markets where brazing is used, notably office furniture and low-cost bicycles, which have square or tubular steel frames, brazing faces strong competition from welding. Welding, however, is a realistic alternative only where considerable investment has been made in automated welding equipment because joints have to be welded sequentially, whereas brazing can be carried out in batch operations. In response to this market challenge, low-melting-point brazes have to be made much more cheaply, through the total elimination of silver.

Table 2.9 High-melting-point ternary braze alloy families for refractory metals, graphite, and alumina

	Brazing temperature range		Application		
Brazing alloy system	°C	°F	Refractory metals	Graphite	Alumina
Ge-Ti-Zr	1300–1600	2370–2910	X	X	...
Cr-Ti-V	1550–1650	2820–3000	X	X	X
Cr-Ti-Zr	1250–1450	2280–2640	X	...	...
Nb-Ti-Zr	1600–1700	2910–3090	X	X	...
Ta-Ti-Zr	1650–2100	3000–3810	X	X	X

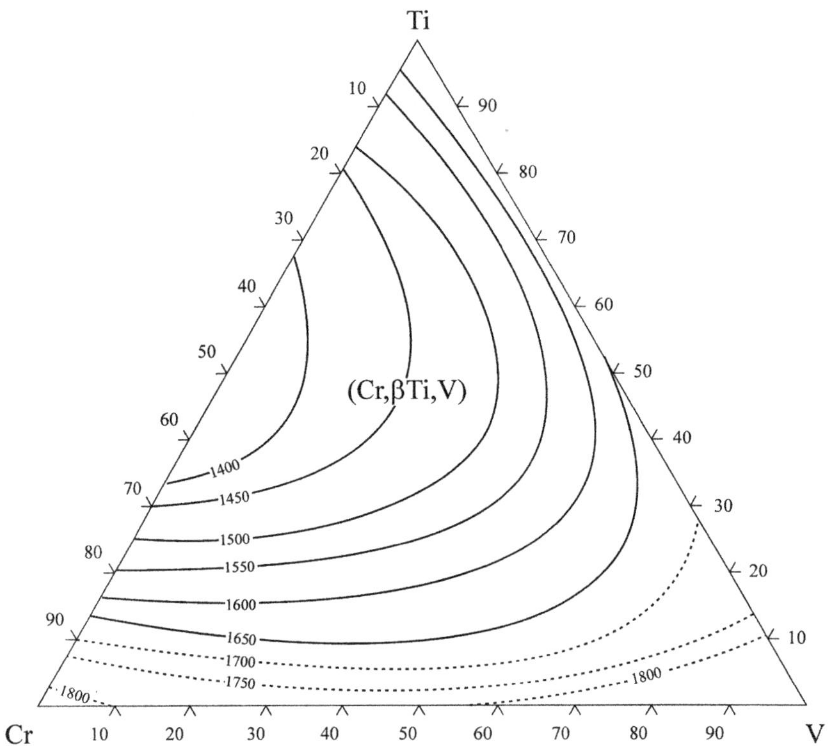

Fig. 2.17 Liquidus surface of the chromium-titanium-vanadium system. Given in wt%. Adapted from Samsonova and Budberg [1965]

They also have to meet a new constraint. In the past, lugs have been used for connecting the steel tubes with the braze used to affix the tubes in the lugs. These lugs fit over the ends of the tubes, help to ensure that the joints are close fitting, and fix the angles of the frame. However, to be effective they must be made to tight tolerance and their use also limits the options on frame geometry available to customers. To reduce cost and weight and also compete with welding processes, lugs are being dispensed with, but the adoption of this practice has the consequence of reducing the precision to which the joint clearances can be controlled. Therefore, the brazing alloys used in tubular frame joining must possess good gap-filling properties. In general, clearances of up to 0.5 mm (20 mils) must be tolerated.

In addressing these requirements, two new silver-free families of brazing alloys have been developed for use at temperatures between 800 and 900 °C (1470 and 1650° F). It has been demonstrated that these brazes are compatible with high-tensile-strength steel tubing. The new brazing alloys are capable of producing joints that meet the tensile and fatigue strengths required of steel tubular assemblies [Jacobson et al. 2002; Eagles, Mitchell, and Rebbeck 1995]. These alloys are based on the copper-manganese-zinc and copper-manganese-silver ternary systems. Manganese lowers the melting temperature of high-copper alloys. The silver content can be reduced to a small percentage (<10%) of the alloy, in the case of the copper-manganese-silver alloys, or dispensed with completely, as in the case of the copper-manganese-zinc brazes. Manganese and zinc are also very inexpensive compared with silver, with manganese being even less expensive than zinc. A small percentage of tin (typically, 5%) may be added to reduce substantially the solidus temperature of copper-manganese-silver alloys, as shown in Table 2.10, but it has been established that tin cannot be used to substantially substitute for the more expensive silver content, without embrittling the braze and joints made with it to steel (shown subsequently).

The two copper-manganese-zinc alloys that have been developed are 70Cu-10Mn-20Zn and 54Cu-4Mn-35Zn-6Ni-1Si. Both compositions lie outside of the brittle range in the ternary copper-manganese-zinc field, as shown in Fig. 2.19. Nickel and silicon are known to promote wetting by copper brazes, especially on oxidized steel surfaces, while the latter element also confers fluidity. However, these extra additions raise the melting point of the alloy and, to compensate, the zinc content is raised. The nickel enters into solid solution with copper and manganese as does the small silicon fraction. The constituent phases of these two alloys, together with their melting points and selected mechanical properties, are indicated in Table 2.11.

Brazing with these alloys can be carried out in air using air-propane torches and a conventional brazing flux containing a fluoroaluminate constituent appropriate to the heating conditions required. Lugless joints made between mild steel tubes using the 54Cu-35Zn-6Ni-4Mn-1Si alloy and a reference silver-base brazing alloy (44Ag-30Cu-26Zn) are shown in Fig. 2.20(a) and (b). The tube ends were profiled to provide a maximum gap of 0.5 mm (20 mils) in the saddle of the joint. As indicated in the photographs, all the brazed joints were filled and there were no holes in the joints.

The tensile strength of joints made with the copper-manganese-zinc brazing alloys compare favorably with those made with the conventional silver-base brazing alloy (44Ag-30Cu-26Zn), which has been used widely with tubular assem-

Table 2.10 Composition and melting range of some low-silver brazing alloys

Composition	Melting range °C	Melting range °F
65Cu-30Mn-5Ag	823–866	1513–1591
62.1Cu-30Mn-7.9Ag	810–853	1490–1567
60Cu-30Mn-5Ag-5Sn	745–822	1373–1512
57.1Cu-30Mn-7.5Ag-5Sn	742–822	1368–1512

Adapted from Eagles, Mitchell and Rebbeck [1995]

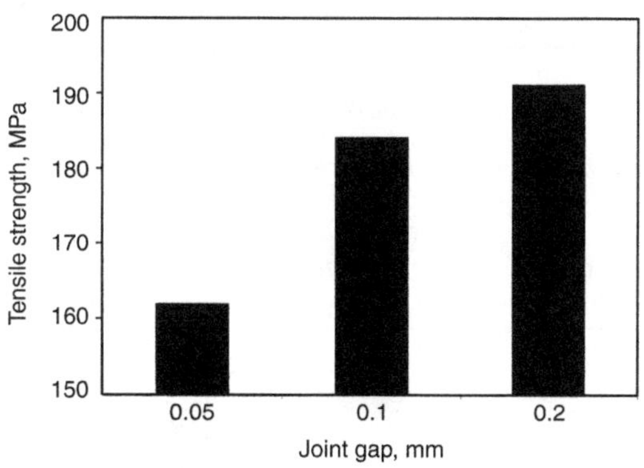

Fig. 2.18 Tensile strength of butt joints made with low-silver brazing alloys between mild steel components as a function of the joint clearance

blies. Both the copper-manganese-zinc alloys and silver-base brazing alloy satisfy the DIN 79100-2000-04 fatigue standard set for bicycle frames in Europe, as can be seen from Fig 2.21.

The copper-manganese-tin alloy system has also been investigated for potentially useful silver-free, low-melting-point brazes [Chatterjee and Mingxi 1990]. While the tensile strengths of alloys containing between 5 and 15% manganese and 15 and 20% tin are reasonable, being in the range 245 to 313 MPa (35.5–45.4 ksi), joints made to mild steel tubular components have been shown to be brittle and tend to fail when brazed assemblies are subjected to mechanical shock as, for example, when they are struck against a hard surface. On the other hand, joints made to copper-base materials have been judged satisfactory. These copper-manganese-tin alloys have the additional disadvantage of not being amenable to fabrication as wire or foil, except by rapid solidification, and even then the foils are somewhat brittle.

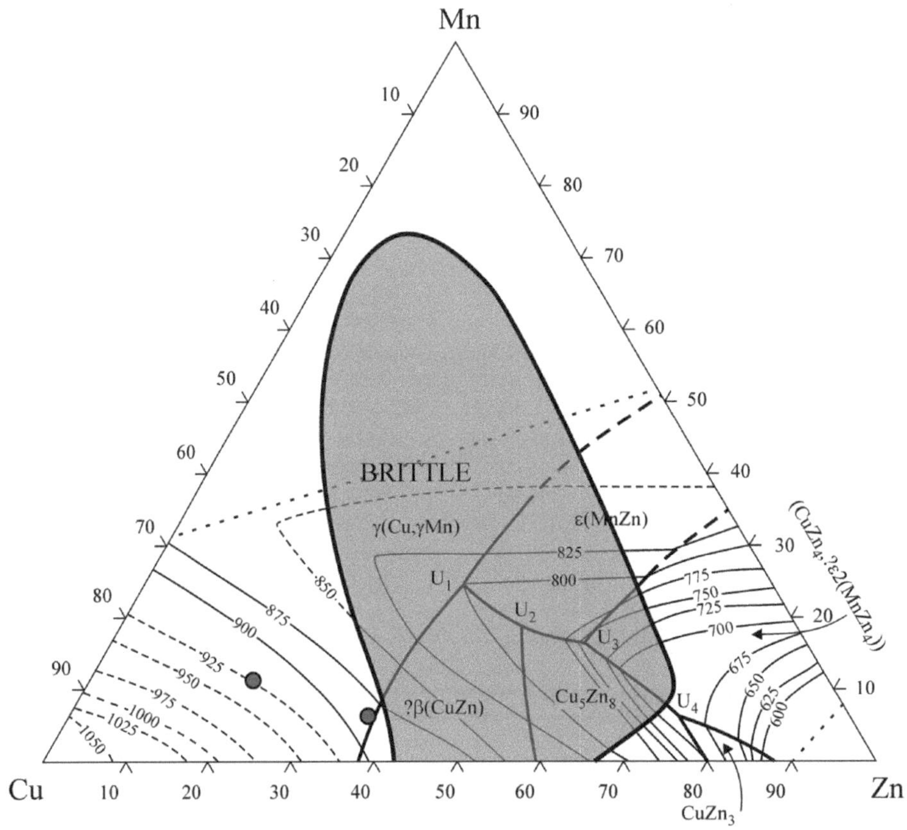

Fig. 2.19 Liquidus surface of the copper-manganese-zinc ternary alloy phase diagram with the compositions of the 70Cu-20Zn-10Mn and 54Cu-35Zn-6Ni-4Mn-1Si brazes, shown as wt%. The light shaded area is the approximate location of brittle alloys. The two darker dots are braze compositions.

Table 2.11 Silver-free brazing alloys based on copper-manganese-zinc alloys: their constitution and selected mechanical properties [Jacobson et al. 2002]

	Melting range			Tensile strength of brazed joints between mild steel tubes (0.5 mm, or 20 mils gap)		Alloy hardness,
Alloy composition	°C	°F	Phases present	MPa	ksi	HV
70Cu-10Mn-20Zn	799–925	1470–1697	(Cu,Mn) + CuZn	270	39.1	160
54Cu-4Mn-35Zn-6Ni-1Si	850–930 (approximate)	1562–1706 (approximate)	(Cu,Mn,Ni,Si) + CuZn	285	41.3	>180

A silver-free replacement for self-fluxing copper-silver-phosphorous brazes, represented by Cu-15Ag-5P, is Cu-9.2Sn-6.4P-5.8Ni. It has closely similar application and service temperatures, reportedly flows well, and forms good joints to copper components. It is prepared by rapid solidification casting technology [Rabinkin 1998]

2.1.13 Aluminum Brazes

Aluminum-bearing brazes are used primarily for joining components of aluminum and its alloys [Altschuller et al. 1990]. With aluminum melting at 660 °C (1220 °F), compatible brazes have among the lowest melting points of all brazing alloys.

Joints made to aluminum components tend to be more susceptible to corrosion than similar joints between other common metals because aluminum has a highly negative electrode potential. In consequence, brazed joints to aluminum and its alloys are more susceptible to galvanic corrosion than joints involving other common base metals when relatively more noble constituents are present. The necessary emphasis placed on corrosion prevention has resulted in the adoption of selected aluminum- and zinc-base alloys as the only satisfactory brazes for joining aluminum alloy components.

The degree of galvanic corrosion that can occur is a function of the difference between the electrode potential of two metals when they are electrically coupled by an electrolyte, such as water. The more negatively biased metal is the one that corrodes. Table 2.12 lists the standard electrode potentials of several common metals with reference to hydrogen and shows that aluminum is usually the metal subject to chemical attack. However, it should be borne in mind that in the majority of filler alloys, the phases present are not pure metals but solid solutions and compounds of various sorts, the electrode potentials of which are often not known and cannot be predicted readily from the constituent elements. Even for elemental metals, the data in Table 2.12, which are obtained under closely specified conditions, can serve only as a rough guide because measured electrode potentials depend on such factors as the nature and strength of the electrolyte and the surface condition and topography of the metal electrodes.

Susceptibility to galvanic corrosion provides one of the main criteria for aluminum being the major constituent of brazes used for joining the same metal. Indeed, all of the commercially

Table 2.12 Electrode potential of selected elements at 25 °C (77 °F)

Element	Electrode potential, V
Gold	+1.50
Silver	+0.80
Copper	+0.34
Hydrogen	0.00
Lead	−0.13
Tin	−0.14
Nickel	−0.25
Cadmium	−0.40
Iron	−0.44
Zinc	−0.74
Silicon	−1.30
Aluminum	−1.66
Magnesium	−2.37

(a) (b)

Fig. 2.20 Lugless joints made between mild steel tubes using (a) the 54Cu-35Zn-6Ni-4Mn-1Si brazing alloy and (b) the reference 44Ag-30Cu-26Zn brazing alloy

available aluminum-bearing brazes are based on the Al-13Si eutectic composition alloy, which melts at 577 °C (1071 °F). The aluminum-silicon phase diagram is presented in Fig. 2.22. Besides depressing the melting point of aluminum, silicon also confers some fluidity, and it is one of the few elements that does not enhance significantly the corrosion of the parent materials. The most commonly used aluminum braze is the Al-12Si alloy, which is just slightly hypoeutectic so as to make it marginally more malleable.

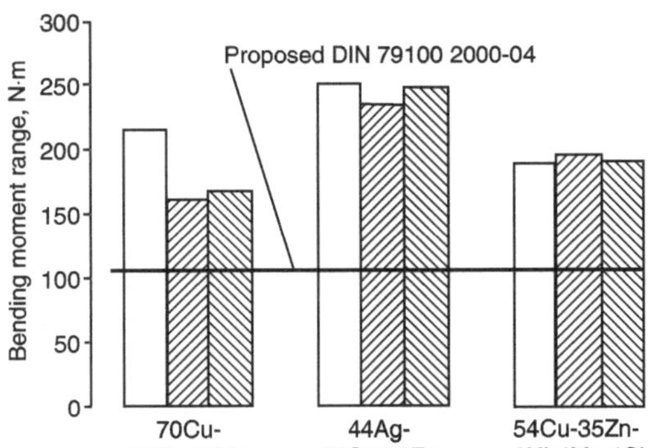

Fig. 2.21 Bending moment to failure at 10^5 cycles for three batches of tubular steel assemblies fabricated with a maximum joint gap of 0.5 mm, made with the two copper-manganese-zinc brazes and the reference silver-base braze (44Ag-30Cu-26Zn)

Aluminum-silicon binary alloys are available in the form of wire, foil, and cladded sheet. The clad material comprises a sheet of a suitable aluminum engineering alloy, coated on one or both sides with filler metal. Each cladding constitutes typically up to 15% by thickness of the total thickness of the sheet. Clad materials are particularly suited for fluxless brazing processes: the joint interface between two mated components that have been roll clad will liquefy on heating above the solidus temperature, which helps to displace the alumina layer and other surface films present.

Aluminum-germanium binary alloys are used as brazes for niche applications where the high cost of germanium and those of preform preparation can be accommodated. Alloy of the eutectic composition is brittle, unless prepared by rapid solidification so that the alloy is endowed with a highly refined grain structure and germanium precipitation is suppressed. [Illgen et al. 1991]. However, the low melting point of this alloy (420 °C, or 788 °F) and its low susceptibility to corrosion make it attractive for applications requiring a filler metal with a melting point in the range 350 to 550 °C (660–1020 °F),

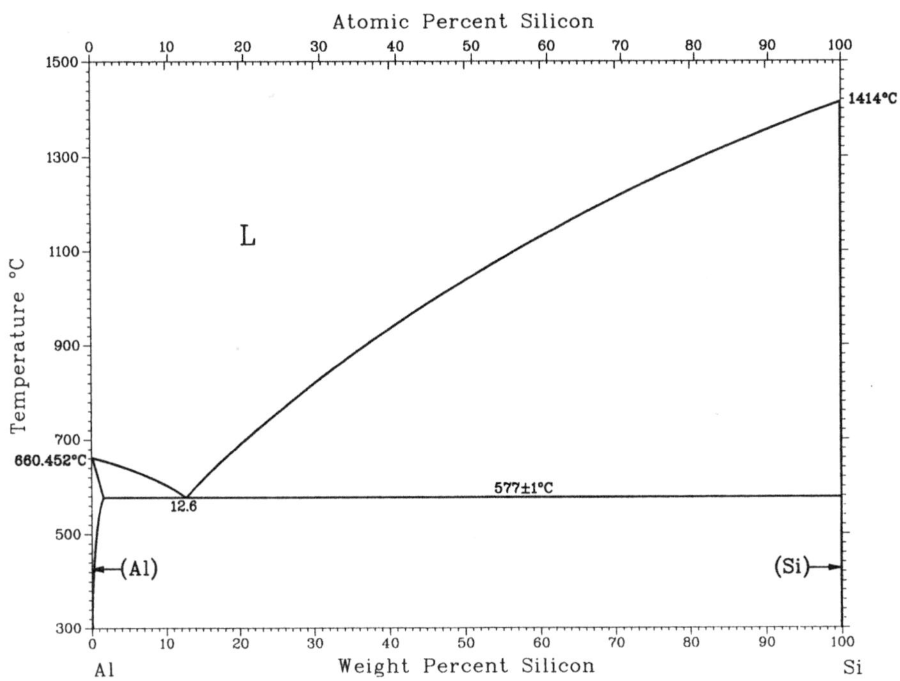

Fig. 2.22 Aluminum-silicon phase diagram

where there is a dearth of available alloys. An example is silicon chip die attachment in the fabrication of semiconductor components [Muller and Ruhlicke 1991].

Lower-melting-point brazes based on aluminum-silicon can be achieved by silver, copper, magnesium, and zinc additions due to the existence of ternary eutectic points in all four Al-Si-X alloy systems (where X = silver, copper, magnesium, and zinc). The aluminum-silver and aluminum-copper eutectic alloys cannot be used as brazes because their microstructures contain a high volume fraction of the brittle and relatively noble intermetallic compounds $AgAl_2$ and Al_2Cu, respectively. The embrittling phases can be sidestepped by preparing the braze as a foil of aluminum-silicon clad with a thin layer (a few microns, or fractions of a mil) of silver or copper. The cladding can be accomplished readily by electroplating. Provided the assembly is held at temperature for several hours, the silver and copper will diffuse completely out of the joint and into the components, whereupon their concentration is sufficiently low so that it no longer presents a problem. Strictly speaking, the process is one of diffusion brazing (see Chapter 6). The requirement to maintain the parts at the joining temperature for an extended period (typically hours) means that this process is suitable only for a restricted number of aluminum casting alloys. Joint strengths exceeding 225 MPa (326 ksi) have been achieved by this method, involving casting alloy A356.0 [Niemann and Wille 1978]. Zinc is extremely volatile at 500 °C (930 °F) and, if present in the alloy as a significant proportion (>10%), this will liberally coat the workpieces and exposed surfaces within the furnace in a layer of this metal. Magnesium is also volatile, but less so, and at least that metal has the benefit of aiding wetting by physically displacing oxide layers as it volatilizes, as well as gettering oxygen from the surrounding atmosphere [Ambrose and Nicholas 1986; Miller 1969]. A selection of representative aluminum filler metals suitable for brazing aluminum alloys are listed in Table 2.13.

Several problems are associated with the use of standard aluminum-base brazes with aluminum alloy components, which represent their main area of application:

- Their melting points are very close to those of most aluminum engineering alloys. For many of these alloys, the melting points of the available brazes can actually exceed the solidus temperatures of the engineering alloys. For this reason, most wrought and cast aluminum alloys cannot be joined by brazing.
- Aluminum has high solubility in these brazes, resulting in significant erosion of the parent metal. Consequently, thin-walled components less than about 0.5 mm (20 mils) thick cannot be joined easily by this means. The alloying increases the melting point of the filler and tends to impede lateral flow and fillet formation by the molten braze.
- They exhibit poor wetting on aluminum alloys when used without fluxes. This is a consequence of the high reactivity of aluminum with oxygen in the atmosphere and the refractory nature of the alumina that forms. By applying special procedures, fluxless brazing of aluminum is possible. This topic is discussed in Chapter 3, section 3.4.3.
- Heavy fluxing is generally required. The fluxing agents react extensively with the aluminum-base parent materials. This results in the formation of large quantities of corrosive flux residues that are difficult to remove. Fluxes for aluminum brazing are considered in Chapter 3, section 3.2.2.
- Most engineering alloys of aluminum rely on precipitation hardening for their boosted mechanical properties (i.e., hardening through the presence of a finely divided second phase in the material). The temperatures required for brazing with the available filler metals

Table 2.13 Typical aluminum-bearing filler metals

Composition, wt%				Melting range			
Al	Si	Cu	Mg	°C	°F	AA number	AWS designation
88.0	10.0	...	2.0	555–585	1030–1085	...	...
86.0	10.0	4.0	...	524–585	975–1085	4145	BAlSi-3
87.0	13.0	...	...	577	1071	4047	BAlSi-4
90.0	10.0	...	...	577–595	1071–1100	4045	BAlSi-5
93.0	7.0	...	...	577–612	1071–1134	4343	BAlSi-2
94.75	5.25	...	...	577–632	1065–1170	4043	BAlSi-1

are too high to be compatible with the heat treatment step that precedes precipitation of the second phase. There have been attempts to remedy this situation by developing new brazing alloys with substantially lower melting points. These alloys were further designed for fluxless brazing of aluminum and additional details are provided in the next section.

Notwithstanding these difficulties, fluxed and fluxless brazing of aluminum are industry-standard processes that are practiced routinely and are readily available at modest cost. Non-heat-treatable wrought aluminum alloys that can be brazed are to be found mostly in the Aluminum Association 1xxx, 3xxx, and 5xxx series designations. The heat treatable alloys most commonly brazed are the 6xxx series. Their principal constituents are magnesium and silicon. The 2xxx and 7xxx aluminum alloys are alloyed extensively and therefore have melting points that are too low to be brazeable with standard aluminum-silicon brazes. Notable exceptions are the 7005 and 7072 alloys.

2.1.13.1 Low-Temperature Brazing of Aluminum

As pointed out previously, the temperature requirements for fluxed and fluxless brazing with standard aluminum alloy brazes are not suitable for use with the majority of aluminum engineering alloys, casting alloys, and the high specific strength alloys favored for aerospace applications. This is because these materials either melt or degrade when exposed to temperatures above 550 °C (1020 °F). For precipitation-strengthened alloys, the goal has been to develop a braze that can be used at a process temperature between 480 °C (900 °F) and 540 °C (1000 °F), so that the joining and solution-treatment process steps might be undertaken in a single, combined furnace cycle, with the aging treatment carried out at a lower temperature in a subsequent step. Aluminum-silicon eutectic braze and the Nocolok (Solvay Fluor, Germany) flux both utilize process temperatures closer to 600 °C (1110 °F) (see Chapter 3, section 3.2.2). The development of a braze that is electrochemically compatible with aluminum, so that it is not a source of corrosion, and has a melting point closer to 500 °C (930 °F), could enable up to 90% of all structural and casting grades of aluminum to be joined by brazing. A virtue of making the new brazing process fluxless is that it would obviate the problem of also having to develop a suitable low-temperature flux.

The technical and commercial need for low-temperature aluminum brazes has provided the impetus for much research in this area. To have optimum utility, an aluminum brazing alloy should satisfy the following conditions:

- A brazing temperature below 550 °C (1020 °F), preferably close to 500 °C (930 °F), and a narrow melting range. This implies that the braze should be of eutectic or near-eutectic composition. A narrow melting range is one of the prerequisites for a filler metal to possess high fluidity when molten, especially given that alloying with aluminum components is likely to be extensive and tend to reduce the fluidity.
- The brazing alloy must have good fluidity when molten in order that it can penetrate and fill narrow joint gaps. Brazed assemblies intended for engineering applications need to have mechanically strong joints, which in turn means that they must be narrow and well filled with braze [Sloboda 1961].
- The ability of the braze to be used with simple furnace equipment and without flux are both attractive from the viewpoint of capital and recurring cost. Consideration must be given to the total equipment requirement, including exhaust gas scrubbers if the brazing process liberates volatile species that cannot be vented directly to the atmosphere.
- The brazing alloy needs to be metallurgically compatible with aluminum, particularly for joining thin-walled sections. This means there must be minimal substrate erosion, and no embrittling phases should form as a result of wetting and alloying with the base materials.
- Likewise, the solidified filler metal must exhibit electrochemical compatibility with the base alloys. Galvanic corrosion is a major source of weakness in joints made to aluminum and its alloys. To minimize this problem, the electrode potential of all the phases in the joint should be fairly close to that of aluminum.
- The constituents of the brazing alloy need to be low cost so that the associated low-temperature brazing process can find wide commercial application. For this reason, germanium, which is roughly 400 times the cost of aluminum, has to be ruled out for most applications.

- Finally, the brazing alloy needs to be capable of being processed readily into a number of forms—rod, wire, paste, and so forth—and have long shelf life.

Analysis of these objectives leads to the conclusion that the primary constituent of any new filler metal suitable for brazing aluminum alloys is aluminum itself because this provides the best chance for ensuring metallurgical and electrochemical compatibility. The requirement is, therefore, to identify alloying additions that will depress the melting point sufficiently, without upsetting these other properties. A list of elements that form possible binary low melting point eutectics with aluminum and their limitations is given in Table 2.14.

The literature records attempts to develop simple brazes for aluminum alloys, involving alloying with more unusual elements including calcium [Hagiwara et al. 1988] indium, tin, and yttrium [Werner, Slaughter, and Gurtner 1972]. However, like those listed in Table 2.14, it is clear that no binary alloy will satisfy all the functional requirements, and therefore higher-order alloys must be considered.

Suzuki, Kagayama, and Takeucji [1993] reported a eutectic braze of composition Al-40Zn-4.2Si with a melting point of 535 °C (995 °F). The disadvantage of this filler metal lies in the fact that the vapor pressure of zinc is very high at 500 °C (932 °F), and the high zinc content makes it entirely unsuitable for a vacuum brazing process. Also, high zinc-containing alloys do not have sufficient corrosion resistance for most applications, involving components of aluminum. Kayamoto and colleagues [1994] identified aluminum-germanium-silicon(-magnesium) filler metals that produced acceptable results in terms of brazing efficacy, but the process temperature of 575 °C (1067 °F) gives it little advantage over the established Al-12Si braze, and the costly germanium content represents a decided drawback, as mentioned previously. Furthermore, neither the Al-Zn-Si nor the Al-Ge-Si alloys are entirely satisfactory from the processing perspective, owing to limitations in the brazing conditions of long cycle time and the need to apply a compressive load during the brazing cycle for good joints to be produced. Other ternary alloy combinations with silver, silicon, germanium, copper, manganese, magnesium, and zinc have been examined at some time, but all suffer from one or more of the drawbacks encountered with the binary alloys [Kayamoto et al. 1996]. Although undesirable from the point of view of metallurgical complexity, low-melting-point brazes for aluminum must therefore be based on high-order multicomponent alloys.

The aluminum-silicon eutectic provides a suitable basis from which to develop a multicomponent aluminum braze. The eutectic composition and those close to it meet all of the target requirements, with the exception of melting point. Copper is relatively inexpensive and one of the more effective additions for lowering the melting point of aluminum, to a minimum of 524 °C (975 °F). However, the ternary eutectic, of composition Al-5.2Si-26.7Cu, contains a large volume fraction of the phase Al_2Cu, which renders the alloy brittle and susceptible to corrosion when in contact with aluminum. Yet, hardness measurements, backed by salt-spray testing, revealed that provided the level of copper is restricted to 20%, the alloy is acceptable from the point of view of corrosion resistance for typical ambient service environments, but then other constituents must be added to depress the liquidus to below 550 °C (1020 °F). Figure 2.23 shows the variation in liquidus temperature and hardness for aluminum-copper-silicon alloys as a function of copper content. It can be seen that the combined addition of copper and silicon is effective in lowering the melting point to below 530 °C (985 °F) but at the expense of ductility. Not surprisingly, then, the most promising low-melting-point aluminum brazes are based on aluminum-silicon-copper.

Low-melting-point aluminum brazes, which largely fulfill the target criteria set out previously, have been developed, and the recommended compositions are listed in Table 2.15 [Jacobson, Humpston, and Sangha 1996; Chuang et al. 2000]. Two of these brazes are

Table 2.14 Binary eutectic alloys with aluminum and their potential as low-melting-point brazes for aluminum engineering alloys

Element complementing aluminum and its weight percentage in the binary eutectic	Eutectic temperature		Drawback, in relation to use with aluminum engineering alloys
	°C	°F	
71.9 Ag	567	1053	Brittle, corrosion
32.7 Cu	548	1018	Brittle, some corrosion
51.6 Ge	420	788	Brittle, cost
35.6 Mg	450	842	Brittle, volatile
12.6 Si	577	1071	High melting point
94.0 Zn	381	718	Volatile, low melting point

based on the Al-Cu-Si-Ni-(Zn) alloy system, and another pair are based on the Al-Cu-Si-Sn-(Mg) system. The liquidus projection of the aluminum-copper-nickel-silicon pseudoternary system at 20% copper is shown in Fig. 2.24.

Certain minor additions were found to produce a marked improvement in brazing behavior. This is the principal reason for the bismuth, beryllium, zinc, and magnesium additions noted in Table 2.15. Zinc and magnesium are effective wetting agents, principally by virtue of their ability to remove residual oxygen and water vapor from the joint surfaces and from the furnace atmosphere. The benefits of bismuth, beryllium, and strontium have a different cause. As mentioned in connection with rapidly solidified nickel-bearing brazes discussed in section 2.1.10, it is well known that wetting and spreading by molten filler metals is enhanced if they possess a highly refined microstructure prior to melting. Elements that produce grain refinement in aluminum alloys are well established and include strontium [Hellawell 1979]. On the other hand, small concentrations of bismuth and beryllium, in combination with tin or zinc destabilize native coatings of alumina on aluminum surfaces, either chemically or physically [Schultze and Schoer 1973; Jacobson, Humpston and Sangha 1996]. A useful side effect of these subpercentage additions is that they further depress the liquidus and solidus temperatures of the braze by 2 to 3 °C (3–4 °F).

Of the four alloys listed in Table 2.15, possibly the nearest to commercial exploitation is the first, namely, Al-20Cu-5Si-2Ni-0.01Bi-0.01Be-0.01Sr. An important consideration in this regard is the ability to co-roll a plate of standard Al-12Si brazing alloy, which already contains the requisite minor additions of bismuth, beryllium, and strontium, sandwiched between two plates of standard Cu-20Ni alloy, or vice versa, to a ductile foil of exactly the desired composition of the low-melting-point braze, because both alloys are themselves ductile. Furthermore, these two alloys are already produced in volume quantities as brazes in their own right. On heating the composite foil to the brazing temperature, the constituents interdiffuse to form the quaternary alloy in situ. A further description of these so-called trifoil preforms is given in Chapter 4, section 4.1.5.

Formation of thin foils of the other compositions listed in Table 2.15 requires hot rolling, with the attendant cost implications and problems of oxide formation because the alloys need to be heated to 250 to 370 °C (480–700 °F) to achieve sufficient ductility for this process to be accomplished satisfactorily, as can be seen from Fig. 2.25.

At least the first two brazes listed in Table 2.15 form joints of acceptable mechanical integ-

Table 2.15 Low-melting-point aluminum braze compositions, based on the Al-20Cu-Si base alloy

Braze alloy composition	Melting range °C	Melting range °F
Al-20Cu-5Si-2Ni-0.01Bi-0.01Be-0.01Sr	515–535	959–995
Al-22.3Cu-9.5Zn-4.5Si-1.2Ni-0.01Bi-0.01Be-0.01Sr	495–505	923–941
Al-20Cu-7Si-2Sn	505–525	941–977
Al-20Cu-7Si-2Sn-1Mg	501–522	934–972

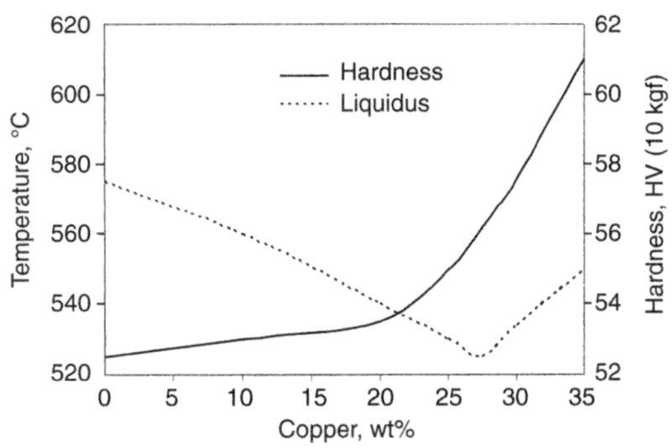

Fig. 2.23 Variation in liquidus temperature and hardness for aluminum-12.6 wt% silicon alloys as a function of copper content

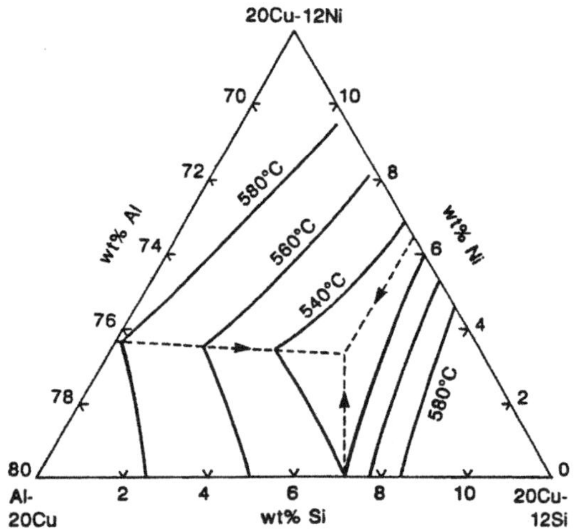

Fig. 2.24 Liquidus projection of the aluminum-copper-silicon-nickel quaternary system at 20% copper

rity to a variety of aluminum engineering alloys, with simple lap joints usually failing in the parent material when stressed in shear. Brazed joints have been successfully made to 3003, 5252, 5083, 6061, 6010, 6013, 6082, 7075, 7475, and 8090 wrought alloys and the common casting alloys 356 and 361 (Table 2.16) [Heine and Sahm 1993]. It has also been established that the joints are not a source of weakness in round butt joints subject to bending fatigue. In common with conventional aluminum brazes, these lower-melting-point brazes exhibit high liquid-to-solid contraction on solidification. Joint filling in parallel gaps is aided therefore by the application of compressive stress during the heating cycle. The applied stress need be only 0.3 to 0.5 MPa (44–73 psi) and, in any event, should not exceed 2 MPa (300 psi) in order to prevent distortion of the parts being brazed because these will obviously soften as they are heated to the brazing temperature [Gempler 1976; Humpston, Jacobson, and Sangha 1995]. Provided the copper concentration is restricted to about 20%, the pitting corrosion potential of the brazes is in the region of 440 mV, compared with 470 to 530 mV for wrought aluminum alloys. As a result, the extent of corrosion is limited to minor pitting of the substrate at the rate of 5 μg/m^2/h, with no accelerated crevice corrosion. The relevant test was 5,000 hour salt-spray testing in accordance with DIN 50021. This rate and extent of corrosion is comparable to that between existing aircraft rivets and skins and is adequate for many situations, particularly if additional protection is provided by anodizing or painting.

Table 2.16 Strength of brazed joints to aluminum parent materials using low-melting-point aluminum brazes, measured in shear

	Shear strength	
Base alloy	MPa	ksi
3001	75	11
3003	70	10
6013	142	21
6061	196	28
6082	141	20
7475	206	30

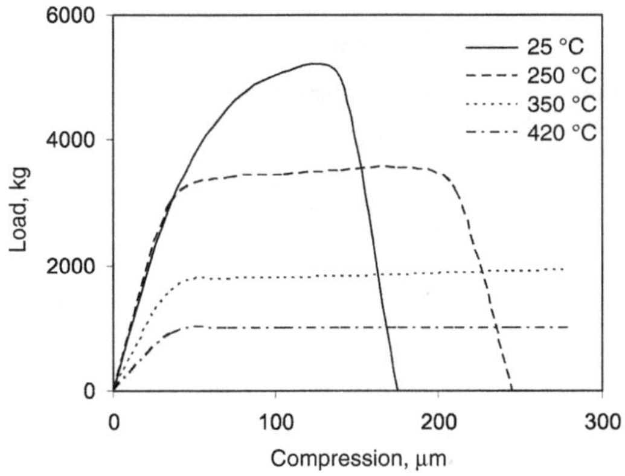

Fig. 2.25 Thermomechanical working behavior of the Al-20Cu-5Si-2Ni braze, which shows it has favorable characteristics for hot working if heated to between 250 and 370 °C (480 and 700 °F)

2.1.13.2 Aluminum Alloy Brazing of Other Materials

Aluminum-silicon-base brazes are commonly used in the semiconductor industry to join to silicon power devices. As the semiconductor must be packaged in a hermetic enclosure containing a dry nitrogen atmosphere, there is obviously little opportunity for corrosion to occur, and the favorable melting point of aluminum brazes can be exploited advantageously [Humpston et al. 1992]. Besides, the electrode potential of the silicon device is similar to that of the silicon phase in the braze. Figure 2.26 shows a microsection between a silicon power device brazed to a molybdenum heat sink using the Al-12Si alloy, and Fig. 2.27 shows a cross section of a joint made using the same braze to copper-metallized components.

Aluminum alloy brazes are also satisfactory for joining metals such as stainless steels, molybdenum, tungsten, and some copper alloys, provided the different electrode potentials of

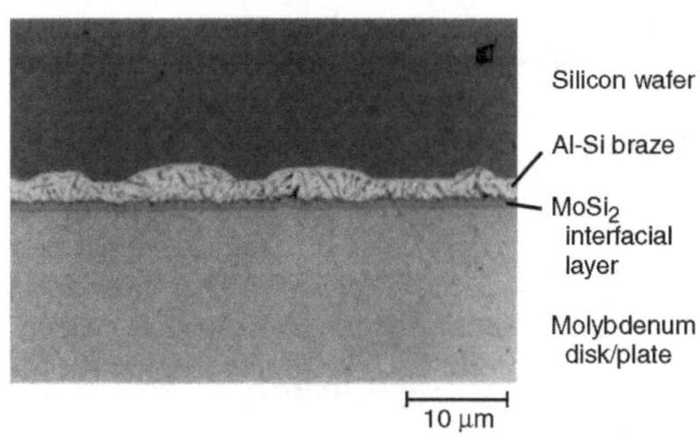

Fig. 2.26 Cross section through a joint made between a silicon wafer and a molybdenum disk using the Al-12Si braze and a peak process temperature of 680 °C (1256 °F), maintained for 30 min. The braze has wet and dissolved the silicon wafer in a nonuniform manner that is detrimental to the electrical performance of devices. 1000×

these metals are recognized and appropriate measures are taken to minimize the risk of corrosion of the joint. This usually means that the service environment must be noncorrosive, which generally implies using some degree of environmental protection.

2.2 Effect of Impurities

During the development of a joining process in the laboratory, conditions tend to be strictly controlled, including the purity of the materials used. In production, standards are often less stringent. Yet, relatively low levels of metallic and nonmetallic impurities in joints, either present in the filler alloy or infiltrated from the parent material during the joining process, can have an effect on properties. Manufacturers supply brazes of suitable purity for most applications, but the joint designer must always be alert to the possibility of minor changes in composition resulting from alloying with the expected constituents of the components and also their impurities.

Among nonmetallic elements, hydrogen, sulfur, and phosphorus are known to be responsible for brittle joints, the problem being exacerbated by the rapid diffusion of these elements into most engineering metals and alloys, so that they can penetrate deeply. Sulfur and phosphorus are especially deleterious to the mechanical integrity of joints made with nickel-base brazes. Sulfur can enter joints from lubricating oils, grease, and paint residues left on surfaces. The reaction of sulfur with nickel leads to the formation of the brittle nickel sulphide, Ni_3S_2, which concentrates in grain boundaries, so that failure occurs by intergranular embrittlement. Likewise, phosphorus, which can enter joints from electroplatings—the element commonly added to nickel electrolytes as a brightening agent—is another source of embrittlement, through the formation of nickel phosphide, Ni_3P. It is for this reason that the self-fluxing copper-(silver)-phosphorous brazing alloys are not generally recommended for use with components of nickel-base alloys, and, for similar reasons, those based on iron (see section 2.2.1).

Likewise, boron, silicon, and carbon also have a detrimental effect on brazed joints made with nickel brazes. These elements not only produce intergranular embrittlement like sulphur and phosphorus, but also enhance erosion (dissolution) of the base metal parts. Furthermore, the presence of carbon in nickel-base brazing alloys in levels greater than about 0.1 wt% can compromise the corrosion resistance of joints [Schwartz 2003].

While the presence of impurities will often have unfavorable consequences, some impurities are highly beneficial and their presence can result in a favorable improvement in the mechanical or physical properties of a joint, or a boost to wetting or spreading behavior of a braze. The effects are often not predictable. Further discussion of deleterious and also of beneficial impurities is continued in the two sections that follow.

2.2.1 Examples of Deleterious Impurities

The embrittlement of brazed joints by impurity elements is illustrated spectacularly by the deleterious effect of even trace quantities of aluminum and phosphorus on joints made between mild steel and brass with the once widely used silver-copper-zinc-cadmium brazing alloys. This effect is not observed in similar joints made between all-steel or all-brass components.

Boughton and Sloboda [1970] made a study to quantify this problem and establish its cause. A series of V-notch Charpy impact test pieces were prepared by brazing together two equal lengths of steel bar with beveled ends, at which the joint was made. The braze contained controlled fractions of either aluminum or phosphorus, and the joining operation was carried out under fixed conditions (temperature, time, heat-

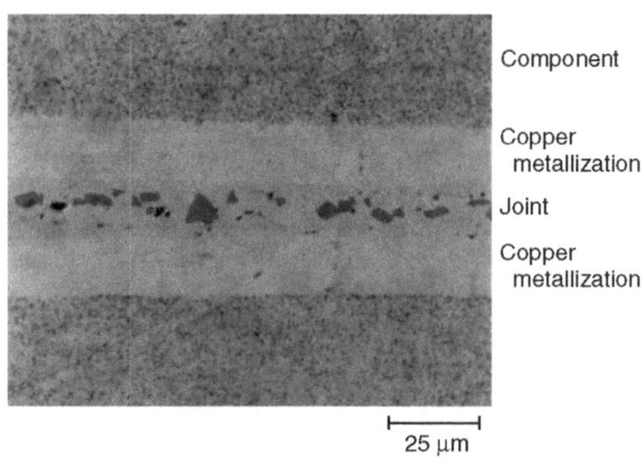

Fig. 2.27 Micrograph of a joint made using Al-12Si braze between components that have been metallized with copper. The joint microstructure comprises silicon in a matrix of the intermetallic compound Al_2Cu. 420×

ing rate, fluxing, cooling rate, etc.). The test pieces were then cleaned and subjected to an impact bending test. From the data, a graph was plotted of the energy absorbed by the test piece versus impurity concentration, reproduced in Fig. 2.28, which shows that for both aluminum and phosphorus there is a critical concentration threshold above which the joints are completely embrittled, and the impact bending strength plummets by nearly two orders of magnitude.

Although the critical impurity concentration is substantially higher than the maximum permitted in the brazing alloy by Standards' specifications, the problem can arise in practice with brazes that conform to these specifications because aluminum and phosphorus are sometimes present in brass, as well as in the braze. On wetting of brass by the molten braze, these elements will dissolve into the molten filler alloy. Elemental analysis obtained in an electron microscopic examination of the fracture surfaces has shown that the cause of the weakness is in the formation of brittle iron-base phases containing aluminum and phosphorus at the braze/steel interface. This finding accounts for the embrittlement problem being evident only in brazed joints between brass and steel. It can be prevented by tightening the impurity limits in brass components when used in such assemblies.

There are other occasional instances when brazing alloys complying with the Standards' specifications contain minor constituents that are not compatible with the parent materials. A case in point is the Ni-14Cr-5Fe-5Si-3B-0.7C brazing alloy (B-Ni1) that observes the prescription set by the Standards to have a carbon content not exceeding 0.7%. When this alloy is used to join martensitic stainless steel, cracks are observed in the fillet surface that extend inward to a depth of about 0.25 mm (0.01 in). Such defects could be catastrophic to the fatigue life of the assembly. These cracks are thought to be associated with the volumetric strain associated with the formation of martensite during cooling of the parent material from the brazing temperature. The problem can be circumvented by conducting an isothermal stress relief treatment just above the martensitic transformation temperature, but this greatly extends the process cycle time. Investigators found that by reducing the carbon content of the braze to 0.1%, and preferably below 0.03%, the modulus of the filler metal was lowered sufficiently to eliminate all instances of fillet cracking [Eng, Ryan, and Doyle 1977].

In vacuum brazing it is common practice to specify brazes with far higher purity than when conducting brazing using fluxes or reactive gas atmospheres. Four and even five nine (i.e., 99.99% or 99.999%) purity multicomponent brazes are available commercially. It is found that a tight restriction on minor impurities generally results in more consistent brazing behavior and superior quality joints that can make the additional cost of high-purity brazes cost effective.

2.2.2 Examples of Beneficial Impurities

Most manufacturers' catalogs contain modified variants of standard brazing alloys for applications in which more rigorous joint performance is required. The additions are often small quantities of elements that are more refractory than the principal constituents. They are made to bolster the mechanical properties and corrosion resistance or wetting behavior of the basic filler metal. The beneficial role of nickel and manganese in improving the mechanical properties and corrosion resistance of silver-copper-zinc and silver-copper-zinc-cadmium brazing alloys is mentioned in sections 2.1.5 and 2.1.6, respectively. The improvements to mechanical strength arise principally from a combination of grain refinement and solution strengthening. Enhanced resistance to oxidation and chemical attack arise from segregation of these minor con-

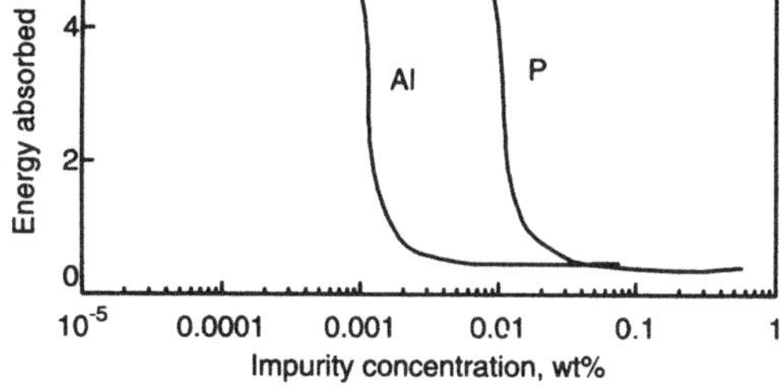

Fig. 2.28 Effect of impurity elements on the impact strength of joints made in mild steel using an Ag-Cu-Zn-Cd filler alloy. Adapted from Boughton and Sloboda [1970]

stituents to the surface of the fillets and exposed grain boundaries where they favorably alter the stability of the native oxide on the solidified braze.

The range of parent materials that a braze will wet satisfactorily can be enhanced by including in the filler metal a constituent that has a higher chemical affinity toward that particular parent material than do the other constituents. Therefore, silver-base brazes containing nickel can be used to join carbide cutting-tips to tool shanks, while manganese additions facilitate wetting of cast iron. In the second case, the improved wetting is due to the ability of manganese to dissolve carbon. The commercial range of these active brazing alloys, often denoted by the suffix ABA, has been augmented considerably by manufacturers taking advantage of the capabilities of rapid solidification enabling them to produce flexible foil and wire with small percentages of elements such as titanium, silicon, aluminum, molybdenum, and vanadium incorporated in the braze. A representative selection of such alloys marketed by one supplier is shown in Table 2.17. The addition of a few percents of titanium to silver-base brazes, for example, considerably enhances their ability to wet engineering ceramics. The subject of active brazing alloys is discussed in further detail in Chapter 7, section 7.2.

A prime example of the effect of impurity elements in changing the wetting and spreading behaviors of brazes is provided by aluminum-bearing brazes when these need to be used without an appropriate fluxing agent. Foils and other preforms of aluminum-rich brazes tend to produce poor wetting and spreading over the joint surfaces. However, by making small additions of elements such as antimony, barium, bismuth, or strontium, wetting can be improved considerably, as explained in section 2.1.13.1. The beneficial effect of some of these minor additions on the surface tension of molten aluminum is shown in Fig. 2.29. The concentration of the individual additions is restricted to a maximum of about 0.3% in order to avoid perceptibly altering the bulk metallurgical characteristics of the braze.

For more conventional brazing alloys, there is a trend of adding trace quantities of cerium, which is reflected in the technical literature. This element boosts the fluidity of many brazes and is a particularly useful addition to quaternary or higher-order compositions, where other additions have been made for functional reasons that tend to compromise the spreading ability of the braze. Much of this deficiency can be made good by a small addition of the highly reactive rare-earth element, cerium. Mischmetal, a low cost mixture of rare earths, has been used at a concentration of 0.08% as a substitute for boron in nickel-base brazes to improve wetting and braze flow. Boron penetrates the grain boundaries of many heat resistant structural alloys, degrading their mechanical properties, and hence its replacement by mischmetal. Lithium, when added as a minor constituent to silver-base brazes, aids wetting because it is effective in reducing the oxides of many elements to metal. The resulting lithium oxide dross is displaced readily by the molten braze. Further development of brazing alloys containing small proportions of lithium is to be expected.

Certain constituents present in parent materials can also benefit brazeability by helping to undermine what are normally resilient oxides, as pointed out by Eustathopoulos, Nicholas and Drevet [1999, 356]. As examples, they cite trace amounts of carbon in stainless steels, which enable the normally tenacious chromium oxide (Cr_2O_3) surface film to be reduced in a vacuum and permit fluxless brazing, and the presence of

Table 2.17 A selection of commercially available active brazing alloys

		Liquidus		Solidus	
Alloy composition (active elements are listed last)(a)	Proprietary name	°C	°F	°C	°F
97.5Au-0.75Ni-1.75V	Gold-ABA-V	1090	1994	1045	1913
96.4Au-3Ni-0.6Ti	Gold-ABA	1030	1886	1003	1837
92.75Cu-3Si-2Al-2.25Ti	Copper-ABA	1024	1875	958	1756
82Au-15.5Ni-1.75V-0.75Mo	Nioro-ABA	960	1760	940	1725
92.75Ag-5Cu-1Al-1.25Ti	Silver-ABA	912	1673	860	1580
68.8Ag-26.7Cu-4.5Ti	Ticusil	900	1650	780	1435
63Ag-35.25Cu-1.75Ti	Cusil-ABA	815	1500	780	1435
63Ag-34.25Cu-1Sn-1.75Ti	Cusin-1-ABA	805	1483	775	1427
59Ag-27.25Cu-12.5In-1.25Ti	Incusil-ABA	715	1319	605	1121

(a) Impurities with a vapor pressure $<10^{-7}$ mm Hg at 500 °C are limited by the manufacturer to a total of 0.05%. All other metallic impurities with a higher vapor pressure at 500 °C are limited to 0.002% each.

molybdenum, which can be responsible for destabilization of the surface oxide on the steel due to the oxidization of the molybdenum and subsequent volatilization.

2.3 Application of Phase Diagrams to Brazing

The selection of a braze for a particular application is often made on the basis of just the melting point and mechanical properties of the braze and its ability to wet the parent materials. The braze is regarded as a uniform layer of metal that simply bridges the gap between the components and binds them together. If only life were that simple! In reality, the formation of the desired metallic bond between the braze and a component requires a degree of alloying between them. The ensuing metallurgical reactions usually lead to a heterogeneity of phases in the joint. To complicate matters further, kinetic factors tend to accentuate the development of this nonuniformity. Such inhomogeneities often determine the quality and overall characteristics of joints, such as their mechanical properties, the ease and extent of braze spreading, the nature of any fillets formed, and so on.

Metallurgical reactions do not cease once the joint has been made but continue to proceed, to a greater or lesser extent, during the service life of the assembly. The rate-controlling step for reaction between two solid metals is the diffusion of atoms between the reacting phases. The relative position of the product of the reaction and the reacting phases will be governed largely by the diffusion coefficients of the participating metals. For individual metals it has been established empirically that the rate of diffusion, D, increases rapidly with absolute temperature, T, following an exponential relationship:

$$D = D_0 \exp^{-Q/kT}$$

where k and D_0 are constants, k being the Boltzmann constant, and D_0 an experimentally determined factor for each combination of reacting phases that may vary with concentration. Q is the activation energy for diffusion, which, to a first approximation, is proportional to the melting point, T_m, of the particular metal [Birchenall 1959]. The rate of reaction will therefore be dependent on the homologous temperature defined as the ratio of T/T_m and will be more pronounced for low-melting-point brazes that see service at elevated temperatures. For example, the use of brazed superalloy and ceramic assemblies at

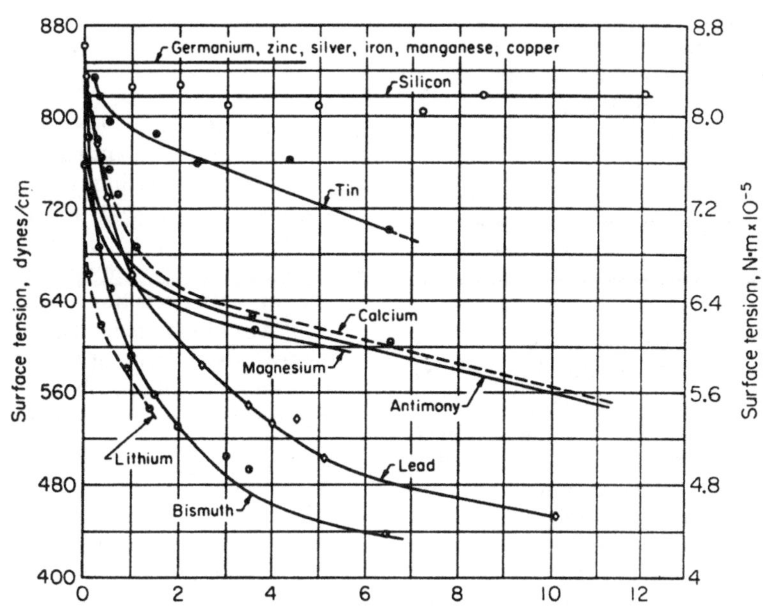

Fig. 2.29 Reduction in the surface tension of molten aluminum produced by various alloying additions. Adapted from Korol'kov [1956]

temperatures that are, in some cases, within 0.9 T_m may result in metallurgical changes that are comparable to those that are more commonly observed in soldered joints.

For a proper understanding of metallurgical reactions between brazes and parent materials, it is essential to have some grasp of the subject of alloy constitution. The "constitution" of an alloy refers to features such as its composition, melting range, range of phase stability, solubility limits, and related parameters that can be deduced from the phase diagram of the system in which the alloy appears. In the following sections, some attention is given to highlighting the value of phase diagrams and suggesting how this source of information might be tapped.

Although the available literature on phase diagrams may appear to be reasonably comprehensive, it is worth bearing in mind that reliable diagrams exist for roughly only 50% of binary combinations, 5% of ternary systems, and 0.5% of quaternary mixtures. A compendium of authoritative alloy phase diagrams is being prepared under the auspices of the International Programme for Alloy Phase Diagram Data (IPAPD) and the first volumes have appeared already. This work is ongoing and updates are to be found in the *Bulletin of Alloy Phase Diagrams* and *Journal of Alloy Phase Equilibria.* On a periodic basis, these publications include a cumulative index that lists all phase diagram evaluations published by members of IPAPD. It is worthwhile consulting the most recent phase diagram available because older diagrams may contain significant errors or omissions.

If the appropriate phase diagram for elucidating specific braze/substrate reactions and joint microstructures is not available in the literature, experimental techniques and a methodology for elucidating gaps in phase diagrams are accessible (for example, see Humpston and Jacobson [1993], Chapter 3.3). Note that all of the phase diagrams in this book are defined in weight percentages of the constituent elements because this is more appropriate to brazing technology than the atomic percentage scale. But, the relative proportion of the elements in intermetallic compounds, such as Cu_3P, refers to atomic weights. General equations for converting atomic to weight percent of constituents in alloys, and vice versa, are given in Appendix A2.1.

The fundamentals of alloy phase diagrams are covered in many metallurgical textbooks and are not repeated here. Readers with little background in this field are referred to some of the publications listed in the Preface, which provide an excellent introduction to the subject. Here, it will suffice to state that a phase diagram is a representation of the thermodynamic stability of phases as a function of composition with respect to particular thermodynamic variables such as temperature or, less commonly, pressure. What is important to remember is that the information provided by the diagrams relates to essentially equilibrium conditions. The phase diagram provides information about the ultimate balance of phases within the joint and those that are likely to be encountered during the progression toward equilibrium. A joined assembly in which the braze and abutting components are different materials is never in true compositional equilibrium as long as the joint remains distinct. In most practical contexts, the composition of a joint will tend toward equilibrium over most of its width and, therefore, phase diagrams are applicable to an assessment of its constitution. However, at the interfaces of the joint with the parent materials, marked compositional gradients will exist, which can cause a significant deviation from equilibrium. These variations will be exacerbated by any temperature gradients that develop during the process cycle and are manifested as the appearance of different phases in those regions. Nevertheless, even here, phase diagrams can assist in the elucidation of the metallurgical reactions and the resulting phases, as shown in the following section.

Phase diagrams can provide the following practical information:

- The melting temperature of the "virgin" braze and of the abutting components
- The probable freezing range of the braze following alloying with the components and hence the remelt temperature of the joint
- Whether the braze remains homogeneous in the joint after reaction with the components and, if not homogeneous, the phases that are likely to be present, or which may form subsequently, together with their elemental compositions and melting temperatures.

Phase diagrams do not reveal:

- The rate of reactions that might occur between the braze and the components and their variation with time and temperature. This applies both when the braze is molten and when it is solid during service.
- The spatial distribution and morphology of phases in the joint, although occasionally it

is possible to deduce whether new phases are likely to form as interfacial layers or will be dispersed throughout the solidified braze. Examples are given in section 2.3.2.

- The wetting characteristics of a particular braze/parent materials combination, in relation to surfaces, even when these are "perfectly" clean. In practice, wetting is likely to be influenced heavily by the oxides, impurities, and residues that are inevitably present on component and braze surfaces but that are extraneous to the alloy phase diagram.
- Physical properties of joints, in particular, the mechanical and corrosion characteristics. However, it is often possible to predict the likely range of certain physical properties by comparison with other known alloy systems.

The simplest diagrams that are encountered in a joining context are those relating to binary alloys where, for example, the braze is a pure metal being used to join components of another metal. This situation is represented by the use of pure copper to braze iron pipes.

High-order alloy systems are naturally more complex and are less well documented, as noted earlier. However, for a given joining process, only a very limited portion of the phase diagram is required and, if this is unavailable, it is often possible to experimentally determine the necessary data, as mentioned previously.

Recently, an exciting breakthrough appears to have been achieved in a method that is able to predict, with a claimed accuracy exceeding 99%, whether compound formation will occur for any binary, ternary, or quaternary system. This predictive tool should be of great assistance in reducing the time and effort to establish the phase relationships in alloy systems, particularly those of higher order that are often necessary to encompass brazing processes [Villars et al. 2001]. This work has also proved that properties of alloys can be determined quantitatively from the property parameters of the constituent elements so that, once additional information retrieval methods are automated, the selection of materials for specific applications will be facilitated greatly. It is hoped that many new materials with exciting property combinations will be discovered using this new analytical approach.

The value of alloy phase diagrams for understanding and optimizing brazing processes can be appreciated best by describing a few specific examples. Discussion of braze/component combinations of progressively increasing metallurgical complexity is presented in the following sections.

2.3.1 Examples Drawn from Binary Alloy Systems

Example 1: A Binary Braze with Complete Solubility in the Solid State, Used to Join Components of One of the Constituent Metals with No New Phase Formation

Consider a gold-nickel braze used with nickel components. The gold-nickel phase diagram is presented in Fig. 2.14, which shows this binary alloy system to possess a minimum melting point of 995 °C (1751 °F), at 18% nickel. On either side of this composition the liquidus and solidus separate, so that all compositions other than pure gold and pure nickel melt over a range of temperature. Within the melting range, the alloy is partly liquid and partly solid. On cooling below the solidus temperature, an alloy in this system exists as a single-phase solid, but as the temperature is lowered further and equilibrium maintained, this phase separates into two: one gold-rich and the other nickel-rich. The temperature of the solid-state decomposition varies with composition, reaching a maximum of 810 °C (1490 °F), which corresponds to an alloy containing 41.7% nickel.

The 18% nickel composition, corresponding to the minimum melting point (technically referred to as a congruent transformation), is normally used for brazing alloys because it completely melts at a unique temperature. The benefits of this property are shared by eutectic alloys and are explained in the next example, which relates to a eutectic alloy braze. On solidification of an alloy with the minimum melting point composition, the molten alloy, *L,* transforms to a single solid phase, *S,* at a unique temperature, in this case, 995 °C (1751 °F). This transformation may be written as:

$$L \leftrightarrow S$$

The formation of a single solid phase distinguishes this type of transformation from a eutectic, where two or more solid phases are formed simultaneously.

For the purposes of this example, it will be assumed that an alloy of composition Au-50Ni is used as the braze. The gold-nickel phase diagram shows that the Au-50Ni alloy has a solidus temperature of 1000 °C (1830 °F) and a liquidus temperature of 1200 °C (2190 °F). Thus, the melting range of an alloy of this composition is 1000 to 1200 °C (1830 to 2190 °F), and the brazing operation must be carried out at a temperature above 1200 °C (2190 °F) in order to obtain any significant flow by the filler metal.

On heating the braze above its liquidus temperature, the molten filler will wet and simultaneously alloy with the nickel substrate. If the reaction is allowed to proceed further, the braze will rapidly dissolve nickel up to an equilibrium value, determined by the intersection of a line drawn on the phase diagram at the process temperature with the liquidus curve, as indicated in Fig. 2.30. Thus, at 1350 °C (2460 °F), the dissolution of nickel by the Au-50Ni braze changes its composition to approximately Au-70Ni.

Because some of the surface region of the nickel component is dissolved by the braze, this process is commonly referred to as dissolution or erosion. As a consequence of this alloying process, the liquidus and solidus temperatures of the filler will change, as is evident from Fig. 2.30, and the microstructure of the solidified braze will be modified accordingly. The practical manifestation of the increase in the melting point of the braze is a reduction in its fluidity and also in the driving force for spreading (see Chapter 1, section 1.2.1). Heavy erosion of the parent materials is therefore generally undesirable. If erosion needs to be minimized, the filler metal must be chosen such that alloying with the parent metal causes the liquidus to rise in temperature rapidly as the composition changes toward that of the parent metal. Under these conditions, isothermal solidification at the joining temperature will occur, which can be used as a safety valve to limit the extent of interalloying. At the same time, the joining operation should be carried out at the lowest practicable process temperature and within the shortest possible time in order that the concentration of the parent material in the filler alloy remains low and the equilibrium composition is not achieved.

The phase diagram represented in Fig. 2.30 shows that, above the liquidus temperature, the braze exists as a single homogeneous liquid phase, and immediately below the solidus temperature it comprises a homogeneous solid phase. Between these two temperatures, the braze is a pasty, two-phase mixture of solid and liquid, the proportions of which are given by the lever rule. Referring to the enlarged portion of the gold-nickel phase diagram shown in Fig. 2.31, at 1100 °C (2010 °F), the weight frac-

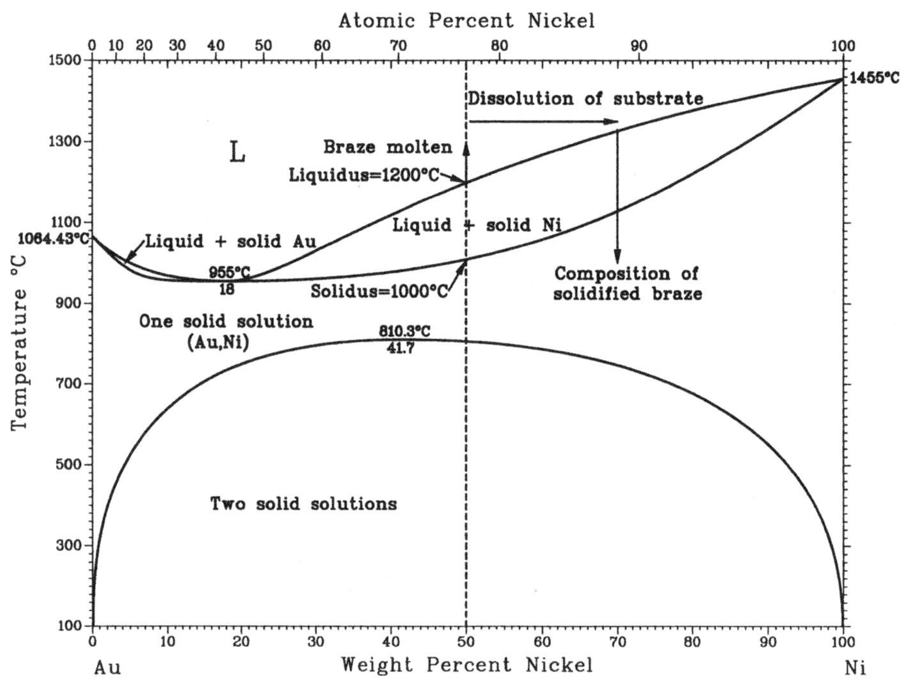

Fig. 2.30 Gold-nickel phase diagram. The erosion of a nickel substrate by a gold-nickel braze and the associated change to the composition of the filler metal are indicated.

tion of braze that does not melt under equilibrium conditions is:

$$\% \text{ solid} = \left(\frac{OX}{XY}\right) \cdot 100$$

The remainder of the braze will have melted, i.e.:

$$\% \text{ liquid} = \left(\frac{OY}{XY}\right) \cdot 100$$

where X is the composition of the liquid phase, Y is the composition of the solid phase, and O is the composition of the alloy. By way of an example, for an alloy of composition Au-50Ni, at 1100 °C (2010 °F), X = Au-38Ni and Y = Au-66Ni. Therefore, at this temperature, the percentage of solid will be (38–50)/(38–66) · 100 = 43%; the percentage of liquid will be 57%.

Although the alloy remains close to compositional equilibrium during melting, in practice it does not do so during solidification. The resolidified braze will tend to develop local concentration gradients of a type known as *coring*. Coring occurs because the compositions of the liquid and solid phases change as the alloy cools through the two-phase region.

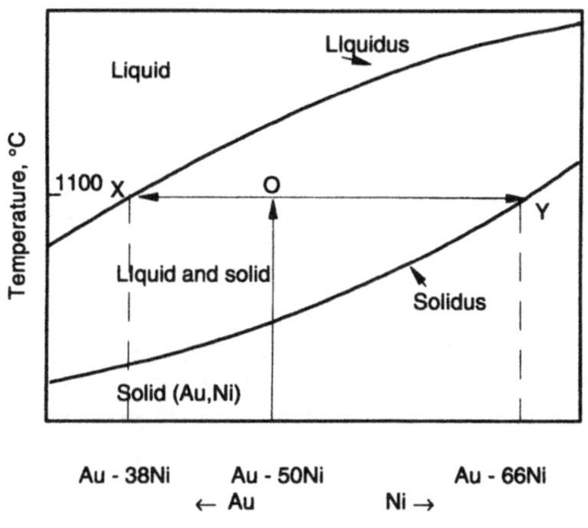

Fig. 2.31 Application of the lever rule to the gold-nickel system

Using the lever rule as discussed, the proportions and composition of liquid and solid can be calculated at various temperatures between the liquidus temperature, at which the alloy is completely liquid, and the solidus, when it has completely solidified. Data determined at five temperatures, corresponding to an aggregate composition of Au-70Ni, is presented in Table 2.18. Here it can be seen that the first liquid to solidify is rich in nickel, with respect to the nominal composition of the alloy, while the last liquid to solidify is rich in gold. If cooling can be controlled to a suitably slow rate, the solid fraction at each stage will continually and uniformly adjust by diffusion to the composition that is indicated on the phase diagram. Under these conditions, when solidification is complete, the alloy will be homogeneous. However, in most practical situations, the rate of solidification will be faster than diffusion can act to homogenize the alloy, and coring will result. The cored microstructure will normally persist on cooling to room temperature.

Being the result of nonequilibrium cooling, coring tends to broaden the temperature range over which an alloy melts, when reheated, and it is usually undesirable in virgin filler metals. Coring can be removed by isothermal annealing of the stock alloy at elevated temperature but below the solidus temperature. This heat treatment will furnish the necessary activation energy for diffusion to bring the alloy toward its equilibrium state and hence toward compositional uniformity. Annealing is most effective when combined with mechanical working that physically remixes material and refines the alloy microstructure.

Example 2: A Binary Eutectic Composition Braze Used with Components of One of the Constituent Metals with No New Phase Formation

A representative example of this type of reaction is a silver-copper braze used to join cop-

Table 2.18 Application of the lever rule to solidification of a Au-70Ni alloy

Temperature		Liquid		Solid	
°C	°F	Fraction, %	Composition	Fraction, %	Composition
1320	2410	100	Au-70Ni	0	. . .
1275	2325	62	Au-61Ni	38	Au-84Ni
1230	2245	36	Au-54Ni	64	Au-78Ni
1185	2165	17	Au-48Ni	83	Au-74Ni
1140	2085	0	. . .	100	Au-70Ni

per components. The silver-copper phase diagram represented in Fig. 2.2 shows that there is a single composition, that of the eutectic, (Ag-28Cu), at which the alloy transforms between a liquid (L) and two solid phases (S_1 and S_2) at a unique temperature, 779 °C (1436 °F), according to the reaction:

$$L \leftrightarrow S_1 + S_2$$

At the eutectic composition, solid alloys form as a mixture of two finely divided phases, one silver-rich and the other copper-rich. For all other compositions, except those of the pure metals, there is a separation between the liquidus and solidus temperatures.

It is seen from the silver-copper phase diagram (Fig. 2.2) that copper is soluble in the Ag-28Cu braze when molten. Thus, at the joining process temperature, the brazing alloy will dissolve copper from the components until the concentration of copper is attained, as described for the gold-nickel braze with nickel components. However, in the case of the eutectic silver-copper braze, the dissolution of copper increases the liquidus temperature of the filler metal in the joint but not its solidus temperature because eutectic transformations are isothermal.

When brazing cycles are short, the braze will not normally dissolve sufficient copper to achieve the equilibrium amount appropriate to the brazing temperature, as indicated on the silver-copper phase diagram (Fig. 2.2). Further dissolution of copper will lead to the formation of a solid, copper-rich phase in the joint as the braze composition enters the two-phase field (liquid plus solid, or $L + S$) of the phase diagram. Owing to the prevailing composition gradient that will normally exist, and also the likely thermal gradient in a practical brazing environment, this solid phase will form preferentially against the copper surface of the component and slow the further dissolution of copper. Hence, the liquidus phase boundary will effectively determine the "equilibrium" composition in brazing processes where the cycle time is sufficiently short. If left at the joining temperature for a sufficiently long time, the silver in the braze would diffuse and disperse throughout the copper components so that the assembly would reach a uniform composition determined by the sum totals of silver and copper in contact. This is actually the basis of the diffusion brazing processes referred to in Chapter 6.

At the commencement of the cooling stage of the process cycle, the molten braze no longer corresponds to the eutectic composition but is rich in copper and, in consequence, now possesses a freezing (i.e., melting) range. On cooling below the liquidus temperature, the excess copper will solidify first, as indicated by the phase diagram. The precipitation tends to occur preferentially at the interface between the components and the braze because this interface tends to be slightly cooler than the volume of the molten braze, as a consequence of heat being lost via the extremities of the assembly, during cooling. This region of the joint is also richer in copper because it is the source of the concentration gradient. Precipitation of the copper-rich phase continues until the temperature and composition of the remaining liquid reach the eutectic point so that final solidification by the molten filler results in the formation of finely divided eutectic. The alloy microstructure will therefore comprise primary dendrites of copper (so-called because they are the first to form on solidification) with the interdendritic spaces filled with the duplex eutectic mixture. The primary copper phase will contain a local concentration gradient as the amount of silver it incorporates varies with temperature; i.e., the primary copper phase is cored.

Alloys of eutectic composition are preferred as filler metals based on the following characteristics:

- Superior spreading behavior when molten. This feature is an immediate consequence of there being no temperature range over which the alloys coexist as solid and liquid, which also applies to the Au-18Ni alloy described in the previous example. Where a pasty mixture can occur, alloying with the materials of the components will diminish the available driving force for spreading, while at the same time, the flow of the partly molten alloy will be impaired due to its high viscosity.
- Superior mechanical properties, arising from the interspersed or duplex character of the eutectic microstructure and the fine grain size. Grain refinement is the only metallurgical process that enhances both the strength and ductility of a metal. The fine triplex microstructure of a ternary eutectic in an aluminum alloy braze can be seen in Fig. 2.32.
- Joining process temperatures can be chosen to be only slightly above the melting point of the alloy, precisely because eutectic composition alloys melt completely at a single temperature, which is usually known with high accuracy.

- A reduced risk of disturbing located components, which can occur easily when the filler appears to be solid but is actually in a pasty state. A rapid liquid-to-solid transformation on cooling, without an intervening pasty stage, minimizes the chance of such an interruption. However, this assumes that alloying of the braze with the component materials does not greatly shift the composition of the braze from the eutectic point. A disturbed joint generally has inferior mechanical properties and the fillets will acquire a rough surface with a frosty appearance.

For these reasons, most brazes are either eutectic compositions or have many of the characteristic features of eutectic alloys. This is certainly true of the silver-copper-zinc family of brazes and virtually all solders.

Example 3: A Binary Eutectic Composition Braze Used with Components of One of the Constituent Metals with Intermetallic Compound Formation

The cobalt-chromium phase diagram is shown in Fig. 2.33. The eutectic composition (Co-42Cr) forms the basis of brazes used to join chromium-based components. An example is Co-19Cr-17Ni-7Si-4W-0.8B-0.3C, which benefits from a melting range occurring at lower temperatures than that of the binary eutectic (1120–1150 °C, or 2050–2100 °F). The reasons for choosing this brazing alloy are the chemical and metallurgical complementarities between the braze and the typical range of high-temperature performance parent materials on which it is used, together with the fact that erosion of the parent material is low and highly predictable. With this combination of component and filler metals, the restricted erosion is a consequence of the formation of the $CoCr_2$ intermetallic compound as a stable layer that separates the molten braze from the remaining chromium in the parent material. This intermetallic phase has a melting point above normal brazing temperatures and it restricts further reaction to a degree determined by temperature and the duration of the brazing process as well as the rate of diffusion of chromium through it. Nevertheless, as seen from earlier examples, diffusion through solids is typically two orders of magnitude slower than through liquids and, therefore, a very thin layer of an interfacial intermetallic compound has a marked effect on the rate of substrate dissolution. Of the other principal constituents of the braze referred to here, nickel is totally soluble in the chromium-cobalt host and can be used to partially substitute for the chromium content, while silicon acts as a melting point depressant, owing to the existence of a low-melting-point cobalt-rich eutectic in the cobalt-silicon system.

While a phase diagram can provide guidance about whether a new phase will form, it cannot normally be used to determine its ultimate distribution and morphology within the joint because this is influenced greatly by a combination of factors. Another piece of important information that cannot be ascertained from equilibrium phase diagrams is the rate of growth of phases.

Example 4: A Binary Peritectic Braze Illustrating Problems Associated with Using a Braze of this Type

The second common type of phase transformation is the peritectic reaction where a liquid, *L*, on cooling, partly solidifies to form a solid phase S_1, and at the peritectic temperature the

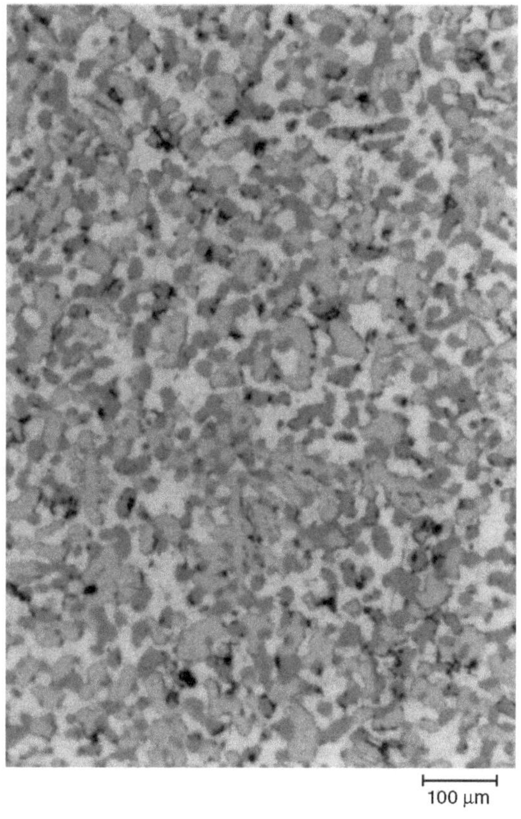

Fig. 2.32 An alloy microstructure characteristic of a (ternary) eutectic transformation. The alloy composition is 35.8Ag-34.2Al-30.0Ge. 100×

remaining liquid reacts with S_1 to form a new solid phase, S_2. This reaction may be written as:

$$L + S_1 \leftrightarrow S_2$$

Remarkably good examples of peritectic reaction occurs in the copper-zinc system. In all, there are five successive peritectic reactions (see Fig. 2.4), of which the most copper-rich is:

$$L + \text{Cu} \leftrightarrow \beta \text{ at } 902\ °\text{C } (1656\ °\text{F})$$

Alloys exhibiting this type of transformation are generally undesirable as brazes because during a peritectic solidification reaction it is not possible to maintain equilibrium conditions. This is due to the fact that diffusion rates in solids are so much slower than they are in liquids, as pointed out previously, so that a nonequilibrium microstructure develops, consisting of islands of the primary solid phase, S_1, completely surrounded by a rim of the second solid phase, S_2. This effect can be mitigated by cooling rapidly through the liquidus/solidus gap and also by selecting alloys with a narrow melting range. This is one of the reasons for limiting the range of copper-zinc alloys that are used as brazes to the copper-rich end of the β-phase field, as indicated in Fig. 2.4.

A quaternary aluminum alloy microstructure exhibiting a peritectic transformation is shown in Fig. 2.34. In such an alloy, liquid that is rich in the lower-melting-point elements will be retained below the peritectic transformation temperature. In consequence, the melting and freezing range of the alloy is widened and the remelt temperature cannot be predicted reliably. Furthermore, the microstructure of the solidified braze will be grossly inhomogeneous and relatively coarse, to the detriment of the mechanical properties of joints made with this alloy.

In higher-order alloys, a number of other types of phase transformation can occur and are generally referred to as *transition reactions*. The majority of these reactions possess features akin to a peritectic transformation and tend to be avoided when selecting alloy compositions as brazes on the same grounds as peritectic alloys.

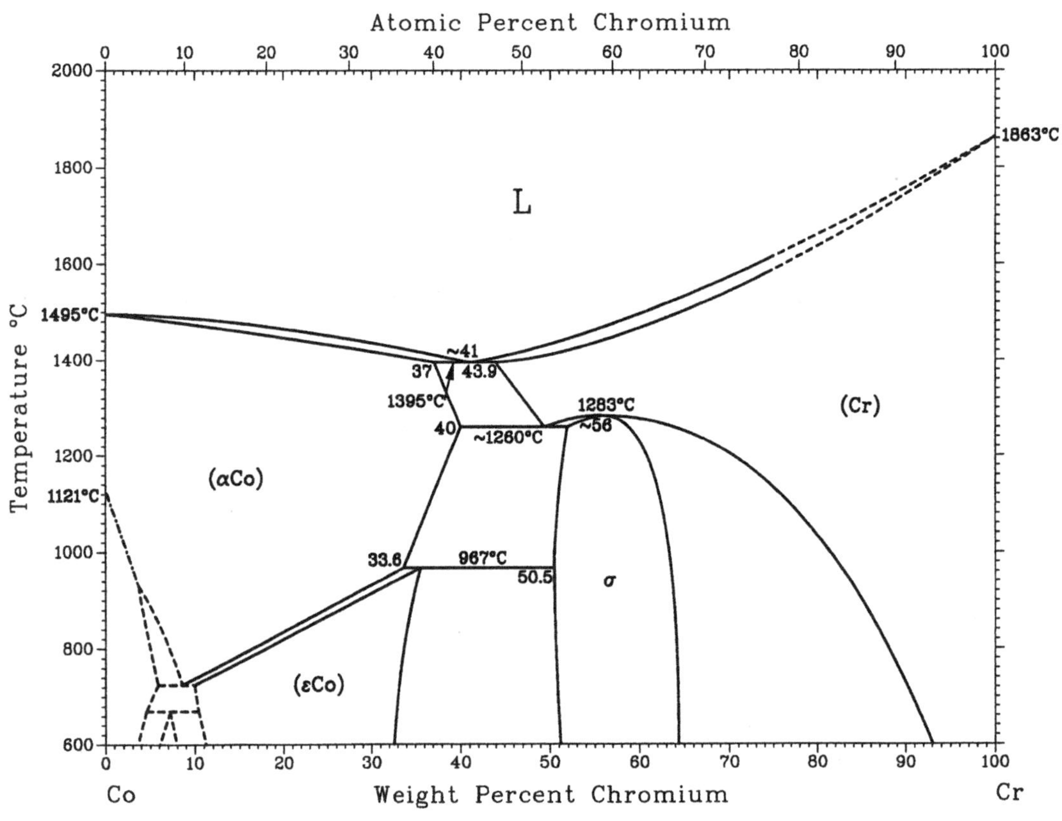

Fig. 2.33 Cobalt-chromium phase diagram. Source: Ishida and Nishizawa [1990]

2.3.2 *Examples Drawn from Ternary Alloy Systems*

It is rare for a brazed joint to be limited to a combination of just two elements, forming a binary alloy system. Usually, the braze is an alloy of at least two metals, while engineering substrates are frequently multicomponent alloys. A ternary alloy system is most usually represented by an equilateral triangle, with each of the vertices corresponding to the three constituent elements. A grid can be drawn on the triangle to provide a linear scale of composition. Temperatures are then represented by a series of isotherms, and the liquidus is mapped on the diagram as a topographical surface viewed in plan. Phase stability as a function of temperature is commonly represented by a figure resembling a binary alloy phase diagram, where either one of the constituents or the ratio of two constituents is held fixed. A single diagram cannot be used to track the solidification sequence because the ensuing composition changes can extend outside the plane of the diagram. For a similar reason, the lever rule cannot be applied to this representation in order to calculate the proportions of phases that exist in equilibrium. However, the lever rule can be used in conjunction with a series of isothermal sections.

Commonly, intermetallic compounds form between the constituents. The volume, distribution, and morphology of these intermetallic phases in a joint can have a pronounced effect on mechanical properties, in particular. From the characteristics of the phase diagram, it is often possible to predict whether the intermetallic compound will tend to form as a continuous interfacial layer against the parent materials or is more likely to be dispersed throughout the joint. In essence, if a line drawn on the phase diagram at the brazing temperature between the braze composition and that of the parent material intersects a new solid phase, that phase will form at the joint interface. If the line passes above a new solid phase, that phase will likely form by precipitation, distributed within the braze, as the brazing alloy solidifies. If the intermetallic compound has poor mechanical properties, then a dispersion of it is preferred because this not only avoids the source of weakness represented by the interfacial intermetallic phase but actually works to advantage by strengthening the braze. Examples of these different situations are described below.

Some care needs to be taken when referring to the intermetallic phases that form in a ternary system because binary phases often have appreciable solubility in the ternary system. For example, in the aluminum-molybdenum-silicon system (Fig. 2.35), to which detailed reference is made in the next section, there exists a continuum of composition between the intermetallic phases $AlMo_3$ and Mo_3Si in the molybdenum-rich corner of the composition triangle. If the latter phase is rewritten as $SiMo_3$, it becomes clear there is considerable scope for variation in the aluminum-to-silicon ratio within a stable phase field, highlighting the fact that the convenient labels adopted from binary alloy practice do not always describe adequately the situation in higher-order alloys. In this particular case, the ternary phase should be written as $(Al,Si)Mo_3$.

Example 5: Complete Intersolubility between a Eutectic Braze and the Metal on the Joint Surfaces

A typical example of intersolubility between a eutectic alloy braze and a metallic surface is provided by a Ag-28Cu eutectic braze used in the fluxless brazing of gold-plated titanium components. For the three component system silver-gold-copper, a ternary phase diagram is required to follow the reaction between the molten filler and the solid components and also to obtain quantitative data on the phases that form. The liquidus surface of the silver-gold-copper system is shown in Fig. 2.13. This surface contains a valley that runs from approximately the center of the diagram, at the 800 °C (1470 °F) isotherm,

Fig. 2.34 An alloy microstructure characteristic of a peritectic transformation. The alloy contains four constituents: aluminum, copper, nickel, and silicon. The primary phase is totally surrounded by a rim of a second phase as a result of the peritectic reaction failing to maintain equilibrium conditions during solidification.

to the eutectic point on the silver-copper binary axis. The minimum in the valley is 767 °C (1413 °F) at a composition of Ag-29Au-26Cu, as mentioned in section 2.1.8.1, which represents the lowest temperature at which liquid can exist in the silver-gold-copper system at atmospheric pressure. Figure 2.36 depicts an isothermal section at 700 °C (1290 °F), where all the phases are solid. These comprise two solid solutions, one combining silver and copper and the other rich in gold, separated in the triangle by a parabolic boundary.

The silver-gold-copper phase diagram forms the basis of many carat gold brazes and hallmarked jewelery alloys. In practice, both of these are higher-order multicomponent alloys. Further details on the metallurgy of jewelery alloys and corresponding brazes can be found in Chapter 5.

A section (or isopleth) through the silver-gold-copper phase diagram at 2% Au is shown in Fig. 2.37.

The composition of each phase in a solidified alloy cannot be read directly from the phase diagram because the orientation of the relevant tie line joining the two conjugate phases is usually not known. A tie line is a line drawn in a phase diagram: it is always straight and ends at points that correspond to two separate phases. These two phases will comprise compositions lying along a tie line in a proportion that is given by the lever rule. A typical tie line, drawn on an isothermal section of the silver-gold-copper phase diagram at 700 °C (1290 °F), is indicated in Fig. 2.36.

For reasons of economy, when gold coatings are used with the Ag-28Cu braze, the thickness of the gold layer will normally represent a maximum of about 2% of the volume of the metal in the joint. In the joining operation involving gold-plated titanium parts, the molten Ag-28Cu braze fuses with the gold layer to form a single solution. The underlying titanium will be wetted by the molten alloy but does not dissolve to any great extent because the process temperature is relatively low. On cooling, the liquid filler solidifies to produce a eutectic microstructure comprising silver- and copper-rich phases, each containing some gold, as indicated by the section through the silver-gold-copper phase diagram at 2% gold, shown in Fig. 2.37. Silver, copper, and gold are among the few metals that do not form brittle phases on alloying with titanium.

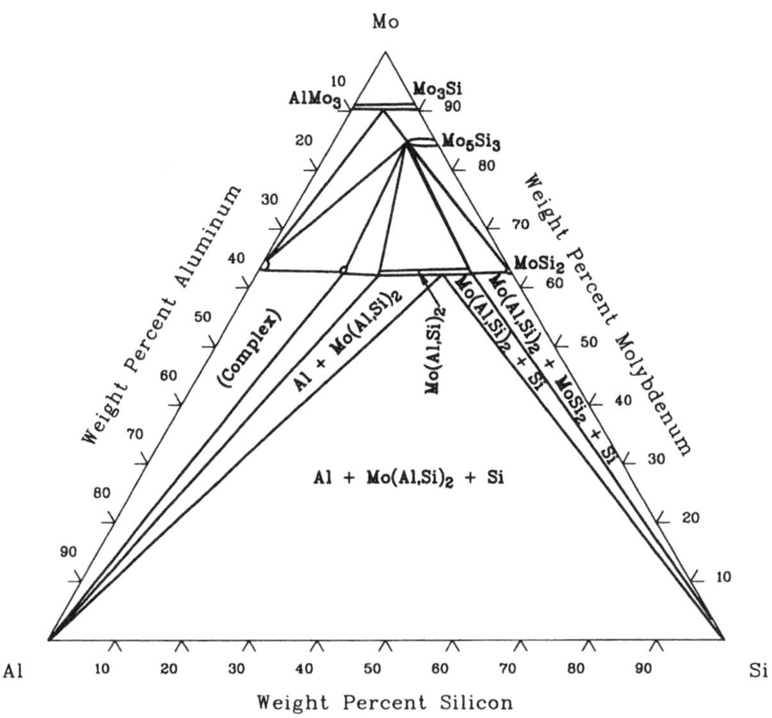

Fig. 2.35 Triangulation of the Al-Mo-Si ternary system at 20 °C (68 °F). Aluminum and silicon enter into ternary equilibrium with the compound $Mo(Al,Si)_2$. Adapted from Brukl, Nowotny, and Benesovsky [1961]

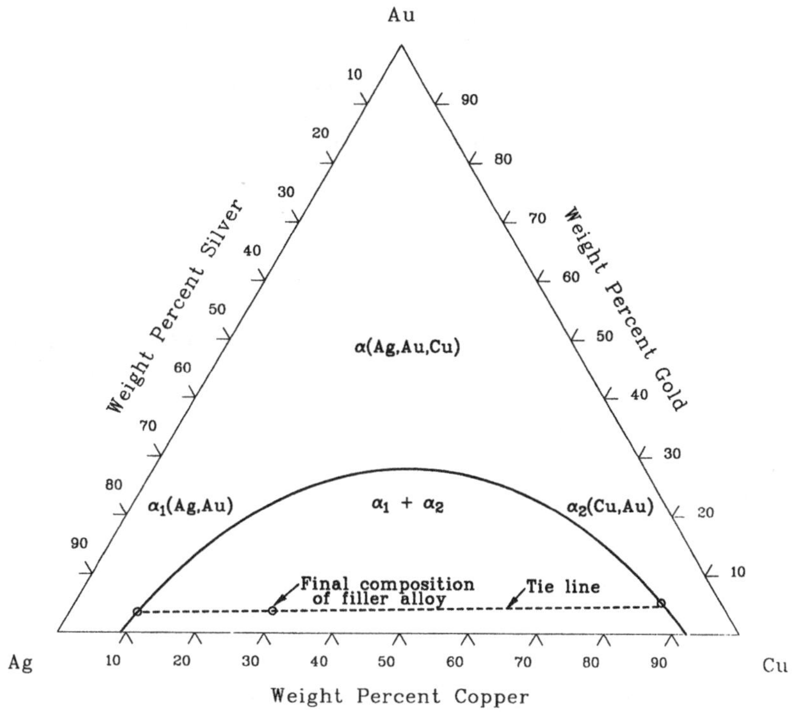

Fig. 2.36 Isothermal projection of the Ag-Au-Cu phase diagram at 700 °C (1290 °F). A hypothetical tie line is shown linking the compositions of the two conjugate phases formed by reaction of the Ag-28Cu braze with a thin gold metallization.

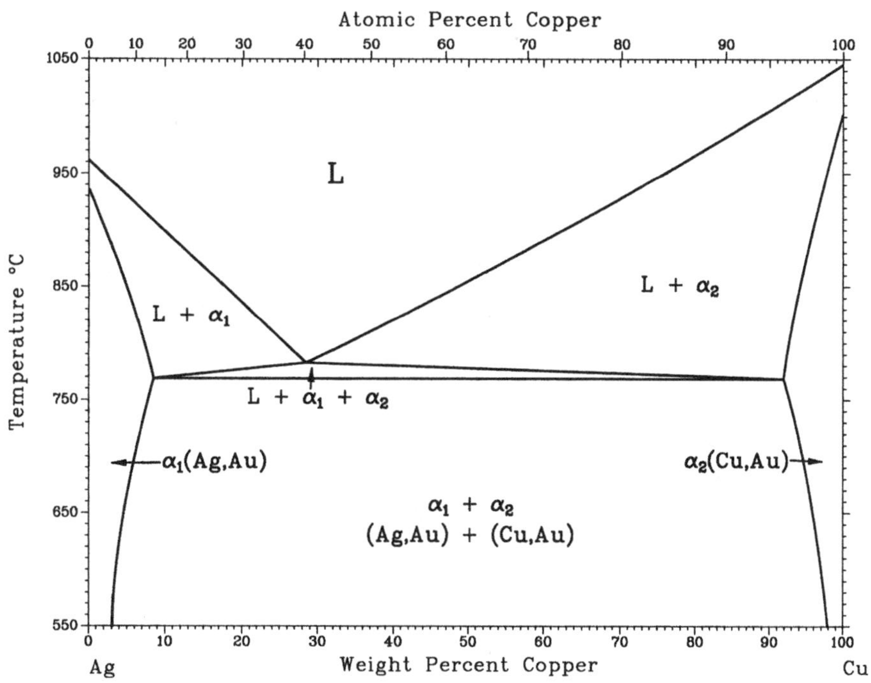

Fig. 2.37 Schematic section through the Ag-Au-Cu ternary system at 2% Au

Providing a noble, but sacrificial layer over the titanium surface enables the joining process to be fluxless. Titanium and its alloys are generally difficult to join because of the readiness with which this metal forms oxides and nitrides. The first step of the plating process is a chemical etch in mixture of nitric and hydrofluoric acids that removes the surface oxide from the titanium surface. Application of the gold plating prevents the oxide from reforming. Stainless steels are sometimes plated with copper or nickel for similar reasons.

A recommended braze for use with many titanium alloys is Ti-20Cu-20Ni, which has a liquidus temperature of approximately 950 °C (1740 °F), because this alloy is similar in composition and properties to many titanium engineering alloys.

Example 6: Interfacial Compound Formation between a Eutectic Braze and a Metallic Component

Brazing of silicon to molybdenum using the aluminum-silicon braze offers a convenient example of interfacial compound formation as a result of alloying between a eutectic alloy braze and a metallic parent material or component. This brazing process is referred to in section 2.1.13.2. The silicon item is a high-power semiconductor device in the form of a disc, and the molybdenum plate is the first part of a heat-sink assembly that has a reasonable match of thermal coefficient of expansion to silicon. Such devices are manufactured in large numbers for traction motor and power conversion applications [Taylor 1992].

Aluminum and silicon form a low-melting-point eutectic at Al-13Si, as shown in Fig. 2.22, so that the slightly aluminum-rich composition that is used for the braze (Al-12Si) is metallurgically compatible with the silicon wafer. An isothermal section at 20 °C (68 °F) of the aluminum-molybdenum-silicon ternary system is shown in Fig. 2.35.

The features to note are:

- Molybdenum is wetted by molten aluminum-silicon brazes and is partly soluble in them.
- The third constituent in the majority of reactions between an aluminum-silicon braze and molybdenum, the other two being the aluminum and silicon phases, is the ternary intermetallic compound $Mo(Al,Si)_2$ [Liu, Shao, and Tsakiroboulos 2000].

Wetting of the molybdenum plate by the brazing alloy results in the growth of a strongly adherent, but reasonably compliant, interfacial layer at the braze/molybdenum interface. This phase is often referred to as *molybdenum disilicide* because the concentration of aluminum is relatively low and it is difficult to pronounce "$Mo(Al,Si)_2$"! The growth kinetics of this phase follow a time/temperature relationship, which takes the form shown in Fig. 2.38. For a sensible process cycle time, the process temperature needs to exceed 650 °C (1200 °F) in order to achieve a continuous interfacial layer. A brazed joint made at a process temperature of 680 °C (1256 °F), sustained for 30 minutes, is shown in Fig. 2.13. The continuous interfacial layer of molybdenum disilicide is evident. The braze has also wet and dissolved silicon from the wafer, as would be expected. The solidified braze is actually now a ternary alloy because it will contain some molybdenum, but the steep slope of the liquidus surface between the ternary eutectic point and molybdenum means that the proportion of molybdenum that dissolves in the liquid braze is extremely small, and a transmission electron microscope is able to only just detect the distributed $Mo(Al,Si)_2$ phase within the solidified microstructure. This is distinct from the very conspicuous interfacial layer of this phase between the braze and the surface of the molybdenum component.

The phase diagram indicates that molybdenum disilicide will form at the braze/molybdenum interface because this compound is solid at the brazing temperature and the prevailing temperature gradient and field of composition favor its stability. If the brazing cycle time is consid-

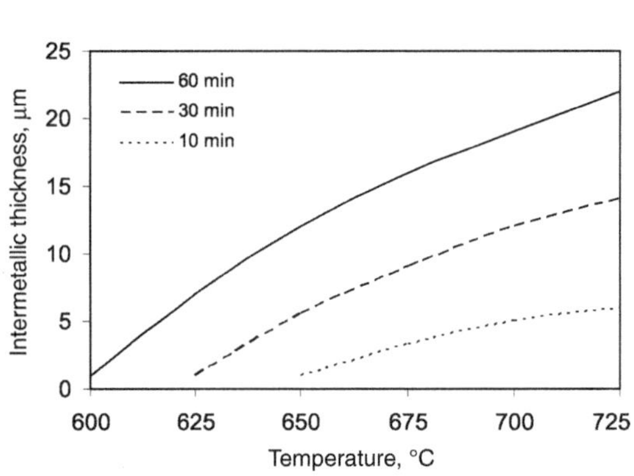

Fig. 2.38 Thickness of the "molybdenum disilicide" layer formed at the interface between Al-12Si braze and molybdenum as a function of process temperature for three different holding times

erably extended, the phase diagram predicts that two additional phases will appear between the molybdenum disilicide and the molybdenum, namely $Mo_5(Al,Si)_3$ and $Mo_3(Al,Si)$.

As noted previously, a phase diagram is unable to furnish any information about the rate of dissolution and growth processes. The rate of formation of interfacial molybdenum-rich intermetallic phases is governed by the rate of interdiffusion of aluminum and silicon through them, which is generally slow. These phases will continue to increase in thickness during elevated temperature service of the component, albeit much more slowly than when the braze is molten. The presence of thick layers of intermetallic phases is generally considered to be detrimental to the mechanical properties of the joints because they usually have inferior physical characteristics, particularly fracture toughness, to both the brazing alloy and the components. However, silicon power devices are not subject to any appreciable mechanical abuse and the interfacial compound is of little consequence to the product function or reliability.

Example 7: Distributed Compound Formation between a Eutectic Braze and a Metallic Component

While the phase diagram can provide guidance about where a new phase will form, it cannot be used to determine their ultimate distribution within the brazed joint, as mentioned earlier. Due to the significant expansion mismatch between silicon and molybdenum, for very large diameter power devices, a backing plate of tungsten is preferred due to its lower coefficient of thermal expansion (CTE), which is 4.5 ppm/K, compared with 5.0 ppm/K for molybdenum, at normal ambient temperatures, and therefore closer to that of silicon. The aluminum-silicon-tungsten phase diagram is wholly analagous to that of aluminum-silicon-molybdenum. However, in this situation, one in which tungsten is wetted by the Al-12Si eutectic braze, the high melting point Si_2W intermetallic phase initially forms at the interface between the molten braze and the tungsten, but it promptly spalls off into the liquid due to intrinsic stress in this layer. The Si_2W phase then becomes redistributed as needles within the body of the braze. It is therefore essential to keep the brazing cycle short so as not to saturate the braze with these hard and embrittling particles.

The mechanical properties of joints containing intermetallic phases can, to some extent, be inferred from a phase diagram, according to whether they are compounds of exact stoichiometric composition (i.e., they are in integral atomic ratios of their constituents), or exist over a range of compositions. Exact stoichiometric compounds tend to form when one of the two elements is strongly metallic in character and the other is significantly less so, in terms of the density of free electrons that bind the atoms of the metal together. Si_2W, Fe_3C and Ni_3B are typical examples of this type of compound. These compounds tend to be hard and brittle. Moreover, because their crystal structures are frequently of low symmetry, that is, they deviate from simple cubic or hexagonal structures, the interfaces of these compounds with other phases tend to be weak. These characteristics are transferred to the joint unless the compounds form as a fine dispersion within the braze. Therefore, their occurrence should be minimized, or even better, avoided, wherever possible.

Compounds that are stable over a range of compositions tend to be more ductile and have crystal structures exhibiting high symmetry (that is, simple cubic or hexagonal), as do most elemental metals. Therefore, they tend to have a benign effect on joint properties. An example of such a compound is Cu_3P, which is a principal constituent of self-fluxing copper-brazes and has a stable composition range of approximately 5% at room temperature. Preforms of this braze are ductile. When these brazes are used with steel components, Fe_3P tends to form at the steel interface with the braze and also within the joints. This phase has no composition width (i.e., it is a line compound) and the resulting joints have low fracture toughness. However, if other elements are incorporated in the braze, the phase Fe_2P can be made to form instead. This intermetallic compound has a composition width of about 1% at room temperature, and the joints have vastly improved mechanical properties. Similarly, copper-beryllium intermetallics have composition widths in double percentage figures and are commonly used as spring materials.

As can be seen from the ternary isothermal section in Fig. 2.35, $Mo(Al,Si)_2$ has some phase width and one might therefore expect it to have a reasonable fracture toughness. By contrast, Si_2W is a line compound and has negligible solubility of aluminum. The lower fracture toughness of this phase has been confirmed by experiment, as discussed previously.

2.3.3 *Complexities Presented by Higher-Order and Nonmetallic Systems*

More often than not, higher-order alloy systems are encountered in joining because both the braze and the parent materials usually each have a minimum of two constituents. Combinations involving five or even larger numbers of elements are not uncommon.

The definition of the plethora of phases that can exist in these higher-order systems represents a daunting task. In order to make the problem more tractable, a reductive approach is often employed. This method usually involves partitioning the multicomponent system into a series of quasi-binary or quasi-ternary alloy systems, each containing different but fixed proportions of the other components and ascertaining these sections of the relevant phase diagrams, in turn. Much care should be exercised in extracting quantitative information from these partial phase diagrams because the tie lines, triangles, quadrilaterals, and so forth that are used with the lever rule to determine the relative proportions of phases present often do not lie in the plane of the selected sections.

Joining to nonmetallic materials presents additional problems that may not be immediately apparent from the relevant phase diagram, assuming that one is available. A common method for joining to materials that contain glass or ceramic phases is to incorporate into the filler highly active metals such as titanium, zirconium, or hafnium that will wet and bond to these materials. Further details are given in Chapter 4, section 4.1.2.2 and Chapter 7, section 7.2. The bonding process relies on the formation of a compound between the active constituent of the filler metal and one of the elements of the nonmetallic material while liberating others. For example, a gold-nickel braze containing titanium will wet the engineering ceramic silicon nitride, Si_3N_4, by forming a surface reaction layer of titanium nitride, leaving some free silicon. But nickel silicide is brittle while the gold present in the braze forms a eutectic (solder) with the free silicon, and this alloy melts at 363 °C (685 °F). The resulting joints are weak and melt well below the temperature expected for the brazed assembly. In the case of silicon nitride, Si_3N_4, brazed with an aluminum-silicon filler alloy, the braze wets the ceramic by dissociating the surface film so it can bond with the silicon, liberating nitrogen in the process. The nitrogen gas will produce voids in the joint unless provision is made for it to escape from the joint gap [Lugscheider and Tillmann, 1991].

The various examples discussed in this section underline the care that must be taken in choosing a filler for use with a particular material. They further demonstrate the need to consider the combination of the brazing alloy and components as an integral materials system in order to gain a satisfactory understanding of the characteristics of brazed joints.

2.4 Depressing the Melting Point of Brazes by Eutectic Alloying

The number of commercially available brazes is finite and it is not uncommon to have an application where it would be desirable either to extend the lower temperature limit on the use of a particular family or to identify filler compositions that melt within a specified range. Examples from recent development activities are new brazes for aluminum engineering alloys having melting points below 530 °C (985 °F), and replacements for lead-tin solder that do not contain lead but which can be used at comparable processing temperatures. Lead-free solders are discussed in the companion volume, *Principles of Soldering*.

For the reasons elaborated in section 2.3.1, eutectic alloys posses several key characteristics that make them a natural choice for filler metals, namely superior spreading behavior when molten, with complete melting occurring at a single temperature that is lower than that of either constituent metal. This property of instantaneous melting enables joining operations to be carried out at temperatures only slightly in excess of the solidification temperature of the braze, and molten eutectic alloys generally possess a high degree of fluidity. Eutectic alloys also possess favorable mechanical properties arising from their well-distributed duplex or higher order microstructure.

In contrast with solders, few of the commonly used brazes are based on eutectic compositions. Among the few notable examples are the self-fluxing silver-copper brazes, which are based on the silver-copper-Cu_3P alloy partial ternary system, represented in Fig. 2.3. The ternary (three component) eutectic alloy in this system has the composition Ag-31Cu-51Cu_3P, and it melts at

646 °C (1195 °F). As is the norm for a eutectic alloy, this temperature is lower than the melting points of all of the three constituent binary eutectics (Ag-Cu, Ag-Cu_3P, and Cu-Cu_3P). What happens, then, to the melting point if further constituents are added? In particular, can it be lowered at will and, if not, how might one determine the limits?

Answers to these questions are provided by the patterns of behavior observed in the development of low-melting-point aluminum brazes and the silver-copper-zinc brazing alloy family. In these cases, there exists sufficient data that can be used to illustrate the principles clearly. Features that are observed consistently enable empirical rules of general applicability to be identified. The two case studies are discussed in turn, and the common characteristics are delineated in the section that follows. Additional examples are discussed in the companion volume, *Principles of Soldering,* Chapter 2, section 2.4.

2.4.1 Silver-Base Brazes

The development of the silver-copper-zinc plus cadmium or tin, plus manganese and/or nickel brazes, through the sequential addition of constituents is discussed in section 2.1. Starting with pure silver, the melting point of the alloy is reduced by the addition of copper, then zinc, followed by cadmium or tin and then with manganese and nickel. This progression is elaborated in Table 2.19 and also diagrammatically in Fig. 2.39. For some of the higher-order alloys, the cited melting point differs from the brazing alloy because here it is the minimum melting temperature in the system that is of interest. Figure 2.39 clearly shows a trend whereby the further reduction in melting point that can be achieved decreases with each new addition ($x(i)$; $i = 1, 2, 3, 4, 5, 6$), there being a total of six constituents in this example. The order of the system, $n = x(1) + \ldots x(i)$, where $1 \leq i \leq 6$.

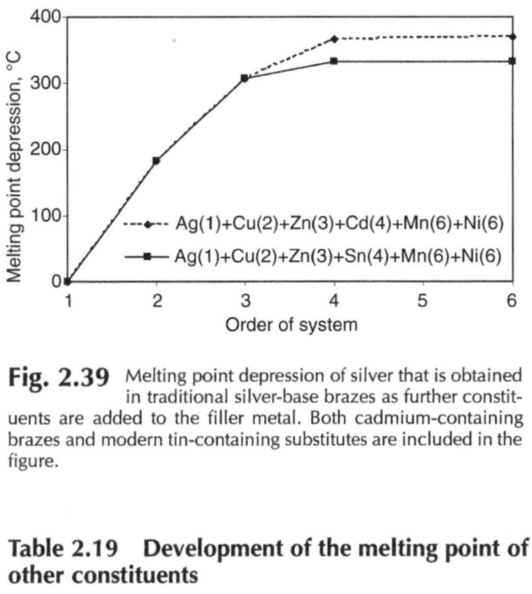

Fig. 2.39 Melting point depression of silver that is obtained in traditional silver-base brazes as further constituents are added to the filler metal. Both cadmium-containing brazes and modern tin-containing substitutes are included in the figure.

2.4.2 Aluminum-Brazing Alloys: New Low-Melting-Point Compositions

The brazing alloys that are used currently for joining aluminum are based on the aluminum-silicon eutectic composition, which has a melting point of 577 °C (1071 °F). For the aluminum alloys that can be brazed, the proximity of the filler liquidus temperature to the melting points of the aluminum-base alloys demands stringent control of temperature during the joining operation. Such control is difficult to achieve with large assemblies and, more generally, in high volume production. For the majority of high-strength aluminum engineering materials, which derive their strength from the mechanism of precipitation hardening, it is desirable that the joining temperature coincides with that at which the precipitation treatment is carried out, typically in the range of 480 to 540 °C (896–1004 °F).

Brazes for aluminum and its alloys are based predominantly on either aluminum-silicon or aluminum-germanium because these elements, when alloyed with aluminum, do not render the

Table 2.19 Development of the melting point of silver-base brazes with sequential additions of other constituents

Order	System	Melting point, °C	Temperature depression, °C
1	Ag	962	0
2	Ag-Cu	779	183
3	Ag-Cu-Zn	655	124
4	Ag-Cu-Zn-Sn	630	25
4	Ag-Cu-Zn-Cd	595	60
6	Ag-Cu-Zn-Sn + Ni, +Mn	629	1
6	Ag-Cu-Zn-Cd +Ni, +Mn	593	2

joint susceptible to galvanic corrosion. Moreover, as explained previously, it is desirable for the selected alloy to be eutectiferous, and this consideration restricts further the choice of alloying constituents. Besides entering into eutectic reaction with aluminum, silicon, and germanium have the further merit of possessing low volatility and therefore being compatible with vacuum brazing.

Whereas silicon has a melting point of 1414 °C (2577 °F) and germanium one of 938 °C (1720 °F), the eutectic temperatures of the aluminum-silicon and aluminum-germanium reactions are 577 °C (1071 °F) and 424 °C (795 °F), respectively, which represent a significant reduction on the melting point of aluminum of 660 °C (1220 °F). In this instance, the melting-point depression resulting from the eutectic alloying correlates with the respective melting temperatures of silicon and germanium. However, the melting point of the alloying element on its own provides no certain guide to the extent of melting point depression that it will produce. As an extreme example, an alloy of the gold-silicon eutectic composition melts at a mere 363 °C (685 °F), which is over 200 °C (390 °F) below that of the aluminum-silicon eutectic, notwithstanding the fact that the melting point of gold (1064 °C, or 1947 °F) is 400 °C (750 °F) higher than that of aluminum.

Table 2.20 shows the effect on the melting point of sequential additions to aluminum-silicon and aluminum-germanium alloys, respectively. The data presented were designed to highlight this effect as well as the relative effectiveness of various alloying elements in changing the melting point, both individually and in combination with others [Jacobson and Humpston 1995].

The quaternary aluminum-copper-nickel-silicon alloy has a eutectic temperature of 518 °C

Table 2.20 Temperature of eutectiferous phase transformations in which one of the participating phases is aluminum, shown in two sequences of progressing alloying

In one of these systems, silicon is a constituent, whereas, in the other, it is germanium instead. It is to be noted that a larger melting-point depression is obtained when germanium is included in the alloy system.

Order	System	Melting point, °C	Temperature depression, −δT, °C	System	Melting point, °C	Temperature depression, −δT, °C
1	Al	660	0	Al	660	0
2	Al-Cu	548	112	Al-Cu	548	112
	Al-Fe	655	5	Al-Ag	566	94
	Al-Mn	659	1	Al-Ge	424	236
	Al-Ni	640	20	Al-Ni	640	20
	Al-Si	577	83			
3	Al-Cu-Fe	547	113	Al-Ag-Cu	515	146
	Al-Cu-Mn	547	113	Al-Ag-Ge	416(a)	244
	Al-Cu-Ni	547	113	Al-Cu-Ni	547	113
	Al-Cu-Si	525	135	Al-Ag-Ni	541(a)	119
	Al-Fe-Mn	654	6	Al-Cu-Ge	418(a)	242
	Al-Fe-Ni	638	22	Al-Ge-Ni	424(a)	236
	Al-Fe-Si	576	84			
	Al-Mn-Ni	624(a)	36			
	Al-Mn-Si	573	87			
	Al-Ni-Si	567	93			
4	Al-Cu-Fe-Mn	544(a)	116	Al-Ag-Cu-Ge	415(a)	245
	Al-Cu-Fe-Ni	546(a)	114	Al-Ag-Cu-Ni	505(a)	155
	Al-Cu-Fe-Si	524	136	Al-Ag-Ge-Ni	416(a)	244
	Al-Cu-Mn-Ni	538(a)	122	Al-Cu-Ge-Ni	415(a)	245
	Al-Cu-Mn-Si	519(a)	141			
	Al-Cu-Ni-Si	518(a)	142			
	Al-Fe-Mn-Ni	623(a)	37			
	Al-Fe-Mn-Si	571(a)	89			
	Al-Fe-Ni-Si	564(a)	36			
	Al-Mn-Ni-Si	563(a)	37			
5	Al-Cu-Fe-Mn-Ni	538(a)	122	Al-Ag-Cu-Ge-Ni	414(a)	246
	Al-Ci-Fe-Mn-Si	517(a)	143			
	Al-Cu-Fe-Ni-Si	517(a)	143			
	Al-Cu-Mn-Ni-Si	517(a)	143			
	Al-Fe-Mn-Ni-Si	563(a)	37			
6	Al-Cu-Fe-Mn-Ni-Si	516(a)	144	Al-Cu-Ge-Fe-Mn-Ni	414(a)	246
				Al-Ag-Ge-Fe-Mn-Ni	416(a)	244
7	Al-Ag-Cu-Fe-Mn-Ni-Si	516(a)	144	Al-Ag-Cu-Ge-Fe-Mn-Ni	412(a)	248

(a) Authors' own measurements

(964 °F). The actual composition selected for the brazing alloys is slightly off eutectic, in order to achieve a satisfactory balance between mechanical properties, corrosion resistance, and melting range. It may be appreciated that even lower melting temperatures may be achieved by introducing other elements, but only at the expense of some other important property, including corrosion resistance (adding silver), cost (adding germanium), mechanical properties (most elements), or vapor pressure (adding magnesium and zinc), making these alloys unsuitable as brazes for engineering applications [Jacobson, Humpston and Sangha 1996; Humpston, Jacobson and Sangha 1995].

2.4.3 General Conclusions for Brazes

The melting-point behavior follows the following pattern (which also applies to solders and other eutectiferous alloys; see also the companion volume *Principles of Soldering,* Chapter 2, section 2.4):

- The melting point drops monotonically with the addition of each successive eutectiferous constituent (or others that are judiciously chosen).
- The magnitude of the melting-point depression is dependent on the specific alloying additions. Some elements are more effective than others in lowering the melting point of the alloy and the change is not always related to the melting point of the new constituent.
- The incremental melting-point depression accompanying the addition of each new alloying element becomes progressively smaller so that the melting point tends to an asymptotic minimum.

These features are consistent with elementary thermodynamic and statistical models, as explained in Appendix A2.2, and are generic to eutectic alloying.

From a practical point of view, the implications are clear. When seeking a reduction in the melting point of a pure metal or of an existing alloy, for use in brazing or other applications, there is a trade-off between keeping the number of constituents low and increasing the number of components beyond three or four, with the intention of reducing the melting point further. In most cases, it is preferable to strictly limit the number of constituents to as few as possible and judiciously choose them to optimize the melting point depression, rather than increase them further, which, in any case, is likely to produce only a relatively small further reduction in the melting point. Multicomponent brazes are beset by disadvantages, chief among which is a considerably more complex alloy phase diagram, if it is indeed known; a greater risk of intermetallic compound formation on alloying with engineering parent materials; and often, a harder and less workable alloy.

Due to the cumulative complications associated with increasing the number of constituents in the filler, it is advantageous, where possible, to limit the choice to alloying additions to those that are most effective at depressing the melting point. Generally, these are elements that have low melting points and very limited solid solubility in the host metal. Other constituents can be added to adjust key properties of brazes (e.g., resistance to corrosion, grain refinement, and spreading behavior), but their concentration should be minimized (which normally means less than 5 wt%), in order not to upset the useful characteristics of the brazing alloy or its melting range.

Appendix A2.1: Conversion between Weight and Atomic Fraction of Constituents of Alloys

In an alloy containing N constituents, conversion from weight to atomic fraction of constituent n may be made using the equation:

$$\text{at.\% } n = \frac{\dfrac{P_n}{A_n}}{\displaystyle\sum_{i=1}^{N} \frac{P_i}{A_i}} \times 100$$

where P is the weight percentage of the constituent denoted by the subscript, A is the atomic weight of the constituent denoted by the subscript, subscript n refers to constituent n, and subscript i refers to each constituent in turn.

Similarly, in an alloy containing N constituents, conversion from atomic to weight fraction of constituent n may be made using the equation:

$$\text{wt\% } n = \frac{P'_n A_n}{\sum_{i=1}^{N} P_i A_i} \times 100$$

where P' is the atomic percentage of the constituent denoted by the subscript, A is the atomic weight of the constituent denoted by the subscript, subscript n refers to constituent n, and subscript i refers to each constituent in turn.

Appendix A2.2: Theoretical Modeling of Eutectic Alloying

The laws of thermodynamics account for a lowering of the melting point, when a substance, B, is added to a pure solvent, A, by an amount given by Raoult's Law in the form of the Clausius-Clapeyron equation. At the liquidus line representing the equilibrium between a solid solution of B in A and a liquid solution of *B* in A, the Clausius-Clapeyron equation takes the form:

$$\frac{\Delta H_{\text{m}}^{(1)}}{R}\left(\frac{1}{T_{\text{m}}^{(2)}} - \frac{1}{T_{\text{m}}^{(1)}}\right) = -\ln\left(\frac{x_1}{x_1^*}\right)$$

where x_1 is the mole fraction of component A of the liquidus composition at temperature $T_{\text{m}}^{(2)}$ and x_1^* is this mole fraction at the limit of solid solubility at the same temperature. $T_{\text{m}}^{(1)}$ is the melting temperature of pure A, and $\Delta H_{\text{m}}^{(1)}$ is its latent heat of fusion. R is the universal gas constant.

The latent heat ΔH_{m} (or enthalpy of fusion) of a metal is proportional to its melting point. This is because entropies of fusion (ΔS_{m}) have similar values for all metals (at roughly 10 $\text{JK}^{-1} \cdot \text{mol}^{-1}$) and to a first approximation, in the absence of any phase changes:

$$\Delta H_{\text{m}} - T_{\text{m}}\Delta S_{\text{m}} = 0 \qquad \text{Eq A2.1}$$

that is,

$$\Delta H_{\text{m}} \propto T_{\text{m}}$$

Equation A2.1 implies, therefore, that the attainable melting-point depression is determined by the melting point of the addition; the lower the melting point of the addition, the lower will be the melting point of the resulting eutectic (assuming that one exists).

The effect on melting point of multiple alloying additions, all with similar melting points, can be deduced as follows. As a simple approximation, Eq A2.1 can be extended stepwise to multicomponent alloys, where, for example, the binary alloy A + B may be considered as the solution "matrix" AB of a pseudobinary alloy AB + C. Equation A4.1 may be rewritten in the form:

$$\frac{\Delta H_{\text{m}}^{(2)}}{R}\left(\frac{1}{T_{\text{m}}^{(3)}} - \frac{1}{T_{\text{m}}^{(2)}}\right) = -\ln\left(\frac{x_2}{x_2^*}\right)$$

with the following conditions satisfied:

$$\Delta H_{\text{m}}^{(2)} < \Delta H_{\text{m}}^{(1)}$$

$$T_{\text{m}}^{(3)} < T_{\text{m}}^{(2)} < T_{\text{m}}^{(1)}$$

$$\frac{x_2}{x_2^*} < \frac{x_1}{x_1^*}$$

The relationship $x_2/x_2^* < x_1/x_1^*$ implies that for the third component C, the attainable liquidus depression $T_{\text{m}}^{(3)} - T_{\text{m}}^{(2)}$ is, in most cases, significantly smaller than it is for the second component B, equal to $T_{\text{m}}^{(2)} - T_{\text{m}}^{(1)}$. If this sequential procedure is applied to additional elements, the general formula for the ith constituent and the corresponding liquidus temperature, $T_{\text{m}}^{(i)}$, is obtained:

$$\frac{\Delta H_{\text{m}}^{(i-1)}}{R}\left(\frac{1}{T_{\text{m}}^{(i)}} - \frac{1}{T_{\text{m}}^{(i-1)}}\right) = -\ln\left(\frac{x_{i=1}}{x_{i-1}^*}\right) \qquad \text{Eq A2.2}$$

This expression is a member of a series of decreasing size for increasing value of i. The melting-point depression is maximized for a large difference between the concentration of the "matrix" x_{i-1} and its solid solubility limit x_{i-1}^*.

Despite the fact that this model considerably oversimplifies the reality, it accounts for the two principal common features of the experimental results on eutectic alloys:

- The progressive reduction in melting temperature as the number of alloying constituents is increased

- The asymptotic narrowing of the melting-point depression of the alloy with the introduction of each additional constituent

Composition-specific thermochemical data on multicomponent alloys (ΔH_m, x_i) are needed in order to apply Eq A2.2 in calculating the liquidus temperature depression through progressive alloying. This information is mostly unavailable for ternary and higher-order systems. However, from our knowledge of thermochemical data for pure metals and binary alloys, it can be inferred that the respective values will differ widely from one metal to another, and therefore large variations in temperature drop are to be expected between different alloying additions.

The physical picture of this behavior may be understood more clearly in terms of the entropy changes accompanying progressive alloying. At the microscopic level, a material system may be viewed as an ensemble of atoms or molecules and, on this basis, entropy provides a measure of the degree of atomic or molecular disorder in the system, according to the relationship:

$$S = k \ln \Omega$$

where S is entropy, k is the Boltzmann constant, and Ω is the degree of disorder, as measured by the number of different distributions available to the atoms or molecules in the system.

In the simplest case, where the volume of the material is shared by i different species of atom, representing different constituents of an alloy, each present in amounts $N_1, N_2, N_3, \ldots, N_i$, such that:

$$N_1 + N_2 + N_3 + \ \ldots\ N_i = \mathrm{N}$$

then, assuming that the different species of atom are equally interchangeable, the number of ways, Ω, in which all the atoms may be arranged amongst the N available sites is:

$$\Omega = \frac{N!}{N_1!N_2!N_3! \ \ldots\ N_i!}$$

and

$$S = k \ln \frac{N!}{N_1!N_2!N_3! \ \ldots\ N_i!}$$

As the number of constituents increases, while keeping the total number of atoms, N, constant, there is a tendency for the entropy to increase as the number of constituents in the material is increased, as follows:

If $N_i \gg 10$ for all values of i, then Stirling's Formula can be applied, whereby:

$$\ln N_i! \approx N_i \ln N_{\mathrm{I}} - N_i$$

and the change in entropy in increasing the number of constituents from 2 (binary system) to 3 (ternary system) is:

$$\Delta S_{2,3} = kN[\ln 3 - \ln 2] = R[\ln 3 - \ln 2]$$

in the particular case where $N_1 = N_2 = \frac{1}{2}N$ for the binary system and $N_1 = N_2 = N_3 = \frac{1}{3}N$ for the ternary system.

Then, on increasing the number of constituents from 3 to 4 (quaternary system), the entropy rises further by an increment:

$$\Delta S_{3,4} = R[\ln 4 - \ln 3] < \Delta S_{2,3}$$

where $N_1 = N_2 = N_3 = N_4 = \frac{1}{4}N$ for the quarternary system.

In general:

$$\Delta S_{i-1,i} = R[\ln i - \ln (i - 1)]$$

where $N_1 = N_2 = N_3 = \ldots = N_{i-1} = N_i = (1/i)N$ and $i \geq 2$.

The pattern is established in which each successive addition increases the entropy of the system overall, but by progressively smaller amounts, as shown by the data in Tables 2.19 and 2.20. In other words, each additional constituent has a relatively smaller effect on the degree of disorder of the system, as one might intuitively expect.

Entropy, S, is related to the Gibbs free energy, G, by the relationship:

$$\left(\frac{dG}{dT}\right)_P = -S$$

Therefore, as the entropy increases, the depression of the Gibbs free energy of a system as a function of temperature increases. This in turn will tend to depress its melting point, although the actual relationship will be governed by the specific free energies of the constituents in the molten and solid states and of the solution they form. In general terms, the picture provided by this elementary expression is consistent with that furnished by Raoult's law and the Clausius-Clapeyron equation.

At first sight, this model may appear inappropriate for a eutectic alloy. However, the special case of a eutectic alloy with a low degree of intersolubility of the pure metal constituents in the solid approximates reasonably well, insofar as each phase is tantamount to a pure constituent and is well dispersed throughout the alloy. This model, therefore, serves as a crude, but nevertheless graphical, illustration of the physical effect of progressive eutectic alloying.

REFERENCES

- *Alloy Phase Diagrams,* 1992. *ASM Handbook,* Vol 3, ASM International
- Altschuller, B. et al., Ed., 1990. *Aluminum Brazing Handbook,* 4th ed., The Aluminum Association Inc.
- Ambrose, J.C. and Nicholas, M.G., 1986. Alloys for Vacuum Brazing Aluminum, *Brazing Soldering,* Vol 11 (No. 3), p 374–379
- Birchenall, C.E., 1959. *Physical Metallurgy,* Metallurgy and Metallurgical Engineering Series, McGraw-Hill
- Boughton, J.D. and Sloboda, M.H., 1970. Embrittling Effects of Trace Quantities of Aluminum and Phosphorus on Brazed Joints in Steel, *Weld. Met. Fabr.,* Vol 8, p 335–339
- Bowen, R.C. and Peterson, D.M., 1987. A Comparison of Rapid Solidification Cast Versus Conventional Die Attach Soft Solders, *IEEE Trans. on Components Hybrids and Manufacturing Technology,* Vol 10 (No. 3), p 341–345
- Brukl, C., Nowotny, H., and Benesovsky, F., 1961. Untersuchungen in der Dreistoffsystemen: V-Al-Si, Nb-Al-Si, Cr-Al-Si Mo-Al-Si bzw. Cr(Mo)-Al-Si, *Monatsh. Chem.,* Vol 92, p 967–980
- Canonico, D.A., Cole, N.C., and Slaughter, G.M., 1977. Direct Brazing of Ceramics, Graphite and Refractory Metals, *Weld. J. Res. Suppl.,* Vol 56 (No. 8), p 31s–38s
- Chatterjee, S.K. and Mingxi, Z., 1990. Tin-Containing Brazing Alloys, *Weld. J. Res. Suppl.,* Vol 69 (No. 10), p 37s–42s
- Chuang, T.H. et al., 2000. Development of a Low Melting Point Filler Metal for Brazing Aluminum Alloys, *Metall. Mater. Trans. A,* Vol 31A, p 2239–2245
- DeCristofaro, N. and Bose, D., 1986. Brazing and Soldering with Rapidly Solidified Filler Metals, *Proc. Conf. Rapidly Solidified Materials,* February 3–5 (San Diego, CA), p 415–424
- DeCristofaro, N. and Henschel, C., 1978. Metglass Brazing Foil, *Weld. J. Res. Suppl.,* Vol 57 (No. 7), p 33s–38s
- Dev, S.C. et al., 1992. A Copper-Based Brazing Alloy for Electronics Industries, *J. Mater. Sci.,* Vol 27 (No. 24), p 6646–6652
- Eagles, A.M., Mitchell, S.C., and Rebbeck, M., 1995. Investigation of the Suitability of New High Manganese, Low-Silver Braze Alloys for Joining Steels, *British Association of Brazing and Soldering,* Vol 13, p 8–12
- Eng, R.D., Ryan, E.J., and Doyle, J.R., 1977. Nickel-Base Brazing Filler Metals for Aircraft Gas Turbine Application, *Weld. J. Res. Suppl.,* Vol 56 (No. 10), p 15s–19s
- Eustathopoulos, N., Nicholas, M.G., and Drevet, B., 1999. *Wettability at High Temperatures,* Pergamon
- Gempler, E.B., 1976. Parameters Evaluated in Long Cycle Aluminum Vacuum Brazing, *Weld. J. Res. Suppl.,* Vol 55 (No. 10), p 293s–301s
- Hagiwara, M. et al., 1988. Aluminum Brazing Material for use in Aluminum Heat Exchanger, U.S. Patent 4,781,888
- Heine, B. and Sahm, K.F., 1993. Flussmittelfrei Hartloten Luftfahrtrelevanter Aluminumlegierurgen mit Niedrigschmelzendem Lot, *Schweissen Schneiden,* Vol 45, p 429–430 (in German)
- Hellawell, A., 1979. The Growth and Structure of Eutectics with Silicon and Germanium, *Progress in Materials Science,* Chalmers, A., Christian, J.W., and Massalski, T.B., Ed., Pergamon Press
- Hosking, F.M. et al., 2000. Microstructural and Mechancial Characterisation of Actively Brazed Alumina Tensile Specimens, *Weld. J. Res. Suppl.,* Vol 81 (No. 8), p 222s–230s
- Humpston, G. et al. 1992. Recent Developments in Silicon/Heat-Sink Assemblies for High Power Applications, *GEC J. Res.,* Vol 9, p 67–78
- Humpston, G. and Jacobson, D.M., 1993. *Principles of Soldering and Brazing,* ASM International
- Humpston, G., Jacobson, D.M., and Sangha, S.P.S., 1995. New Filler Metals and Processes for Fluxless Brazing of Aluminum Engineering Alloys, *Mater. Sci. Technol.,* Vol 11, p 1161–1167

- Illgen, L. et al., 1991. Preparation of Ductile Al-Ge Soldering Foils by PFC Technique, *Mater. Sci. Eng.* A, Vol 133, p 738–741
- Ishida K. and Nishizawa, T., 1990. Co-Cr (Cobalt-Chromium), *Binary Alloy Phase Diagrams,* 2nd ed., Massalski, T.D., Ed., ASM International, p 1180
- Jacobson, D.M., et al., 2002, The Development of New Silver-Free Brazing Alloys for Steel Tubular Assembly, *Weld. J. Res. Suppl.,* Vol 81 (No. 8), p 149s–155s
- Jacobson, D.M. and Humpston, G., 1995. Lowering the Melting Point of Solders and Brazes by Eutectic Alloying, *GEC J. Res.,* Vol 12, p 112–121
- Jacobson, D.M., Humpston, G., and Sangha, S.P.S., 1996. A New Low Melting Point Aluminum Braze, *Weld. J. Res. Suppl.,* Vol 75 (No. 8), p 243s–249s
- Johnson, A., 1990. Evidence for Au-Si Bonding in Liquid Gold-Silicon Alloys from Electrical Resistivity Measurements, *Solid State Commun.,* Vol 76 (No. 6), p 733–775
- Kang, S. and Kim, H.J., 1995. Design of High Temperature Brazing Alloys for Ceramic-Metal Joints, *Weld. J. Res. Suppl.,* Vol 74 (No. 9), p 289s–295s
- Kayamoto, J.H. et al., 1994. *Proc. Conf. Workshop of the Japanese Welding Society,* Vol 12, p 495–501 (in Japanese)
- Kayamoto, T. et al., 1996. Vacuum Brazing 2017 Aluminum Alloy Using Al-Ge-Cu Based Fillers, *J. Jap. Weld. Soc.,* Vol 14 (No. 1), p 63–68
- Korol'kov, A.M. 1956. Effect of Added Elements on the Surface Tension of Aluminum at 700 to 740°C, *Otdelenie Teknicheskik Nauk,* Vol 2, p 35–42
- Lee, K.J. and Nash, P., 1990. Ni-P (Nickel-Phosphorus), *Binary Alloy Phase Diagrams,* 2nd ed., Massalski, T.D., Ed., ASM International, p 2835
- Liu, Y., Shao, G., and Tsakiropoulos, P., 2000. Thermodynamic Reassessment of the Mo-Si and Al-Mo-Si Systems, *Intermetallics,* Vol 8 (No. 8), p 953–962
- Lugscheider, E. and Schittny, Th., 1988. Wide Gap Brazing of Stainless Steel and Carbon Steel, *Brazing Soldering,* Vol 14 (No. 1), p 27–29
- Lugscheider, E. and Tillmann, W., 1991. Development of New Active Filler Metals for Joining Silicon Carbide and Nitride, Paper 11, *Proc. Conf. British Association for Brazing and Soldering,* September 3–5 (Stratford-upon-Avon, U.K.)
- Miller, C.J. 1969. Fluxless Aluminum Brazing furnace, U.S. Patent 3,460,816
- Miller, V.R. and Schwaneke, A.E., 1978. Interfacial Compositions of Silver Filler Metals on Copper, Brass and Steel, *Weld. J. Res. Suppl.,* Vol 57 (No. 10), p 303s–310s
- Mizuhara, H. and Mally, K., 1985. Ceramic-to-Metal Joining with Active Brazing Filler Metal, *Weld. J. Res. Suppl.,* Vol 64 (No. 10), p 27s–32s
- Muller, W. and Ruhlicke, D., 1991. Microcrystalline Aluminum-Germanium Solder Foil for Chip Bonding, *Metallwissenschaft und Technik,* Vol 45 (No. 1), p 41–45
- Nicholas, M.G., 1998. *Joining Processes: Introduction to Brazing and Diffusion Bonding,* Kluwer Academic
- Niemann, J.T. and Wille, G.W., 1978. Fluxless Diffusion Brazing of Aluminum Castings, *Weld. J. Res. Suppl.,* Vol 57 (No. 10), p 285s–291s
- Petzow, G. and Aldinger, F., 1968. Nonvariante Gleichgewichte und Schmelzraume im System Silber-Kupfer-Zink-Kadmium, *Z. Metallkd.,* Vol 59 (No. 2), p 145–153; Vol 59 (No. 5), p 390–395; Vol 59 (No. 7), p 583–589
- Rabinkin, A., 1998. Stability to Aging of Copper-to-Copper Joints Brazed with Metglas MBF-2005 and BCuP-5 Filler Metals, *Weld. J. Res. Suppl.,* Vol 77 (No. 10), p 29s–30s
- Rabinkin, A. and Liebermann, H.H., 1993. Brazing and Soldering with Rapidly Solidified Alloys, *Rapid Solidified Alloys, Processes, Structures, Properties and Applications,* Liebermann, H.H., Ed., Dekker
- Rabinkin, A., Wenski, E., and Ribaudo, A., 1998. Brazing Stainless Steel Using a New MBF-Series of Ni-Cr-B-S-Amorphous Brazing Foils, *Weld. J. Res. Suppl.,* Vol 77 (No. 2), p 66s–75s
- Samsonova, N.N. and Budberg, P.D., 1995. Cr-Ti-V, *Handbook of Ternary Alloy Phase Diagrams,* Vol 7, Villars, Prince, and Okamoto, Ed., ASM International, p 9261
- Schultze, W. and Schoer, H., 1973. Fluxless Brazing of Aluminum Using Protective Gas, *Weld. J. Res. Suppl.,* Vol 52 (No. 10), p 644–651
- Schwartz, M.M., 2003. *Brazing,* ASM International

- Sloboda, M.H., 1961. Design and Strength of Brazed Joints, *Weld. Met. Fabr.,* Vol 6, p 291–296
- Sloboda, M.H., 1971. Industrial Gold Brazing Alloys, *Gold Bull.,* Vol 4 (No. 1), p 2–8
- Stephens, J.J. and Greulich, F.A., 1995. Elevated Temperature Creep and Fracture Properties of the 62Cu-35Au-3Ni Braze Alloy, *Met. Mater. Trans. A,* Vol 26 (No. 6), p 1471–1482
- Suzuki, K., Kagayama, M., and Takeucji, Y., 1993. *J. Jap. Institute Light Met.,* Vol 43 (No. 10), p 533–538 (in Japanese)
- Taylor, P.D., 1992. *Thyristor Design and Realisation,* John Wiley and Sons
- Timmins, P.F., 1994. The Development of Non-Toxic Ag-Based Brazing Alloys, *Weld. J. Res. Suppl.,* Vol 73 (No. 9), p 31s–33s
- Tuah-Poku, I., Dollar, M., and Massalski, T.B., 1988. A Study of the Transient Liquid Phase Bonding Process Applied to a Ag/Cu/Ag Sandwich Joint, *Metall. Trans. A,* Vol 19A (No. 3), p 675–686
- Villars, P. et al., 2001. Binary, Ternary and Quaternary Compounds Former/Nonformer Prediction via Mendeleev Number, *J. Alloy. Compd.,* Vol 317–318, April 12, p 26–38
- Werner, J., Slaughter, G.M., and Gurtner, F.B., 1972. Development of Filler Metals and Procedures for Vacuum Brazing of Aluminum, *Weld. J. Res. Suppl.,* Vol 53 (No. 2), p 64s–70s

CHAPTER 3

The Joining Environment

WHEN CONSIDERING the metallurgical aspects of brazing in Chapter 2, it is assumed that components and the filler were perfectly clean and remained so throughout the process cycle, enabling the constituents to interact freely so that the filler metal can wet and spread over the component surfaces. However, this situation represents the ideal case because oxides and other nonmetallic species are usually present on surfaces that have been exposed to ambient atmospheres and these will interfere with or inhibit wetting and alloying. As a general guide, if the most refractory constituent of a parent metal is present at a concentration of above one atomic percent (see Chapter 2, Appendix A2.1 for formulae for conversion between weight and atomic weight percentages), it will be the major component in the surface film and will impede wetting, unless reduced, dissolved, or displaced. Any oxygen or moisture present in the joining environment will exacerbate this effect further, particularly as the kinetics of oxidation reactions are highly temperature dependent. Thus, the nature and quality of joints depend not only on alloying reactions but also on the processing environment—in particular on whether the surroundings are oxidizing, reducing, or neutral. The term *surroundings* refers to both the gas atmosphere itself and any chemicals, such as fluxes, that are in the vicinity of the workpiece.

The relative ease with which some common engineering materials can be brazed is given in Table 3.1. Most nonmetallic materials are not wetted by most conventional brazes, even when these have clean surfaces. Where wetting does occur, the contact angle between the molten braze and the parent material is often high and thus the braze does not spread over the component surfaces. For nonmetals, this situation cannot be remedied with the help of chemical fluxes because these are unable to change the physical properties of the intrinsic materials that govern the wetting characteristics, as explained in Chapter 1, section 1.2. Wetting and spreading of a braze on nonmetals can be induced by incorporating within the braze highly active elements, such as titanium, that react chemically with the base materials to form interfacial compounds that the braze can wet. Although the manner in which reactive fillers promote wetting is different from that of chemical fluxes, they can also be used to promote wetting of oxidized metal surfaces and thereby provide an effective fluxing action. The fluxing mechanism, in this case, involves the reduction by the active constituent of the braze of oxides of less refractory metals in the parent material, thereby creating a metallic surface on which wetting by the braze can proceed in a conventional manner through alloying. Active brazes are described in Chapter 7, section 7.2. Sometimes, and particularly for higher-melting-point brazes, active metals are the majority constituents of the braze.

Materials used in joining, whether brazes, fluxes, or atmospheres, are becoming increasingly subjected to restrictions on the grounds of health, safety, and pollution concerns. These regulations can limit the choice of materials and process that are deemed acceptable for industrial use. For example, some grades of free-machining steels and leaded brass contain lead globules in the microstructure and cannot be brazed satisfactorily. Attempts to do so tend to cause volatilization of some of the lead and contamination

Table 3.1 Relative ease of brazing some common engineering materials

Material	Degree of difficulty
Precious metals, copper, nickel, cobalt, carbon steels	Easy
Aluminum, tungsten, molybdenum, carbides, stainless steels, cast iron	Fair
Titanium, zirconium, beryllium, graphite, ceramics	Difficult

of the furnace lining and furniture with a material classified as hazardous!

3.1 Joining Atmospheres

Many types of assembly demand brazing under a protective atmosphere, including assemblies intended for service in a vacuum environment, which must be free from volatile contaminants and parent metal components that are disfigured by oxide scale. The categories of joining atmospheres that are available and their interrelationships are shown in Fig. 3.1. Generally, fluxes are needed only when carrying out the joining operation in air or other oxidizing environments.

Two distinct types of atmosphere are used for brazing:

- Chemically inert gas atmospheres (e.g., argon, nitrogen, helium, vacuum). These function by excluding oxygen and other gaseous elements that might react with the components to form surface films and inhibit flowing of and wetting by the braze.
- Chemically active atmospheres, both gases and fluxes, which are designed to react with surface films present on the components and/or the filler metal during the joining cycle and remove them in the process. These atmospheres may either decompose surface films (as does hydrogen when acting on certain oxide or sulfide layers, for example) or react with the films to produce compounds that can be displaced by the molten filler metal. An example of the latter is magnesium vapor that is introduced during the furnace brazing of aluminum. The vapor reacts with the alumina surface coating to form a complex aluminum-magnesium oxide spinel, which is broken up readily by molten filler metals (see section 3.4.3). Brazing fluxes often function by dissolving oxides.

Controlled gas atmospheres require a confining vessel and this invariably means a furnace of some type. Furnace joining also offers other advantages:

- The process may be easily automated for either batch or continuous production because the heating conditions can be accurately controlled and reproduced without the need for much operator skill.
- Furnace joining offers uniform heating of the components of almost any geometry and is suitable for parts that are likely to distort if heated locally.
- The atmospheric protection afforded leads to economies in the use of flux and post-process operations, such as cleaning and the removal of flux residues.

Against this must be considered the following potential disadvantages:

- Capital costs of the equipment, including the associated gas atmosphere handling or

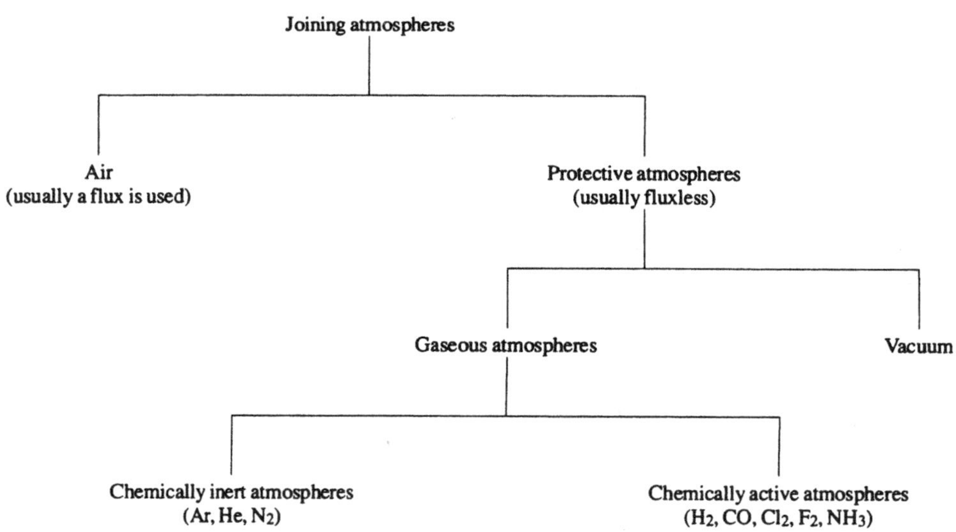

Fig. 3.1 Interrelationship of joining atmospheres

vacuum system, may be significant in relation to the processing costs.

- Recurring costs, particularly those arising from the consumption of gas atmospheres used for processing and maintenance of vacuum pumps
- The entire assembly is heated during the process cycle, which can result in a loss of mechanical properties, even to components that are divorced from the joint area(s).
- The range of permissible parent materials and brazes tends to be restricted to elements and chemicals of low volatility to avoid contamination of the furnace. For a similar reason, most fluxes are undesirable.

Certain metals are not compatible with standard gas atmospheres (oxygen, nitrogen, hydrogen, and carbon-containing gases). Hydrogen atmospheres can cause hydrogen embrittlement of some metals, including titanium, zirconium, niobium (columbium), and tantalum. The hydrogen diffuses into the components and combines with the metal to form hydrides. This lowers the fracture toughness as well as the strain rate transition between ductile and brittle fracture. Many steels are effected adversely by hydrogen, but a short low-temperature (100 to 200 °C, or 200 to 400 °F) bake in air or nitrogen is usually sufficient to diffuse the dissolved hydrogen back out. Oxygen-containing copper alloys can respond unfavorably to hydrogen atmospheres owing to the internal formation of high-pressure steam. The product is a completely useless metal sponge. Hydrogen can combine with graphite that may be a part of internal furnace accessories such as sensors or fixtures to form methane, which will carburize steel. Carbon monoxide reduces the oxides of iron, nickel, cobalt, and copper. However, it is poisonous and must be suitably vented and the workplace monitored continuously for leaks.

Carbon monoxide/dioxide mixes are often used for brazing steels to provide a stable reducing atmosphere that inhibits decarburization.

Nitrogen atmospheres cannot be used when the parent materials and filler metals contain elements susceptible to nitriding, namely chromium, molybdenum, titanium, zirconium, and boron. Boron combines with nitrogen to form boron nitride. Ultimately, this redistributes to form as a black film on the components and the furnace furniture, and it prevents wetting and spreading by the braze. Because the nitriding reaction obeys normal time- and temperature-dependent kinetics, if the heating rate is sufficiently fast, it is possible to get good results by maintaining the nitrogen level in the atmosphere at a low level ($<$0.5 vol%), but process control must be rigorous.

Where the furnace atmosphere is derived from burnt fuel gas, care should be taken to ensure that the source hydrocarbon is free of sulfur. Nickel alloys are embrittled by even small quantities of sulfur, stemming from the formation of nickel sulfide at grain boundaries. Residual contamination from machining and other metalworking lubricants is another source of sulfur. Some brasses are intolerant of ammonia, as are many stainless steels. Ammonia is sometimes "cracked" to yield a nitrogen/hydrogen mix, the advantage being that ammonia can be obtained easily in liquid form, facilitating storage of large quantities of process gas. Any residual ammonia in the furnace atmosphere can result in stress cracking of brass and nitriding of stainless steel. Therefore, the requirements of each component in an assembly must be assessed individually, together with the heating method and other materials in the vicinity, and the atmosphere chosen to suit the ensemble.

When carrying out a brazing operation in a controlled atmosphere, one must take into account the material of the furnace lining and furniture. For example, if the furnace contains items made of carbon steel and the furnace load includes a low-carbon stainless steel, carburization of the stainless steel can occur if the furnace atmosphere contains sufficient moisture. Water vapor reacts with carbon to form carbon monoxide, and this can result in a redistribution of carbon from the furnace furniture onto the workpieces.

3.1.1 Atmospheres and Reduction of Oxide Films

A principal process requirement for successful brazing is to ensure that the joint surfaces are free from oxides and other films that can inhibit wetting by the molten braze and the formation of strong metallic bonds. The ability to remove a layer of oxide from a given metal depends on the ease of either physically detaching the film from the underlying metal or of chemically separating the oxygen ions from the metallic ions present in the oxide, that is, the strength of the relevant molecular bonds. Chemical reduction of metal oxide by atmospheres is considered first.

Chemical thermodynamics can be used to determine the propensity for a metal to spontaneously oxidize or, conversely for an oxide to dissociate. A measure of the strength of a metal-to-oxygen chemical bond is given by the change in the Gibbs free energy that occurs when that metal reacts to form the oxide, as detailed in Appendix A3.1 at the end of this chapter. Here, it is noted that the Gibbs free energy, *G*, is an important thermodynamic function in chemistry because incremental changes in its value involve only incremental changes in pressure, *P*, and temperature, *T*, for reversible reactions:

$$dG = VdP - SdT$$

(see Appendix A3.1 for a definition of symbols)

Chemical reactions, such as oxidation and reduction, which are reversible, can take place at constant pressure and temperature, so that the Gibbs free energy of the material system does not then change in the course of the reaction. Table 3.2 shows the Gibbs free energy of formation of oxides for a selection of metals at room temperature. This formation energy is sometimes referred to reciprocally as the dissociation potential of the oxide. The least stable metal oxides are those of the noble metals, gold, silver, and members of the platinum group. These metals are therefore the most readily brazed, while the refractory metals and the light metals, notably aluminum, beryllium, and magnesium, have particularly stable oxides so that these metals are the most difficult to join.

Other factors need to be considered in connection with oxide reduction. In particular, many metals form different oxides of varying stability—for example, copper oxidizes to form cuprous oxide (Cu_2O) and cupric oxide (CuO). Furthermore, oxides formed on alloy surfaces are not generally pure metal oxides but rather compound or other forms of mixed oxide. Because iron and chromium can have isomorphous oxides, Fe_2O_3 and Cr_2O_3, a solid solution oxide, $(Fe,Cr)_2O_3$, is formed on chrome steels over a certain range of temperatures. This mixed oxide is more difficult to reduce than Fe_2O_3 but is easier than Cr_2O_3. Many alloys are covered by oxides of nonuniform composition and structure, adding further complexity to the subject. This is particularly true of brazes that are almost inevitably multicomponent alloys.

Oxide reduction (or disruption) can be aided by the presence of certain minor constituents in the parent materials, as mentioned in Chapter 2, section 2.2. Such complexities make it difficult to achieve a comprehensive theoretical understanding of oxide removal in brazing processes. In its present state of development, chemical thermodynamics is not even able to predict accurately conditions under which dissociation of oxides will occur but can provide only a semiquantitative indication, particularly when the kinetics of reaction are taken into account. Therefore, the thermodynamic principles for analyzing oxide reduction is considered here only for pure metals.

3.1.2 Thermodynamic Aspects of Oxide Reduction

All chemical reactions are reversible, including oxidation reactions. In general, the oxidation of any metal can be described by an equation of the form:

$$nM + [m/2]O_2 \leftrightarrow M_nO_m \qquad \text{(Eq 3.1)}$$

The reaction will proceed spontaneously in either direction—namely, oxidation of the metal, or, conversely, reduction of the oxide, if it is energetically favorable to do so. A condensed treatment of relevant thermodynamic functions and their relationships, which has by necessity required a degree of oversimplification, is given in Appendix A3.1. A more rigorous treatment is given in standard textbooks on thermodynamics.

Table 3.2 Comparative values for free energies of formation of metal oxides of common braze constituents and selected metals at room temperature (25 °C, or 77 °F)

The more negative the value, the more stable the oxide.

Element	Common oxide	Free energy of formation at 25 °C (77 °F), kJ/mol (rounded values)
Gold	Au_2O_3	+50
Silver	Ag_2O	−10
Copper	CuO	−250
	Cu_2O	−300
Nickel	NiO	−430
Iron	FeO	−490
	Fe_2O_3	−500
	Fe_3O_4	−510
Tin	SnO	−510
	SnO_2	−515
Zinc	ZnO	−625
Chromium	Cr_2O_3	−700
Silicon	SiO_2	−860
Titanium	TiO_2	−900
Aluminum	Al_2O_3	−1050
Magnesium	MgO	−1140

In Appendix A3.1, it is shown that the thermodynamic relationship between the driving force, which is represented by the Gibbs free energy change (ΔG) per molecule of oxygen produced in the oxidation reaction, temperature and the dissociation pressure of the oxide in question [$P_{O_2}^{M}$], is:

$$\Delta G_{(2/m)M_nO_m} = RT \ln P_{O_2}^{M}$$

(see Eq 3A.6), where, R is the universal gas constant and T is temperature in Kelvin.

As pointed out in Appendix A3.1, derivation of this relationship required some simplifying assumptions, including insolubility of oxygen in the metal, but it is substantially valid in most situations involving pure metals.

The free energy change (ΔG) for oxidation reactions involving a series of metals can be charted on a diagram as a function of temperature, as shown in Fig. 3.2. This representation is known as an Ellingham diagram. The Ellingham diagram can be used to determine whether, in principle, an atmosphere is capable of reducing surface oxides, although it does not provide any indication of the kinetics of the reactions. The use of the Ellingham diagram in brazing practice is described subsequently.

It may be noted that the slope of the curves on the Ellingham diagram, for solid metals at atmospheric pressure, are largely identical since the slope is a measure of the entropy increment $\Delta S(T)$ for the designated reaction, as defined in Appendix A3.1, according to the relationship:

$$\left[\frac{\Delta G(T)}{T}\right]_P = -\Delta S(T)$$

The free energy curves are essentially straight lines below and above the melting point. The discontinuous changes in slope that occur at the melting point are due to changes in entropy between the solid and liquid states.

At any given temperature, the smaller is the equilibrium partial pressure of oxygen in the metal oxide, the stronger is the bond between the oxide and the parent metal, that is, the greater is the stability of the oxide. The partial pressure of a gas in an atmosphere is defined in Chapter 1, section 1.3.2.5. Thus, the tendency for the ox-

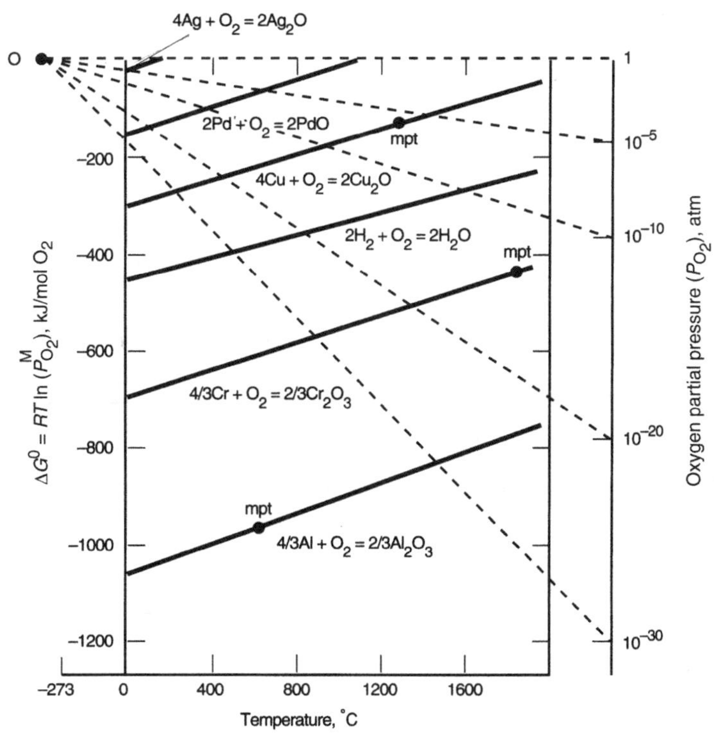

Fig. 3.2 Simplified Ellingham diagram showing the free-energy change for oxidation of several metals. Oxide stability is reduced by elevated temperature and decreased oxygen partial pressure. Each dashed line corresponds to the Gibbs free-energy change as a function of temperature, relating to a particular oxygen partial pressure. mpt, melting point

ide to decompose will be greater the lower the oxygen content of the atmosphere and the higher the temperature. Expressed mathematically, at a given temperature: If $P^{M}_{O_2} < P^{A}_{O_2}$, where $P^{M}_{O_2}$ is the dissociation pressure of the oxide and $P^{A}_{O_2}$ is the partial pressure of oxygen in the atmosphere, the oxide is stable. However, if $P^{M}_{O_2} > P^{A}_{O_2}$, the oxide will spontaneously reduce to the metal.

As shown on the Ellingham diagram, the dissociation pressure $P^{M}_{O_2}$ increases with rising temperature for all metals. Thus, for an atmosphere containing a given partial pressure of oxygen, there exists a critical temperature, T_c, at which the boundary condition will apply, whereby:

$$P^{M}_{O_2} = P^{A}_{O_2}$$

and the oxide will commence to dissociate spontaneously, according to thermodynamic theory.

This condition can, in principle, be achieved by varying either the oxygen partial pressure or the temperature, or both, as explained in the following section. However, there are practical limits to this, notably when the temperature required is so high that the material melts or some physical property degrades irreversibly, or if the partial pressure of oxygen needs to be so low that it is practically unattainable. The farther down the diagram a particular metal-oxygen reaction curve lies, the more stable is the oxide and the more difficult it is to reduce (i.e., higher temperatures and atmospheres of lower oxygen content are required to effect reduction).

When applying the Ellingham diagram to brazing, the following points should be borne in mind:

- The oxides present on the surfaces are assumed to be of pure composition, corresponding to those represented on the diagram.
- The diagram can be used only to establish whether the reduction of the oxide in question is thermodynamically possible; it does not provide any information about the rates of oxide removal. Even when conditions are favorable for reduction of the oxide, the rate might be too slow to make a process based on atmosphere control alone economically viable.
- The diagram indicates only the conditions under which oxide is spontaneously reduced to metal. In practice, it is often found that oxide removal will occur at higher partial oxygen pressures than the limiting value required to satisfy thermodynamic criteria, as pointed out in the section that follows.

3.1.3 Practical Application of the Ellingham Diagram

3.1.3.1 Brazing in Inert Atmospheres and Vacuum

As seen in the simplified Ellingham diagram shown in Fig. 3.2, silver oxide (Ag_2O) and palladium oxide (PdO) can be reduced to metal below their melting points by heating in air to just above 180 and 920 °C (356 and 1688 °F), respectively, the temperature at which $P^{M}_{O_2} = P^{A}_{O_2}$ for the relevant oxidation/reduction reaction. In practice, silver oxide is not considered to dissociate until it is heated to at least 190 °C (375 °F): the excess temperature is required to drive the reaction at a reasonable rate; moreover, the oxide is seldom pure.

For many metals, heating alone in air is inadequate to reduce the oxide because the components are degraded or even melt before the critical temperature, T_c (at which the oxide will spontaneously decompose), is reached. Moreover, the rate of oxidation roughly doubles with each 25 °C (45 °F) rise in temperature. Thus, stable oxides become progressively thicker and tenacious, and consequently more difficult to remove, over the time interval that the component is being heated to the critical temperature. Excessive oxidation can damage component surfaces, particularly if the film spalls off locally, because the rate of oxidation will be nonuniform over the surface, producing an unsightly finish. For these reasons, it is usual practice to heat the components in a suitable inert atmosphere or vacuum, which will both protect the surfaces from further oxidation and reduce the partial oxygen pressure and hence the critical temperature.

The conditions of temperature and oxygen partial pressure required to spontaneously reduce a metal oxide can be deduced from the Ellingham diagram. Reduction will occur when the free energy curve for metal-oxide formation lies above the oxygen partial pressure curve at the temperature of interest, that is, when the oxygen pressure in the atmosphere is less than that which will cause the metal under consideration to oxidize. These curves are marked on Fig. 3.2 as dashed lines originating from point *O* and intersecting the oxygen partial pressure side scale. Thus, the critical temperature for the reduction of PdO decreases from 920 °C (1688 °F) in pure oxygen at atmospheric pressure to 380 °C (715 °F) if the oxygen partial pressure is decreased to 10^{-10} atm (10^{-2} mPa). It can be seen from the more detailed Ellingham diagram given in Fig. 3.3 that oxide reduction in vacuum is practicable only for copper, palladium, silver, iron, and nickel, under realistic process conditions. For metals having oxidizing reaction curves that are located below the 10^{-10} atmospheres (10^{-2} mPa) oxygen partial pressure curve, such as chromium and aluminum, it will be energetically favorable for the metal to oxidize by reaction

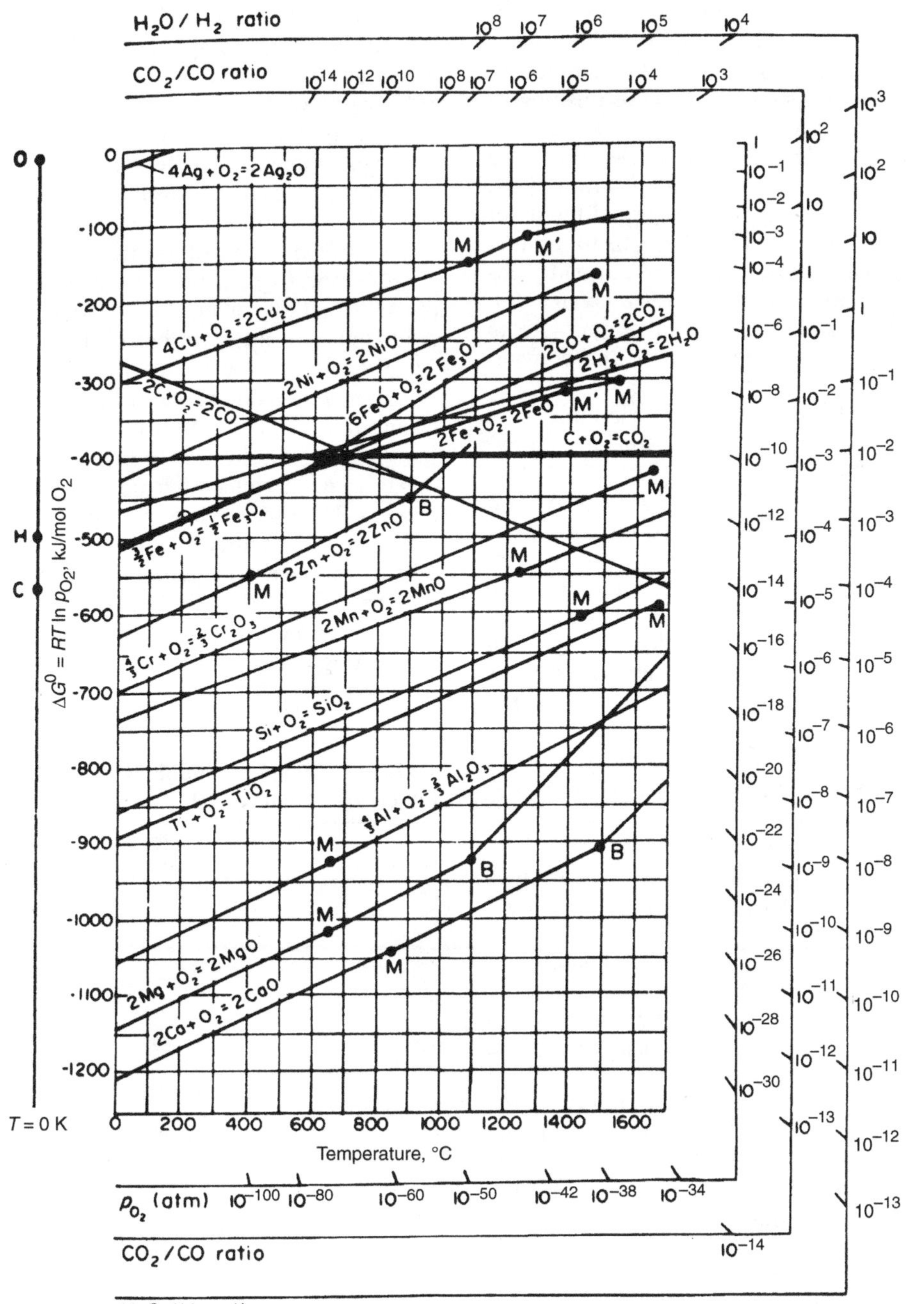

Fig. 3.3 Ellingham diagram for selected oxides. *M*, melting point of metal; *B*, boiling point of metal; *M′*, melting point of oxide

with the residual oxygen and any water vapor present in the furnace atmosphere in most industrial plants. The oxygen partial pressure in a vacuum furnace can be reduced substantially below the gas pressure of the vacuum by repeatedly pumping out and backfilling the chamber with a dry, oxygen-free gas (see Chapter 1, section 1.3.2.5). Care must be taken to ensure that the inlet system is completely leak tight, otherwise, some oxygen will be bled into the furnace and this will impair or even nullify the benefit of the inert atmosphere. A periodic flushing of the chamber with the inert gas will also serve to prevent any buildup of oxygen released in the dissociation of oxides during the heating cycle. However, a partial oxygen pressure of the order of 10^{-10} atm (10^{-2} mPa) is about the minimum, which can be achieved using high-quality industrial equipment. Note that it is convenient to use the atmosphere as the unit of pressure in thermodynamic calculations, and this convention is applied to Ellingham diagrams.

For metals having oxidizing reaction curves that are located below the 10^{-10} atm (10^{-2} mPa) oxygen partial pressure curve (that is, a line joining the point O on the $T = 0\ K$ axis on the left, to the 10^{-10} atm value on the partial oxygen pressure, $P_{O_2} = P^A_{O_2}$, scale on the right side of the Ellingham diagram, as shown in Fig. 3.2 and 3.3), it will be energetically favorable for the metal to oxidize by reaction with the residual oxygen and any water vapor present in the furnace atmosphere. The metal, in this case, includes many common braze constituents. Therefore, industrial quality vacuum and inert gas atmospheres are incapable of preventing degradation of most brazes during normal heating cycles. Obviously, an atmosphere that is largely free of oxygen and water vapor will slow further oxidation greatly, but cannot prevent or reverse it.

While the removal of some oxide coatings by the reduction of the partial oxygen pressure would appear to be practically impossible, components covered with these oxides are capable of being vacuum brazed by methods that will be described. Examples are chromium oxide (Cr_2O_3) and alumina (Al_2O_3). The dissociation of Cr_2O_3 at 1000 °C (1830 °F) requires a partial oxygen pressure of less than 10^{-17} atm (10^{-9} mPa), while it requires the partial pressure of oxygen to be less than about 10^{-50} atm (10^{-42} mPa) to reduce a film of Al_2O_3 to the metal at 700 °C (1290 °F).

As mentioned previously, care must be taken to select an atmosphere that is inert toward all the metals in the assembly being joined. Vacuum can degrade certain materials, notably brass, even at brazing temperatures, due to the loss of zinc through volatilization, a consequence of the high vapor pressure of this element. Likewise, manganese-containing brazes are unstable in high vacuum at temperatures much above 750 °C (1380 °F) and are not recommended for use under these conditions. Table 3.3 lists the boiling/sublimination temperatures of selected elements at 10^{-10} atm (10^{-2} mPa). For metals to be joined under reduced pressure, the process temperature must be considerably lower than the boiling/sublimation temperature (by a factor of $<\frac{1}{2}$ in K/K), if volatilization is not to be significant.

The effectiveness of using the process temperature and oxygen partial pressure to control oxide reduction, or at least prevent significant oxidation, is limited further by the presence of adsorbed water vapor on the walls of the vacuum chamber and on other free surfaces inside it. The desorption of water vapor effectively increases the oxygen partial pressure in the chamber, and this has a deleterious effect on the oxide removal process. Therefore, it is good practice to heat the walls of the chamber to promote desorption, while simultaneously removing the vapor from the chamber by alternately pumping out and/or flushing with dry, inert gas before commencing the heating cycle.

Table 3.3 Boiling/sublimination temperature of selected elements at 10^{-10} atm (10^{-2} mPa)

	Boiling/sublimation temperature	
Element	°C	°F
Cd	100	212
Zn	150	302
Mg	210	410
Sb	300	572
Bi	350	662
In	525	977
Mn	550	1022
Ag	630	1166
Al	725	1337
Sn	730	1346
Cu	780	1436
Cr	800	1472
Au	880	1616
Pd	905	1661
Fe	950	1742
Co	1020	1868
Ni	1025	1877
Ti	1130	2066
Mo	1680	3056
W	2230	4046

Values are rounded. Note the high position of tin and the low position of manganese and zinc in the table in relation to their melting points.

A simple qualitative indication of the oxygen content of the atmosphere in a brazing furnace can be obtained by including a thin foil of titanium (100 μm, or 4 mils thickness) with the furnace load. Any oxygen present will progressively color tint and then deeply discolor the titanium, while the ductility of the foil—in terms of its ability to be bent round a pencil without fracture—will also decline. Foil test pieces can be kept to act as visual reference standards of acceptable and inadequate quality of the furnace atmosphere.

Large-scale industrial processes requiring nitrogen gas often rely on cryogenic liquid nitrogen for several reasons—not least of these is the ease of convenience of delivery and storage. Furthermore, nitrogen boiled off from a cryogenic tank containing the liquified gas possesses lower levels of oxygen and water vapor (typically <2 ppm combined) than all but the purest grades of bottled nitrogen. It is also relatively inexpensive, being comparable in price per liter to bottled mineral water. Owing to increasingly stringent environmental legislation, joining in inert atmospheres is gaining in popularity. In commercial systems, the nitrogen ambient contains less typically than 10 ppm of other gaseous constituents. The running costs associated with the large volumes of nitrogen that are required to achieve this quality of atmosphere are offset by the ability to dispense with post-joining treatments because reduced quantities of fluxes and cleaning fluids are required and thereby reduce the associated health and environmental problems accordingly. Figure 3.4 shows an aluminum component assembled by fluxless brazing in a nitrogen flow furnace, that is, a furnace where the portal on the far side from the gas supply is continuously open to air, and so providing easy access to the furnace chamber for volume manufacturing. The fluxless brazing of aluminum is described in further detail in section 3.4.3 of this chapter. Gas atmospheres have the singular advantage of superior thermal transfer on heating and cooling, compared with a vacuum process. Even among different atmospheres there can be appreciable differences in heat transfer characteristics. For example, hydrogen gas has a thermal conductivity seven times that of nitrogen (see Table 3.4).

Table 3.4 Thermal conductivities of brazing atmospheres, relative to air

Soldering atmosphere	Relative thermal conductivity
Carbon dioxide	0.62
Argon	0.68
Nitrogen	0.99
Air	1
Helium	5.8
Hydrogen	6.9

Fig. 3.4 A component made of an aluminum engineering alloy (type 6082), fabricated by fluxless brazing in a nitrogen flow furnace. The brazed joints exist at the interface between the web members and the face plates and also at the intersection of each web member. A similar component is shown partly jigged prior to brazing in Fig. 1.2. Courtesy of BAE Systems Ltd.

For certain applications, inert gases other than nitrogen may be more appropriate. Of these less-common inert gases, argon and carbon dioxide are probably the most widely used. Both can be purchased in high-purity form. Carbon dioxide is often recommended in applications where the atmosphere is confined, but open to air at various portals, because the greater molecular weight of carbon dioxide enables it to displace and exclude air more effectively than does nitrogen [Esquivel and Chavez 1992]. In the presence of graphite furnace furniture, carbon dioxide tends to dissociate at higher brazing temperatures to carbon monoxide and thereby provides a function of oxygen reduction, which can compensate partially for any oxygen ingress. Argon is more expensive than the other two gases, and its use is therefore always confined to joining in closed volumes.

Reference to exothermic and endothermic brazing atmospheres may be found in trade journals and other publications. These are formed from hydrocarbon fuel gas, either natural or synthetic mixtures that are combusted in a retort together with a controlled ratio of air. The retort may contain a catalyst to help tailor the mix of reaction products. In an exothermic burn, the heat liberated in the combustion is sufficient to sustain the reaction so the gas is effectively also preheated. Endothermic burns require additional heat be supplied to the retort. The combusted product will be typically a mixture of nitrogen, hydrogen, methane or ethane, carbon monoxide, carbon dioxide, and water vapor. As the relative proportion of air to fuel gas is increased, the mixture changes from endothermic to exothermic, up to a limiting ratio. This type of gas mixture has the merit that it is relatively inexpensive, because the fuel supply can be obtained from a utility company with no storage facility required and may be made sufficiently reducing to be able to braze carbon steels. Often the atmosphere is referred to by the ratio of air to fuel, e.g., 7:1 exothermic atmosphere.

3.1.3.2 Brazing in Reducing Atmospheres

If the partial oxygen pressure surrounding the workpiece cannot be sufficiently lowered by creating a vacuum or introducing an inert gas so as to effect oxide removal, then a reducing atmosphere might be able to remove the oxide. The three most widely used reducing gases are hydrogen, carbon monoxide, and "cracked" ammonia (that is, ammonia dissociated into nitrogen and hydrogen).

For the case of a hydrogen atmosphere, the boundary condition for oxide reduction is governed by the two reactions:

$$n\mathrm{M} + [m/2]\mathrm{O}_2 \leftrightarrow \mathrm{M}_n\mathrm{O}_m$$

determined by $\Delta G_{\mathrm{M}_n\mathrm{O}_m}$, and

$$2\mathrm{H}_2 + \mathrm{O}_2 \leftrightarrow 2\mathrm{H}_2\mathrm{O}$$

determined by $\Delta G_{2\mathrm{H}_2\mathrm{O}}$. Subtracting these equations gives:

$$n\mathrm{M} + m\mathrm{H}_2\mathrm{O} \leftrightarrow \mathrm{M}_n\mathrm{O}_m + m\mathrm{H}_2$$

determined by $\Delta G_{\mathrm{M}_n\mathrm{O}_m+m\mathrm{H}_2}$, and

$$\Delta G_{\mathrm{M}_n\mathrm{O}_m+m\mathrm{H}_2} = \Delta G_{\mathrm{M}_n\mathrm{O}_m} - \left(\frac{m}{2}\right)\Delta G_{2\mathrm{H}_2\mathrm{O}}$$

Now, as shown in Appendix A3.1:

$$\Delta G_{\mathrm{M}_n\mathrm{O}_m+m\mathrm{H}_2} = -mRT\ \ln\left(\frac{P^{\mathrm{A}}_{\mathrm{H}_2}}{P^{\mathrm{A}}_{\mathrm{H}_2\mathrm{O}}}\right)$$

and

$$\Delta G_{\mathrm{M}_n\mathrm{O}_m} = \frac{m}{2}\ RT\ \ln(P^{\mathrm{M}}_{\mathrm{O}_2}) = \frac{m}{2}\ RT\ \ln(P^{\mathrm{A}}_{\mathrm{O}_2})$$

under the boundary condition for oxidation. Therefore:

$$RT\ \ln(P^{\mathrm{A}}_{\mathrm{O}_2}) = -2RT\ \ln\left(\frac{P^{\mathrm{A}}_{\mathrm{H}_2}}{P^{\mathrm{A}}_{\mathrm{H}_2\mathrm{O}}}\right) + \Delta G_{2\mathrm{H}_2\mathrm{O}} \qquad \text{(Eq 3.2)}$$

The term $RT\ \ln(P^{\mathrm{A}}_{\mathrm{O}_2})$ is the oxygen potential, ΔG_{O_2}, which provides a measure of the "oxidizing strength" of the atmosphere for converting a metal to its oxide via the reaction given in Eq 3.1. According to Eq 3.2, the oxygen potential of the atmosphere can be reduced by increasing the hydrogen content and/or by lowering the fraction of water vapor present, other factors being equal.

If, instead of hydrogen, a reducing atmosphere of carbon monoxide is used, the reduction equation is then:

$$2\mathrm{CO} + \mathrm{O}_2 \leftrightarrow 2\mathrm{CO}_2$$

which, by similar reasoning, leads to the equality:

$$RT\ \ln(P_{O_2}^{A}) = -2RT\ \ln\left(\frac{P_{CO}^{A}}{P_{CO_2}^{A}}\right) + \Delta G_{2CO_2}$$

In this case, the oxygen potential can be reduced by increasing the carbon monoxide fraction of the atmosphere and/or reducing the carbon dioxide content.

Rather than substitute measured data into these equations in order to determine the temperature and concentration of gases in the furnace atmosphere to effect reduction, it is possible to obtain the solution more simply by a graphical procedure using the Ellingham diagram. For convenience to the user, the Ellingham diagram is provided with a series of side scales giving the partial oxygen pressure corresponding to ratios of H_2/H_2O and CO/CO_2, as shown in Fig. 3.3. On the left-hand side is shown an axis for $T = -273$ °C (0 K) with points marked at values of the free energies, ΔG, at this temperature for the hydrogen/oxygen (point *H*), the carbon monoxide/oxygen (point *C*), and other reactions. Each of these points is associated with a side scale that wraps around the other three sides of the Ellingham diagram (see Fig. 3.3).

As an example of the use of the Ellingham diagram, we shall consider the conditions for reduction of chromium oxide by hydrogen. The Gibbs free energy of formation of Cr_2O_3 as a function of temperature is represented by curve *AB* in Fig. 3.5. Values of the free energy of formation of water vapor from the reaction of hydrogen with oxygen are represented by a family of curves diverging from the point *H*, each curve corresponding to a different ratio of the partial pressures $P_{H_2}^{A}/P_{H_2O}^{A}$ and the molar ratio H_2/H_2O in the atmosphere. When curve *AB* crosses *PQ*, belonging to the family of water vapor curves

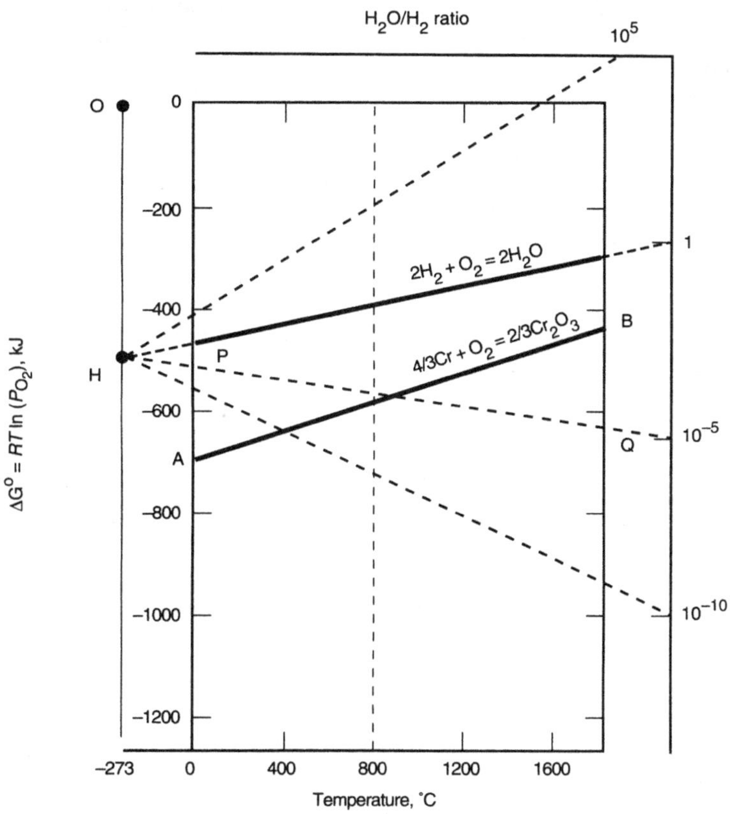

Fig. 3.5 Simplified Ellingham diagram illustrating the graphical method for determining the temperature and H_2O/H_2 ratio that will spontaneously reduce a metal oxide to metal (here, Cr_2O_3 to Cr). The set of dashed lines corresponds to the Gibbs free energy change as a function of temperature for the reaction of hydrogen with oxygen to produce water vapor for different H_2O/H_2 ratios.

and corresponding to a water vapor/hydrogen partial pressure ratio of $1{:}10^{-5}$, the chromium/oxygen and hydrogen/oxygen reactions are in equilibrium because their respective free energies are the same. This means that the oxygen potentials for the two reactions are identical. When curve *AB* lies above *PQ* at a particular temperature, chromium oxide will be reduced spontaneously by the hydrogen to form chromium and water vapor because the latter is more stable than Cr_2O_3. The reverse is true when curve *AB* lies below *PQ*.

Therefore, at any particular temperature, it is possible to use this graphical representation to identify the H_2/H_2O partial pressure ratio that is required to reduce the surface oxide on a metal. For example, a partial pressure ratio of H_2/H_2O greater than or equal to $1{:}10^{-5}$ should reduce Cr_2O_3 to chromium at 800 °C (1470 °F) and above. The calculation of the conditions needed to reduce oxides using carbon monoxide is similar, except that point *C* on the −273 °C (0 K) axis and the CO/CO_2 partial pressure ratio scale are now employed (see Fig. 3.3).

A graphic demonstration of the improvement in wetting by improving the quality of the joining atmosphere is provided by Fig. 3.6, which shows the area over which a pellet of molten copper spreads at 1120 °C (2050 °F) on mild steel, as a function of the H_2/H_2O partial pressure ratio. Note that there is a continuous improvement as the process atmosphere becomes more reducing and not a simple wet/dewet transition that would be expected from a fully literal interpretation of the Ellingham diagram.

In joining technology, it is normal to express the fraction of water vapor in a hydrogen atmosphere in terms of the dew point, a parameter that can be measured directly. By definition, the dew point is the temperature at which water vapor in a given enclosure is saturated. The dew point may be measured simply by observing the onset of condensation of moisture on a polished metal surface as the temperature is lowered. The relationship between dew point and the logarithm of the partial pressure of water vapor is shown in Fig. 3.7. The dryness that may be obtained with conventional drying agents at room temperature is marked on this curve. It is clear that both silica gel and phosphorus pentoxide (P_2O_5) are considerably less effective than chilling the hydrogen to the liquid nitrogen boiling point (−196 °C, or −321 °F) at removing water from a gas supply that is piped into a furnace. Disposable and recyclable molecular sieves are available for most common gases at moderate cost and should be used at furnace inlets as a matter of good practice.

Monitoring of the dew point or oxygen content of a furnace is accomplished readily with modern instrumentation. The same equipment also provides a method for determining the leak tightness of the gas system. In a leak-free system, the oxygen and water vapor content of the gas stream will not change with the flow rate of

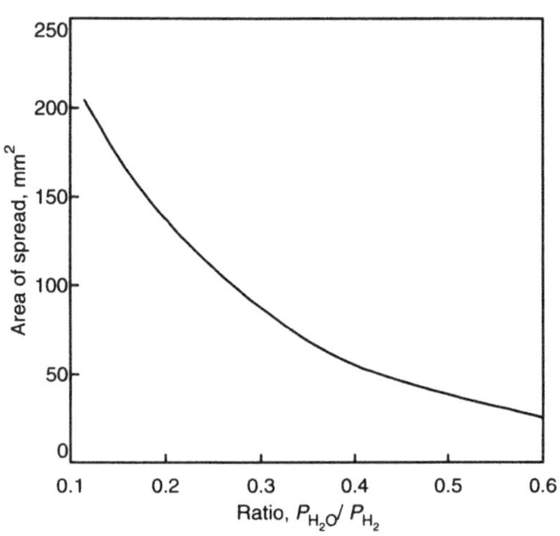

Fig. 3.6 Area of spread by molten copper on mild steel, at 1120 °C (2050 °F), in controlled H_2O/H_2 ratio atmospheres. Source: Bannos [1984]

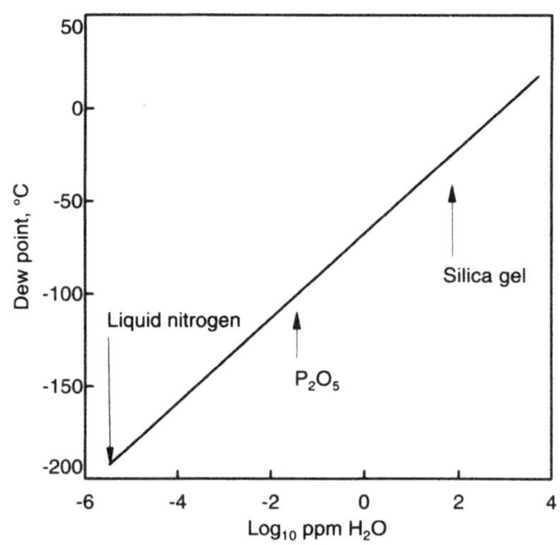

Fig. 3.7 Relationship between dew point and fraction of water vapor in an atmosphere. Drying agents are considerably less effective than low temperatures at reducing the moisture level. Note: vol% = ppm × 10^4

the gas. If the quality of the atmosphere improves markedly as the flow rate is increased, this is a good indication that there is a leak in the system or unsuitable materials have been used. Only a few grades of plastic pipework are effectively impervious to water vapor and these are still greatly inferior to an all-metal gas conveyance system. It should be noted that regulators and flow meters must also be free of organic materials and seals where the highest quality atmospheres are demanded.

Normal chemical reduction is not the only mechanism that can remove surface oxides. It is often found possible to perform flux-free joining at oxygen pressures that are significantly higher than the dissociation pressure of the oxide concerned. For example, the dissociation pressure for titanium dioxide at 1000 °C (1830 °F) is 10^{-30} atm, whereas fluxless brazing and even diffusion bonding of titanium is possible at lower temperatures under moderate vacuum conditions (10^{-9} atm = 0.1 mPa), see Fig. 1.9 in Chapter 1. This is because titanium has a capacity for absorbing large volumes of gases such as hydrogen, nitrogen and oxygen on heating above about 800 °C (1470 °F) and this, in effect, means that its surface remains free of an oxide coating during a typical joining operation [Stubbington 1988]. This characteristic should not be relied on to accommodate a poor-quality atmosphere because the oxidation is detrimental to the mechanical properties of titanium alloys. As another example, at temperatures above about 1000 °C (1830 °F), in moderate vacuum, it appears that the oxide film on the surface of many stainless steels loses its self-repairing ability. This permits wetting by copper-base brazes using a lower-quality atmosphere than would otherwise be expected [Wigley, Sandefur and Lawing 1981].

3.1.3.3 Alternative Atmospheres for Oxide Reduction

Other gases such as chlorine, fluorine, and boron triflouride are more effective than hydrogen and carbon monoxide at removing surface oxides of particular metals, as is clearly indicated on the relevant Ellingham diagrams [Wicks and Block 1963]. Such gases operate partly by converting the oxide to a halide that is volatile at the joining temperature and vaporizes during the heating cycle. These halide atmospheres also chemically attack the underlying metal and physically undermine the oxide, as occurs in the fluxing of aluminum. The use of halide fluxes in aluminum brazing is discussed in section 3.2.2. Unfortunately, these and other chemically corrosive gases tend also to attack furnace linings, furniture, and seals. They also require complex gas handling and exhaust scrubbing equipment in order to comply with health and safety legislation. Consequently, they tend not to be widely used.

3.2 Chemical Fluxes for Brazing

Successful brazing is largely dependent on the ability of the filler to wet and spread on component surfaces. A major barrier to wetting is presented by stable nonmetallic films and coatings on the surfaces, in particular, oxides and carbonaceous residues. Oxide films often endow beneficial attributes to metals, such as corrosion resistance, but their presence on the faying surfaces, and on the surfaces of the brazes, present more than a nonmetallic barrier to wetting and spreading. Oxides are generally poor thermal conductors, compared with metals, and impede heat transfer, thereby exacerbating temperature gradients present and delaying fusion of the braze with the parent metal. Fluxes are chemical agents that are used to remove these layers and thereby promote wetting by the molten filler. In order to be effective in exposing a bare metal surface to the filler, a flux must be capable of fulfilling the following functions:

- Remove oxides and other films that exist on surfaces to be joined by either chemical or physical means. Chemical mechanisms involve reaction of the flux with surface oxides to effect reduction or dissolution. Physical removal of a layer of oxide by a flux is contingent on its ability to weaken adhesion of the oxide film. Normally, the oxide is undermined as a result of penetration by the flux through naturally occurring pores, fissures, and other flaws in the layer, followed by electrolytic action at the interface between the oxide and the parent material.
- Protect the cleaned joint from reoxidation during the joining cycle
- Be displaced by the molten filler as the latter spreads over the faying surfaces.

While molten, fluxes form a thermal blanket around the joint that helps to spread the heat evenly during the heating cycle. The flux also tends to reduce the surface tension between the braze and the joint surfaces, thereby enhancing wetting.

Ideally, the flux should leave no residues or produce residues that are removed easily by, for example, being soluble in water. Fluoborate flux residues are soluble in water, but borosilicate fluxes require a solution containing 5 to 10% sulfuric acid to facilitate their removal. For stubborn deposits, sodium dichromate may be added or a phosphate solution used. Boric acid, a common constituent of many fluxes, is soluble only in hot water so cleaning solutions are often warmed. Fluxes should also be compatible with the filler and substrate materials. For example, so-called "black fluxes," because they contain elemental boron, are not suitable for use with aluminum, magnesium, and titanium components and filler metals owing to the ready formation of borides of these metals.

Chemical fluxes always function while in a gaseous or liquid form, although they are frequently solid at room temperature. If the flux is liquid at the joining temperature, it has to wet the joint surfaces in order to be effective. A flux that is liquid can beneficially help suppress the volatilization of high vapor-pressure constituents of filler metals and thereby improve joint quality. This is particularly true in dip brazing.

Fluxes can be introduced to the joint in a number of ways, some of which are discussed subsequently. Brazing fluxes are usually applied in the form of a powder or paste immediately prior to the heating cycle. The joint is then heated to the required bonding temperature, by which point solid fluxes have become molten, ideally just before the filler metal melts. When designing a joint, allowance must be made for the flux to escape. Flux entrapment at sharp corners of internal features is a common problem.

A flux can also be placed within or adjacent to the joint together with the filler metal as a preform and the assembly heated to the bonding temperature. As a properly chosen flux will melt at a temperature below the melting point of the filler, the molten flux is able to spread over the joint surfaces and clean them before the filler metal melts and displaces the flux.

Another method involves introducing the flux together with the filler into a joint already held at the bonding temperature, in the form of a flux-coated brazing rod. Although this technique is widely practiced because it is fast and convenient, it is not recommended because the heated component surfaces are unprotected until the filler is applied. More aggressive fluxes are then required, which in turn tends to accentuate corrosion and cleaning problems. The reason for the flux being placed on the outside of brazing rods but on the inside of solder wire owes to the relative ratios of flux-to-filler metal required for these two processes. Because brazing is conducted at higher temperatures, reaction rates are faster and so a larger quantity of flux is required to protect the parent materials and fillers against oxidation during the joining process. However, at least one manufacturer sells preforms of brazing alloy in which the brazing flux is internal. The reduced quantity of flux means that they function best in furnace brazing with a controlled atmosphere. Here, the principal attraction is simplification of preparation prior to brazing.

Fluxes can be applied to the faying surfaces, together with the filler metal, prior to the heating cycle, in the form of tapes, pastes, and creams, which are normally proprietary formulations. They comprise mixtures of the filler metal, which is present as a powder of a prescribed grain size range together with a flux and a water-based or organic binder that is selected to produce the desired viscosity and to dry or burn off without leaving contaminating residues. Polyisobutylene is often used as a binder ingredient because the degree of "stickiness" and viscosity can be adjusted by altering the length of the polymer chain, and when thermally degraded, it does not leave a carbonaceous residue. Other common carrier liquids are petroleum-based and polyethylene-glycol-based. The higher viscosity and longer life after application of these pastes is useful in some manufacturing environments. Pastes and creams are particularly used in automated reflow brazing operations because they can be screen printed or dispensed using syringes. Because of the large surface area of the powdered filler metal in contact with the flux, corrosion is inevitable. Therefore, these products have a finite shelf life. Selected braze pastes also are available in the form of tape, which comprises a clean-burning binder material in the form of sheet that incorporates powdered braze and flux. Braze tapes tend to be fragile and must be handled with care. Most brazing fluxes are creamy white in color and are virtually indistinguishable to the naked eye. As mentioned before, the exceptions are those fluxes that contain elemental boron and, consequently, have a gray-black hue.

Certain fluxes are totally soluble in water and so are applied to the faying surfaces as a liquid. These may also be fed into the fuel gas stream during torch brazing so that fresh flux is being applied continuously to the joint area through the heating cycle.

Several different fluxing mechanisms cover the majority of brazing operations that are encountered. Even these are sufficiently complex not to be understood in detail at the present time. However, fluxing mechanisms can be classified according to whether they remove the nonmetallic surface coating by physical or chemical means.

A flux can chemically remove a surface coating by:

- Dissolving the coating
- Reacting with the coating to form a product that is unstable at the bonding temperature
- Reducing the coating to metal in an exchange reaction, such as occurs when reducing gases are effective in removing oxides

A surface coating can also be physically removed. This usually occurs through:

- Erosion of the underlying metal. In this mechanism the flux does not react with the surface coating itself but is able to percolate through it and react with the underlying metal, thereby causing detachment of the coating.
- Wetting of the coating in a manner that causes it to spall off. This mode of fluxing applies to joining processes where components are subject to the thermal shock that cracks the coating due to the relatively brittle nature of oxide layers. Immersion in molten salts and fluidized bed furnaces are examples of this type of process. The process usually relies on the filler metal wetting the parent material through the fissures and spreading underneath the nonmetallic skin to complete its removal.

Many fluxes function by a combination of mechanisms and for this reason, fluxing action is best illustrated with reference to specific examples.

The requirements on brazing fluxes, in particular, the service temperature, greatly restrict the choice of materials that can be used. Consequently, whereas braze Standards (e.g., EN 1044, 1999) list 93 alloy compositions, the matching flux Standard (EN 1045, 1999) contains only seven entries. The rated working temperature range of a flux generally assumes a brazing time of up to 30 seconds in air. Therefore, a flux with a working temperature close to the liquidus temperature of the braze can be used in a controlled atmosphere for longer times. If large components are to be brazed in air, a flux with a higher working temperature will be necessary to allow for the slower rate of heating. Sometimes, particularly for brazing at high temperature, a dual flux system may be used so that some protection is provided to the workpiece at moderate temperatures and a second, high-working-temperature flux then takes over at the brazing temperature. Molybdenum may be brazed in air using this technique.

3.2.1 Brazing Flux Chemistry

Although brazing flux chemistry is a fairly sophisticated science, a few common ingredients account for the vast proportion of the market. The limitation arises on account of the high temperatures involved in brazing, which totally excludes all aqueous and organic materials from consideration. Instead, glass complexes must be used. These possess low volatility and generally also have a low permeability to air and so can provide the cleaned surfaces of the joint and filler metal with the necessary protection against reoxidation.

At the lower end of the brazing temperature range, (i.e., 600–800 °C, or 1110–1470 °F), the glass carrier is based on borates [$-B_xO_y$], with the ratio of oxygen to boron optimized to provide a balance between viscosity and permeability to oxygen. In general, the higher is the oxygen to boron ratio, the higher is the viscosity, and the permeability is reduced correspondingly. The requirement for the flux to be displaced by the molten braze places a limit on the viscosity that can be tolerated. Above 800 °C (1470 °F), borates alone are too permeable and need to be replaced partly by silicates [Si_xO_y] so that the resulting glass is a borosilicate that has a higher viscosity and is therefore able to protect workpieces to higher temperatures. However, borosilicate residues are largely insoluble in water, necessitating more rigorous cleaning procedures, in contrast with those of the simple borates that can be dissolved in water, making them more convenient to use.

In addition to providing protection against oxidation, borates have the ability to dissolve a limited fraction of oxides from the surfaces of steel and copper components. This accounts for the effectiveness of borax (sodium tetraborate) as a flux at temperatures above about 750 °C

(1380 °F). Adding boric acid or boric oxide to borax lowers the melting point of the flux because it reduces the oxygen to boron ratio. The resulting increased chemical activity means that these fluxes facilitate improved wetting by silver-base and silver-free brazes on carbides and alloys that form refractory oxides though having constituents that include chromium, nickel, and cobalt.

When heated by a torch, sodium salts, including borax, produce a bright yellow glare, which is unpleasant to operators. The glare is reduced substantially by partly or totally substituting the borax with the corresponding potassium salt, potassium tetraborate, albeit with a small cost penalty.

Borates on their own are insufficiently active to clean surfaces of many metals. Therefore, fluxes intended for general use contain a proportion of halides. Glare considerations have favored potassium halides, in particular, potassium fluorides, over the equivalent sodium salts. The improved fluxing action results largely from the greater oxide dissolution ability of the fluorides. Indeed, the dominant mechanism responsible for surface cleaning by the fluoborate type fluxes is believed to be direct dissolution of the oxides, and there is little evidence for any of the other recognized types of cleaning action referred to previously. At the same time, the fluorides reduce the melting point of the flux because they disrupt the cross linkages in the borate network structure [Eustathopoulos, Nicholas and Drevet 1999, p 355–56]. The fluoborate fluxes especially suit the low-melting-point silver-base quaternary alloy brazes, which melt at temperatures down to about 590 °C (1095 °F), because they can be used at comparable temperatures. The fluoride addition does, however, reduce the upper working temperature limit of these fluxes because it increases the permeability to oxygen and reduces their thermal stability, owing to the formation of hydrofluoric acid on heating, which is volatile. Commercial fluoborate fluxes fall into the elemental composition range B:F = 1:0.75 to 1.5; B:K = 1:0.55 to 1.1; F:K = 1:0.55 to 0.8, by atomic ratio. This range is represented in Fig. 3.8.

Wetting agents are not strictly necessary for brazing fluxes because, at the elevated joining temperatures used, organic residues will have decomposed, leaving carbonaceous deposits that will be either eliminated through oxidation or cleaned off the surface by the flux. Nevertheless,

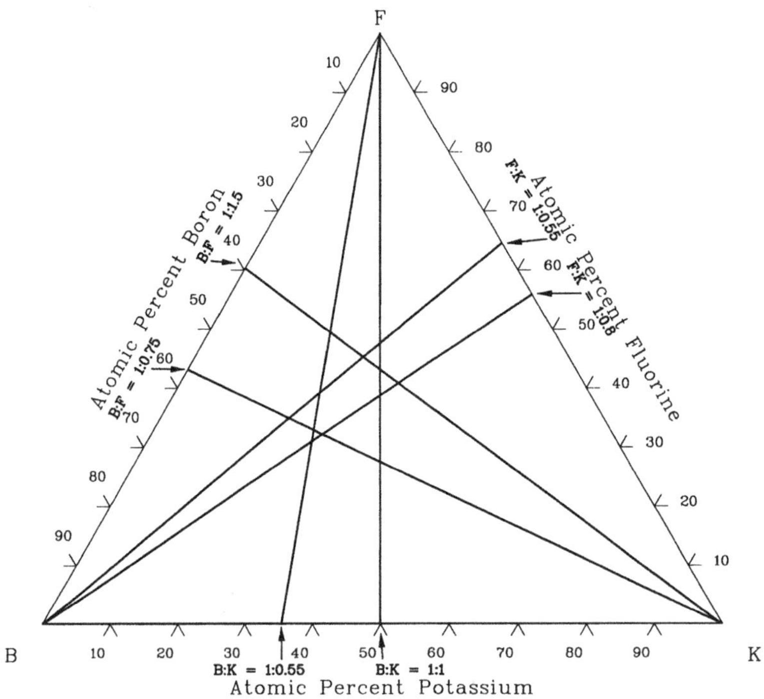

Fig. 3.8 B:K:F atom ratios of common brazing fluxes

wetting and rheological agents are added to flux pastes to produce a smooth consistency, which aids application to the workpiece. Because many brazing fluxes have water-based carriers, it is good practice to thoroughly degrease all components prior to brazing and thereafter handle all items only with gloves or tools that have been similarly cleaned.

Interaction between brazing fluxes, metal oxides, and brazing alloys is often more complex than suggested by the preceding discussion. For example, it is well known that in torch brazing it is easier to make joints that are free of defects to oxidized copper components with silver-copper-zinc brazes than with zinc-free brazes (e.g., silver-copper-tin), under cover of a fluoborate flux. It transpires that copper oxides dissolve in fluoborates slowly so that during the course of a normal brazing cycle, there is insufficient time to remove a thick film of oxide. The flux can dissolve the outer layer of cupric oxide (CuO), but not the bulk of the film, which is predominantly of cuprous oxide (Cu_2O). The flux is, however, effective at removing the oxide skin from the surface of both types of brazing alloy mentioned previously. Removal of this skin permits the brazing alloy to wet the cuprous oxide, which is then removed by a combination of dissolution and reduction by the braze. For this reason, silver-copper brazes that contain an element with a high affinity for oxygen (i.e., zinc and cadmium) wet and spread better on oxidized surfaces of copper and its alloys than brazes without these ingredients. Thus, when brazing copper components in air, the fluxing action amounts to prevention of the formation of scale on the components, but the removal of oxides from the surfaces of the brazing alloy only. In furnace brazing, with a controlled atmosphere and well-prepared components, the extended cycle time and relative thinness of the oxide skin on the copper components means that the flux is able to dissolve all of the cuprous oxide, and the differentiation between zinc-containing and zinc-free brazes is less marked.

Commercial fluxes are proprietary formulations that contain specific ingredients tailored to application requirements and incorporate various subtleties. By way of example, many brazing operators require a flux that will coat a heated rod of the braze when dipped into a tub of the flux powder. This flux-coated rod is then applied to the workpiece, and the brazing operation is carried out using a torch in air. To satisfy this mode of application, which speeds up the joining procedure considerably, fluxes that contain close to 70% of a hydrated potassium fluoborate compound have been formulated. This compound releases sufficient moisture, when heated, to form a sticky paste that will adhere to a metal rod. This example illustrates the finer points affecting user preference and helps to explain why it is best to consult suppliers when considering which flux to use for a particular application. It is also logical that fluxes intended for one combination of parent material and brazing alloy may not be as effective for a different combination, especially if the second brazing alloy melts at a different temperature.

Common brazing fluxes are only suitable for use at temperatures up to about 1200 °C (2200 °F). The higher temperature fluxes comprise a mixture of sodium fluoborate and tetraborate with a significant proportion of silica, boric acid, and elemental boron. Their main use is brazing of copper-, iron-, and nickel-base alloys. Above this temperature, gas atmospheres with sufficient reductive potential are available and the brazes themselves tend to contain active elements (see Chapter 4, section 4.1.2.2 and Chapter 7, section 7.2). The higher the brazing temperature, the more difficult it is to achieve satisfactory joints by local heating methods, particularly by using gas torches. For these reasons, most high-temperature brazing is conducted in closed furnaces under a protective or reducing atmosphere.

Flouride-containing fluxes are corrosive toward skin and fingernails. Also, when heated, hydrofluoric acid and boron-triflouride will be produced. Although the quantities are small, prolonged exposure of furnace equipment and operators to these corrosive vapors is not recommended. For this reason, some of the more common brazing fluxes are available in fluoride-free variants. Those currently available are not as chemically active and, therefore, as effective as their fluoride-containing counterparts, and are only suitable for copper, brass, silverware, and selected iron-base materials.

In dip brazing, flux is used as the process atmosphere. The components to be brazed are appropriately jigged and then immersed in a hot liquid, which at brazing process temperatures is a molten salt. The bath medium usually comprises a mixture of sodium, potassium, and barium chlorides augmented by borax or cryolite as fluxing agents. Carburizing and cyaniding salts are also available, obtained by adding sodium carbonate and sodium cyanide, respectively, so brazing of ferrous components can be combined with these surface treatments. Dip brazing of steels and other iron-base components has largely fallen out of favor. The principal drawback of this process is that it coats everything with flux so that considerable post-joining cleaning is required. Also, it is frequently necessary to preheat the components in a furnace and occasionally cool them in a controlled manner after brazing, as in the case of cast iron, so that the equipment and finishing costs are higher than for a single furnace brazing cycle. Nevertheless, dip brazing is still used occasionally for joining of copper-zinc alloys (brasses) where the hydrostatic pressure of the bath medium suppresses dezincification substantially. Dip brazing is practiced widely for aluminum alloys (see section 3.2.2.1).

3.2.2 Fluxes for Aluminum and Its Alloys

Aluminum forms a natural refractory oxide that is remarkably stable and tenacious. It is mechanically durable, with a hardness that is inferior only to that of diamond, and its high melting point (2050 °C, or 3722 °F) reflects its high degree of physical stability. Alumina is also chemically stable to the extent that it cannot be directly reduced to the metal by aqueous reagents. On exposure to air, a layer of alumina will form almost instantaneously on the surface of aluminum and will grow to an equilibrium thickness of between 2 and 5 nm (0.08 and 0.2 μin.) at ambient temperature. On heating to brazing temperatures, of around 500 to 600 °C (930–1110 °F), the thickness of this surface coating will increase to about 1 μm (40 μin.). Therefore, special fluxes have been formulated for use with aluminum alloys. These fluxes have to be particularly effective in protecting the metal from oxidation.

Aluminum fluxes can be divided into two categories: those that are suitable for use with solders at temperatures below 450 °C (840 °F), and those that can be used at higher temperatures with brazes. A commonality between the aluminum soldering and brazing fluxes is that they all contain halide compounds (with one notable exception). These are highly corrosive, especially in the presence of moisture, including humid atmospheres. Therefore, all flux residues must be removed as completely as possible. The cleaning processes are very laborious and costly, and there is always a danger that some residues will survive the cleaning procedures, resulting in corrosion in the vicinity of the joint. Fluxes used for aluminum soldering are discussed separately in the companion volume *Principles of Soldering*.

3.2.2.1 Liquid Fluxes

There are two principal types of fluxes used for brazing aluminum: chloride- and fluoride-base formulations [The Aluminum Association 1990].

Chloride-Base Fluxes

The active ingredients of these fluxes are chlorides, usually of the alkali earth metals, especially of lithium, which is a particularly effective melting-point depressant. These fluxes physically undermine the oxide by infiltrating naturally occurring cracks in the alumina; upon reaching the metal they proceed to degrade adhesion of the oxide layer and mechanically displace it. The flux residues left on the workpiece surfaces after brazing are highly corrosive and must be removed completely.

Fluoride-Based Fluxes

Many of the well-known fluxes of this type contain a mixed sodium aluminum fluoride that, when molten, can dissolve alumina, but the residues are a source of severe corrosion if left on the surfaces of components. However, by using potassium rather than sodium in the formulation, the flux can be made pH neutral, without compromising its ability to dissolve alumina. The proprietary Nocolok flux comprises a eutectic between K_3AlF_6 and $KAlF_4$, which melts at 562 °C (1044 °F) and just below the eutectic point of aluminum-silicon. The Nocolok (Solvay Fluor, Germany) flux is not hygroscopic and its residues do not corrode aluminum. Therefore, it may be applied to joint surfaces and left there and painted where appropriate. The residue does not spall off during thermal or other forms of stressing [Cooke, Wright and Hirschfield 1978]. The Nocolok process has a reasonably large, although quite specific, process envelope [Ashby 1993]. Some of its key features are:

- *Cleaning:* The flux is water-based and to ensure good coverage on the parts to be brazed, it is essential that they are degreased thoroughly and present a water wettable (hydrophilic) surface. To this end, unless the previous history of the parts is known and controlled, it is recommended that the faying surfaces are degreased, given an aqueous clean, followed by a light etch using a mild alkali solution and a final rinse in demineralized water.

- *Flux loading:* The flux needs to be applied at a loading of 3 to 5 g/m^2 (10 to 16 oz/ft^2). The flux is virtually insoluble in water but, for environmental reasons and those of cost, water provides a good carrier medium. Consequently, the slurry needs to be agitated constantly and great care needs to be taken in the design of the transport system and the spray unit used to apply flux to components. The flux particle size is typically in the range 2 to 6 μm (80–500 μin.) diameter. Prior to brazing, the water is removed by a drying process carried out at 180 to 250 °C (360–480 °F). Special instruments, "flux solids meters" have been devised for measuring the loading of the potassium fluoraluminate brazing flux in an aqueous suspension.
- *Furnace atmosphere:* The atmosphere in the furnace must have a combined oxygen and water vapor level below 500 ppm, which is the maximum recommended for the Nocolok process. Modern furnaces are able to better this by an order of magnitude and at remarkably low gas-flow rates. The process operating temperature range is 575 to 620 °C (1070–1150 °F). The actual temperature used should be just above the melting point of the brazing alloy. The heating rate must be greater than 20 °C (36 °F) per minute. Parts may be removed from the furnace and exposed to air once their temperature has decreased to below 400 °C (750 °F).
- *Exhaust scrubbing:* During the brazing process, small quantities of hydrogen fluoride are produced as a product of the reaction between the flux and residual moisture. Typical concentrations are 20 to 50 ppm of the exhaust gas. An activated alumina scrubber will remove approximately 90% of the hydrogen fluoride generated in the brazing furnace, rendering the discharge able to meet environmental standards (<0.05 mg/m^3, or 5×10^{-8} oz/ft^3 in Europe).
- *Parent metal limitations:* The process cannot be used if the parent metal contains more than 0.5% magnesium. This is because the flux reacts with magnesium to form MgF_2, which in turn prevents further reaction between the flux and the surface oxide. Higher-magnesium-content alloys can be brazed by this process but only if the surface is modified through roll cladding with a magnesium-free alloy that is compatible with the flux. The cladding alloy should also have a melting point that is lower than that of the base alloy.

The Nocolok process is used for the manufacture of heat exchangers in considerable quantities for the automotive industry and other applications. Several dedicated factories exist in which the operation of this process is their sole endeavor.

Several of the fluxes used for the brazing of aluminum and its alloys are associated with the dip method of brazing, also referred to as salt bath brazing. These fluxes are composed of chloride and fluoride mixtures of the alkaline earth elements and of aluminum. The fluxing action is by the physical mechanism of detachment of the alumina layer. This process results in the accumulation of oxide platelets in the flux, which increases its viscosity. This platelet formation has the undesirable feature of enhancing corrosion of the brazed components and of limiting the effectiveness of the flux because platelets that are left partly adhered to the surface of the brazed parts tend to trap flux and obstruct cleaning.

There has been some disagreement about the fluxing mechanism of aluminum using mixtures of chloride-fluoride compounds, although the consensus is that the active constituents in this case are the fluoride constituents, with the chlorides acting essentially as melting-point depressants. Jordan and Milner [1951] claimed that the fluxing mechanism involves a cell action that is dependent on the presence of oxygen. Terrill et al. [1971] have claimed that hydrogen fluoride plays an important role. They have suggested that the fluorides in the fluxes generate hydrogen fluoride by reaction with dissolved moisture in the salt bath and with moisture picked up from the surrounding atmosphere. The hydrogen fluoride reacts with areas of aluminum exposed by crazing of the oxide surface layer, thereby liberating hydrogen, which prevents the aluminum from reoxidizing and, presumably, the gas also helps detach the remaining layer of oxide. Certainly, there is evidence for the evolution of hydrogen fluoride from the flux bath. The chlorides in the salt bath can react with the moisture to form oxychlorides. Oxides and hydrides that react with the fluxes to turn the bath alkaline and render it ineffective are also formed. These reactions progressively exhaust the flux of its activity. For this reason, it is good practice to regularly purge a dip brazing bath of moisture by dipping in aluminum sheets. The aluminum reacts with the moisture to release hydrogen, as mentioned previously, and the gas bubbles to the surface, where it burns with a yellow flame. The disappearance of the flame tends to be used by operators as an indication that the moisture level is sufficiently low to resume brazing.

A third and subtly different method of brazing aluminum alloys, with flux, is also available [Timsit and Janeway 1994]. The process involves coating the parts to be joined with a water-based slurry that carries the flux and also powder of a metal that forms a low-melting-point eutectic with aluminum. Silicon is the obvious first choice. On heating, the flux that is used melts at 562 °C (1044 °F) and removes the oxide layer on the aluminum, enabling the silicon particles to come into intimate contact with the bare metal. When the temperature reaches 577 °C (1071 °F), the silicon removes rapidly into the aluminum, forming a layer of molten braze in situ. Sufficient liquid is drawn into the joint region by capillary action so as to enable a fillet to form. For the process to work, the range of process conditions outlined previously for fluxed brazing of clad parts in the Nocolok process must be observed, as the same flux is used. The particle size of the metal constituent is not critical, 1 to 100 μm (40 μin.–4 mils) being specified as acceptable and surface coverage being between 1 and 80 g/m^2 (3 and 260 oz/ft^2). Silicon is not the only filler metal constituent that can be used in this manner; zinc, copper, and germanium are equally successful because these elements form low-melting-point eutectic alloys with aluminum. However, the lower melting points cannot be used to advantage because the process temperature is fixed by the melting temperature of the flux. Furthermore, the slurry requires modification—to a water/alcohol mix in the case of germanium—owing to the greater surface reactivity of these other metals.

When using aluminum brazes in the form of powder, it is important that the fines <30 μm (1 mil) diameter are removed. The high surface area-to-volume ratio of these particles has been shown to inhibit spreading, because the flux paste needs to be 70 times the mass of the calculated oxygen content in the metal powder in order to achieve good flow of the braze. Figure 3.9 shows a tube-to-plate joint in aluminum made by brazing in air using a commercial flux and a low-melting-point aluminum braze of the type described in Chapter 2, section 2.1.13.1.

3.2.2.2 Gaseous Fluxes

Gaseous fluxing of aluminum can be achieved by using halide vapors as the furnace atmosphere, under conditions that may be approximately determined from the appropriate Ellingham diagram. This approach is based on the reduction reaction:

$$3MHa_2 + Al_2O_3 \rightarrow 2AlHa_3 + 3MO$$

where Ha is the halide of metal M. Chlorides, bromides, and iodides of boron and phosphorus have been found to be particularly effective in this respect. The use of fluorides does not lead to the formation of good fillets, possibly because the aluminum fluoride is not volatile at typical aluminum brazing temperatures and is therefore not displaced completely from the joint region by the molten braze. It has been established by experiment that the main mechanism responsible for the fluxing action in this case is again the physical one, involving the undermining and detachment of the surface oxide by the molten braze, rather than chemical reduction, although a reduction reaction stage is possibly responsible for initiating wetting [Milner 1958].

Furnacing, gas handling, and exhaust scrubbing equipment that is compatible with halide vapor atmospheres and their reaction products incurs a capital premium. Given that satisfactory and more convenient alternatives exist for both fluxed and fluxless brazing of aluminum, gaseous brazing of aluminum with halide and magnesium vapors (described in section 3.4.3) are no longer practiced commercially.

Fig. 3.9 Tube-to-plate joint in aluminum engineering alloy components, torch brazed in air using a low-melting-point aluminum-base filler metal and commercially available flux

3.3 Self-Fluxing Brazes

Certain brazing alloys have been formulated to provide a self-fluxing action during the heating cycle used for bonding (see Chapter 2, section 2.1.2 and 2.1.3). The fluxing agent is an element that has a high affinity for oxygen such as lithium or phosphorus. Brazes containing phosphorus are by far the most numerous members of this family, which includes copper-phosphorus, silver-copper-phosphorus, copper-tin-phosphorus, and copper-nickel-tin-phosphorus alloys, exemplified by those listed in Table 3.5. As can be seen from the table, the addition of tin to the copper-phosphorus alloys depresses the solidus and liquidus temperatures. Silver also lowers the melting temperature and, in addition, enhances significantly the mechanical properties of joints, as pointed out in Chapter 2, section 2.1.3. The maximum continuous service temperature of joints made with the phosphorus-containing alloys is usually restricted to below about 200 °C (390 °F) to avoid selective oxidation of any residual phosphorus and a consequent degradation in joint properties. Long-term exposure of joined assemblies to sulfur-laden atmospheres should be avoided for similar reasons. The principal attractions of these brazing alloys are their relatively low brazing temperatures, simplicity of use as no additional fluxing is required, and their affordability for many applications.

Phosphorus usually represents about 5 to 7% of the self-fluxing alloy composition, and other active elements may be present in even lower proportions. In comparison, the slag contains typically five times that amount, leaving the filler with correspondingly less of the fluxing elements, and these are concentrated predominantly in intermetallic phosphide phases.

The self-fluxing copper-phosphorous brazes are restricted mostly to joining copper-base parent metals and are used widely in plumbing applications. Simple copper-phosphorous brazes produce weak joints when used to join steels due to the formation of a near-continuous interfacial layer of brittle phosphides [Boughton and Solboda 1970]. However, moderate joint strengths can be obtained with nickel-base parent materials and nickel-containing steels. A logical extension of this is self-fluxing brazes containing a small proportion of nickel. The nickel converts the continuous layer of brittle (tetragonal) Fe_3P at the braze/steel interface to discontinuous islands of Fe_2P, which is also a more ductile phase, with a hexagonal close-packed crystal structure [Mottram, Wronski, and Chilton 1986]. By this means, moderate joint strengths that are suitable for less-demanding applications may be achieved. As might be expected, the phosphide phases are not simple binary compounds but are a complex combination of all the phosphide formers in the system.

The self-fluxing ability of the phosphorus-containing brazes toward cuprous oxide films may be demonstrated easily. If a pellet of the brazing alloy is taken and heated slowly in air, a heavy oxide scale will form. As soon as the alloy becomes molten, the scale appears to be dissolved by the alloy, the surface of which becomes bright and shiny. On cooling, the pellet will again reoxidize. However, if the molten braze is quench-cooled, a glassy, semitransparent, film is retained on the surface. The phosphate slag formed under these conditions is deliquescent. The quantity of the slag formed is disproportionately small in comparison with the initial volume of oxide scale. This indicates that only part of the metal oxide is consumed to form the phosphorus-rich flux; the remainder is re-

Table 3.5 Selected self-fluxing alloys based on copper-(silver)-phosphorus and their melting ranges

Composition	Solidus temperature °C	Solidus temperature °F	Liquidus temperature °C	Liquidus temperature °F	AWS designation
92.75Cu-7.25P	710	1310	795	1463	BCuP-2
95Cu-5P	710	1310	925	1697	BCuP-1
87Cu-7Sn-6P	657	1215	687	1269	. . .
85.5Cu-7Sn-1.5Ni-6P	612	1134	682	1260	. . .
75Cu-15Sn-5Ni-5P	600	1112	610	1130	(OKC600)(a)
74.75Cu-18Ag-7.25P	644	1191	644	1191	. . .
80Cu-15Ag-5P	644	1191	805	1481	BCuP-5
86.75Cu-6Ag-7.25P	644	1191	720	1325	BCuP-4
88.25Cu-5Ag-6.75P	644	1191	770	1418	BCuP-7
89Cu-5Ag-6P	644	1191	815	1499	BCuP-3
91Cu-2Ag-7P	644	1191	785	1445	BCuP-6

(a) Alloy patented by Outokumpu, Finland

duced to metallic copper and dissolves back into the braze.

Experiments have revealed that the phosphate slag formed on the surface of the molten braze has only a secondary role in the removal of oxide films from the component surfaces. The fluxing action occurs in the following manner (Fig. 3.10). The molten filler reacts with the oxide on the surface of braze to form a molten slag. This slag, which is rich in the active element (phosphorus or lithium), then floats to the free surface of the filler and protects the joint and the filler from further oxidation. The filler then wets and spreads over the still oxidized component metal surface and removes the oxide skin by both chemical reduction and physical displacement to form the joint. The slag has a limited capability to directly dissolve surface oxides of certain metals, and the process is slow compared with typical torch brazing cycle times, which represent the main area of exploitation of these brazes.

The principal component of the flux is copper metaphosphate, CuP_2O_6, which has a melting point of approximately 620 °C (1150 °F). This compound may be synthesized and used directly as a brazing flux. However, it has limited applicability because the dissolution of most metal oxides results in a sharp increase in melting point, which curtails spreading. Indeed, it is for this reason that there are only a few variants of self-fluxing brazes and a limited range of substrate materials with which they can be used effectively.

Copper-phosphorous brazing alloys have relatively low ductility, making it difficult to prepare preforms. This may be remedied by incorporating in the brazing alloy 1% of silver or chromium, together with 0.1% silicon. Silver and chromium act as grain refiners and impart significant ductility to the alloy, while silicon modifies the Cu_3P phase, which is normally rod-like, and renders it more spheroidal. This improves the fracture toughness, and the combined benefits of these additions greatly ease preform manufacture [Dorofeeva 1993].

Ductile foils of self-fluxing brazes may also be realized by chill-block melt spinning and other rapid solidification casting technologies [Datta, Rabinkin, and Bose 1984]—one example being the Cu-8Ni-8Sn-7P braze. Nickel and tin play a key role in this approach because copper-phosphorous alloys, on their own, have very poor glass-forming ability. The copper-phosphorous binary brazes show no amorphous phases when cooled at industrial rapid solidification rates, in the range 4 to 12% phosphorus. The glass-forming ability is improved greatly for ternary alloys by adding nickel and increased still further for quaternary alloys by adding tin [Bangwei et al. 1993]. The range over which the alloy can be produced in an amorphous phase is then 6 to 8.5% phosphorus. These tin- and nickel-modified variants permit a reduction in the brazing temperature to about 650 °C (1200 °F), although the resulting joints have strengths around only 100 MPa (2 lb/ft^2). Tin-containing self-fluxing brazes are not recommended for joining to steel, even when the braze contains a significant concentration of nickel, because the tin stabilizes the interfacial phosphide layer to a greater extent than nickel can destabilize it [Chatterjee and Mingxi 1990].

The spreading behavior and mechanical properties of rapidly solidified self-fluxing brazes tend to vary significantly with phosphorus con-

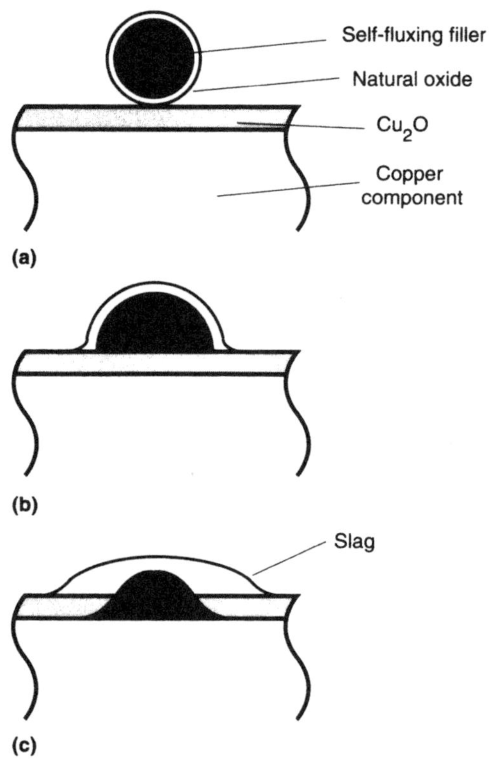

Fig. 3.10 Wetting mechanism of self-fluxing filler metals. (a) Self-fluxing filler applied to copper component. (b) Filler and its oxide melt and wet the oxide film on the component surface. (c) Oxide film on the component dissolves in the molten braze to form a slag that floats to the free surface. The filler then wets and spreads over the clean substrate surface.

centration. This is illustrated for alloys of composition Cu-8Ni-4Sn-*x*P, where *x* is in the range 6 to 8.5 wt% [Bangwei et al. 1993]. As can be seen from Fig. 3.11, spreading of the braze improves with increasing phosphorus content at the expense of mechanical integrity. Adoption of higher brazing temperatures and longer cycle times can compensate partly for the deleterious effect on mechanical properties of a higher level of phosphorus. Elevated brazing temperatures encourage the braze constituents, in particular phosphorus, to diffuse into the parent materials so there are proportionately fewer brittle phases in the solidified joint microstructure.

An interesting variant of fluxless copper-phosphorous brazing alloys is obtained by making small additions of rare earth elements. The rare earth elements are so described because they were originally thought to have a low abundance in the Earth's crust because they were difficult to win from minerals and and even more so to separate from one another. It is now known that lanthanum, cerium, and neodymium are actually more abundant than lead, and vast ore reserves have been found in China and the United States. There are thirty rare earth elements, which is really another name for the elements contained in the lanthanide and actinide series of group three of the periodic table. However, one element of the lanthanide series (promethium) and most of the actinides are transuranium elements, that is, man-made and atomically unstable. Sometimes reference is made to a rare earth called mischmetal. As the name suggests, this is an alloy mixture of the rare earth elements in the proportion of their natural abundance and, therefore, its composition varies with the ore from which it was obtained: monazite, xenotime, or bastnasite. Being a mixture representative of the ore, mischmetal is considerably less expensive than individual rare earths and is therefore used in preference where the collective properties of these metals is all that is required. It is usually given the chemical symbol M and it will typically contain 50% cerium and 30% lanthanum. The general symbol used for rare earth is RE. Rare earth elements have the common attribute that they are extremely reactive toward other metals and most atmospheres.

Addition of rare earth elements to copper-nickel-tin-phosphorus brazes has the effect of improving substantially the mechanical properties of the resulting joints. Figure 3.12 shows the effect of rare earth additions (unspecified with respect to element) on the tensile strength of joints to copper substrates made using Cu-8Ni-8Sn-7P-RE alloys at 700 °C (1290 °F). The optimum concentration of rare earth addition appears to be about 0.2%. The improvement is attributed to refinement of the braze microstructure and appears to be optimum when the nickel and tin contents are 6 and 4%, respectively (Fig. 3.13) [Bangwei et al. 1993]. The microstructure or mechanical properties of joints made when brazing to steel are not reported in that study.

Self-fluxing brazes using lithium typically contain 0.2% of this element. The low concentration means that this element can be added to a much wider variety of brazes to favorably alter their wetting behavior without upsetting other properties. Brittle interfacial phases do not form between lithium and most engineering materials. This accounts for the fact that lithium-contain-

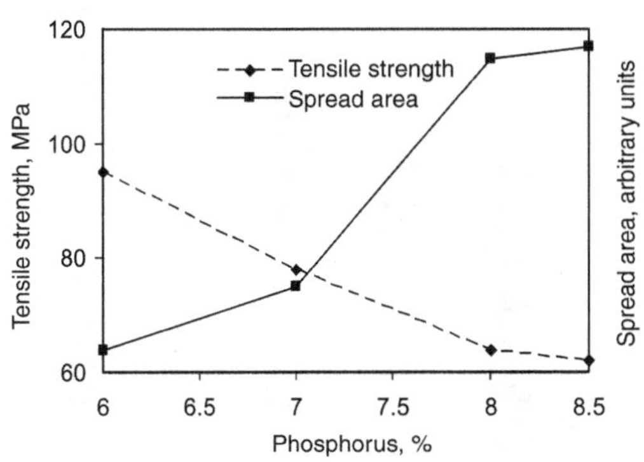

Fig. 3.11 Tensile strength and area of spread data for alloys of Cu-8Ni-4Sn-*x*P ($6 \leq x \leq 8.5$) on copper substrates, heated to 700 °C (1290 °F) for 10 min. Braze spreading and joint mechanical properties have opposing tendencies as a function of phosphorus concentration.

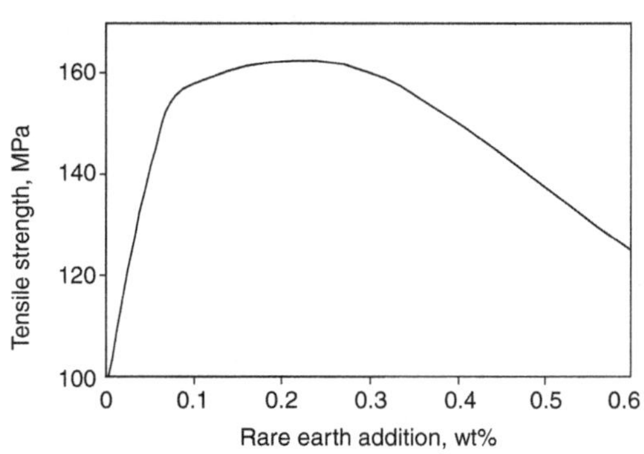

Fig. 3.12 Effect of rare earth additions on the tensile strength of joints made to copper test pieces using Cu-8Sn-7P-6Ni-RE braze by flame heating. The optimum addition appears to be about 0.2 wt%.

ing self-fluxing brazes have been successfully formulated around silver-, copper-, nickel-, and cobalt-base alloys. The lithium in the braze functions in a manner more akin to the rare earth additions described previously than to phosphorus. That is, because lithium is a highly active element, it has the ability to reduce many metal oxides, and it is this characteristic that imparts the apparent self-fluxing ability. However, the lithium oxide skin that forms in consequence is too thin and discontinuous to protect the braze and component surfaces from reoxidation when brazing in air. Also, because the quantity of lithium in the braze is relatively small, there is a risk that it will be exhausted by the time the components and braze have been heated to the process temperature. Consequently, lithium-containing self-fluxing brazes are suitable for use only in high-quality-controlled atmospheres with components and braze preforms that have been cleaned so as to leave only the thinnest possible surface oxide. One benefit of the lithium-containing self-fluxing brazes, as compared with their phosphorous counterparts, is that because the lithium is effectively consumed during the brazing process, the joints are not compromised mechanically by this constituent nor are they subject to any additional restrictions in terms of service temperature or chemical environment.

3.4 Fluxless Brazing

The vast majority of commercial brazing operations conducted at relatively low temperatures (<1000 °C, or 1830 °F) involve the use of chemical fluxes. The application of an appropriate flux considerably widens the tolerance of the brazing operation to an ambient air environment and maintains the brazeability of surfaces through the heating cycle, as explained in section 3.2 of this chapter. The heating cycle when brazing in air using a torch is more rapid than for furnace brazing, and the capital costs tend to be lower. However, there is a penalty to be paid when fluxes are used, in the residues that they leave behind. If flux residues are not removed completely, the chemical contamination can impair the product function, performance, and life.

Highly specified components, particularly those that end up sealed in hermetic enclosures, usually proscribe the use of all volatile nonmetallic materials in their assembly, including fluxes, as a means of guaranteeing service life. Examples of such products are microwave power sources and amplifiers, namely magnetrons and klystrons, which comprise a number of shaped pieces of metal and ceramic components housed in a vacuum enclosure (Fig. 3.14). Magnetrons are produced in large numbers as power sources in microwave cookers, while klystrons are used in radar installations and television stations to generate their broadcast signals. For these and other comparable products, there is the need to use fluxless brazing processes. Apart from excluding corrosive and volatile chemicals, fluxless brazing offers the further advantages of eliminating processing steps, including cleaning after the brazing operation. Fluxless brazing also eliminates a major source of voids in joints, namely, that arising from trapped flux products.

3.4.1 Process Considerations

Key requirements of a successful fluxless brazing process are to eliminate surface contamination from the faying surfaces and to provide sufficient protection of exposed surfaces from oxidation through the heating cycle. At the same time, deleterious contaminants must be excluded from the joining environment. A flux is designed to achieve these conditions and, if it is dispensed with, then other means have to be found to satisfy them.

As might be expected, there is normally a direct correlation between oxide thickness and the quality of a braze joint. While it is fairly straightforward to achieve a nonoxidizing atmosphere of sufficient quality to maintain the brazeability of joint surfaces of many engineering metals and

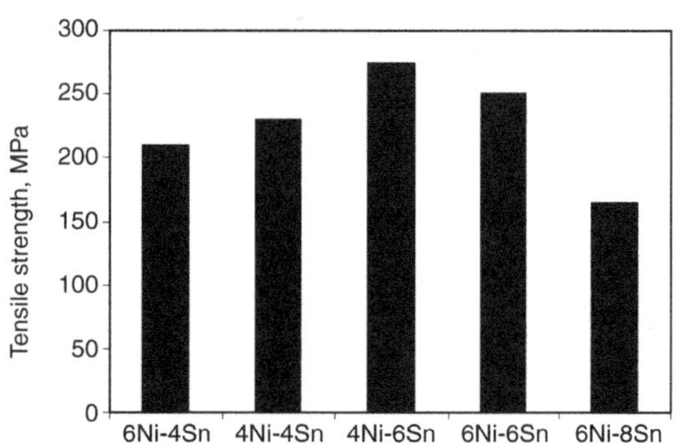

Fig. 3.13 Tensile strength of copper-to-copper joints brazed with amorphous Cu-7P-*x*Ni-*y*Sn-7P-0.2RE braze by flame heating. The respective values of *x* and *y* are given alongside the bars on the chart. Maximum strength is achieved with 4% nickel and 6% tin.

alloys during a rapid heating cycle, initial removal of surface contamination, and particularly the native oxides, is more complex. Tackling this problem first requires knowledge of the nature and thickness of the oxides involved.

3.4.1.1 Oxide Formation and Removal

During fluxless brazing, the prime impediment to wetting and spreading is the presence of oxide films on the surfaces of the parent and filler metals. All base metals are covered with a thin film of oxide through contact with air. In this respect, the noble metals, gold and platinum, are exceptional.

The thickness of the native oxide on a base metal surface depends very much on the history of the particular component. The principal factors governing the thickness of the oxide layer are time of exposure to air, and temperature. The oxidation of some common metals and metallizations at room temperature, after mechanical cleaning of the surface, are indicated in Fig. 3.15. This graph shows that the surface of these base metals will be covered with more than 5 nm (0.2 μin.) of oxide within 5 minutes of the cleaning operation, although the oxide layer could be substantially thicker if the same metal were exposed to certain chemicals or heated in air. Obviously, this surface contamination must be removed or displaced by some means before wetting by the braze can proceed.

The curves represented in Fig. 3.15 should be taken as being indicative only, because in practice the growth rate is affected by a multitude of factors, including surface roughness, residual stresses, and humidity. For elevated temperature, an approximate rule of thumb is that the native oxide thickness on base metals eventually doubles with every 200 °C (360 °F) increase in temperature. Acceleration of oxidation to this higher rate occurs at quite low temperatures, as can be seen from Fig. 3.16. For this reason, it is desirable to heat the joint rapidly and as soon as possible after cleaning of the surfaces, even when carrying out a brazing operation with the benefit of flux, and especially so in its absence. A detailed theoretical treatment of native oxide growth on base metals is given by Martin and Fromm [1977].

The dominant species of oxide that forms on the surface of braze is usually consistent with

Fig. 3.14 An electron gun of a klystron (high-power microwave amplifier) operating in the S-band (2.60–3.95 GHz). Courtesy of TMD Ltd.

that predicted based on classical thermodynamics, namely, the stable oxide of the constituent element that has the highest free energy of oxide formation. The free energies of oxide formation of typical elements used in brazes are listed in Table 3.2. Metal hydroxides are thermally unstable in relation to the oxides, so native surface films are usually not hydrated to any appreciable extent.

3.4.1.2 Self-Dissolution of Braze Oxides

Published work indicates that if a molten metal or alloy is heated above some critical temperature, then dissolution of thin films of surface oxide into the bulk of the melt can occur, the extent depending on the solubility of the oxide. If the heating cycle is conducted in a controlled atmosphere that prevents reoxidation of the free surface, then a fluxless brazing process may be possible under these conditions. Titanium is the only metal where this mechanism is exploited readily in brazing operations on parts of this metal and its alloys, thanks to the fact that above about 925 °C (1700 °F), titanium dissolves oxygen as fast as the oxide can form on the surface (see Chapter 1, section 1.1.7.3). However, the process cycle time must be kept short because internal oxidation of titanium impairs its mechanical properties.

A somewhat similar situation pertains to copper, silver, and possibly iron and nickel. For these metals, the thermally induced removal of the oxides occurs when they are heated in an atmosphere with a low partial oxygen pressure and arises from the thermodynamic instability of the oxides at elevated temperatures, rather than through their dissolution in the respective metals (see section 3.1 in this chapter).

Certain brazes, including those containing phosphorus and lithium described in section 3.3, have an ability to dissolve metal oxides. However, this process results in loss of a constituent from the filler metal with a consequent change in its melting point and intrinsic physical properties. Also, a consequence of the dissolution is usually the formation of dross on the surface of the braze, which impedes spreading and can result in oxide being trapped in the joint gap.

3.4.1.3 Mechanical Removal of Oxides

The native oxides on the surface of the braze and components can, of course, be removed physically by abrasion, through filing, rubbing on abrasive papers, burnishing, and sand blasting. Grit blasting is often used as a surface cleaning technique on multicomponent materials where the phases present have substantially different physical properties and can be removed selectively by this means. For example, graphite inclusions in cast iron and nonmetallic residues on the surface of sand-cast components, both of which tend to inhibit wetting by brazes, can be removed preferentially by grit blasting. The composition and shape of the particles used in the grit-blasting medium are crucial factors to the success of the cleaning operation, because spherical particles provide the least abrasion, while jagged particles are more likely to impregnate the surface of metals. Embedded particles of ceramics such as silica (sand) and alumina will impede wetting by the braze, but metallic blasting media such as Nicroblast, a proprietary Ni-Cr-Fe grit, are less detrimental in this respect [Solomon, Delair, and Thyssen 2003].

The main drawback of all mechanical methods of cleaning surfaces is that native oxide films promptly regrow if the parts are exposed

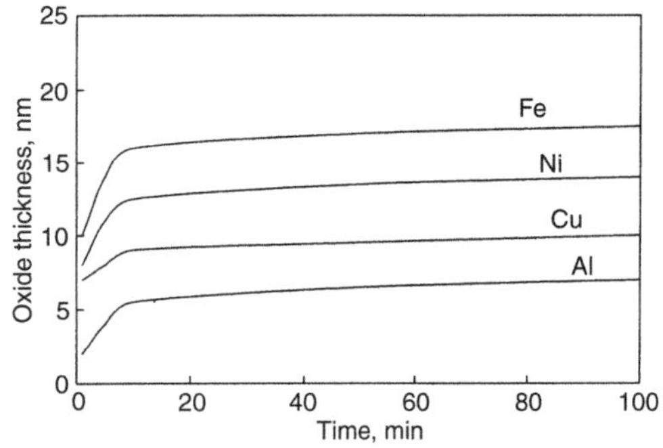

Fig. 3.15 Oxide growth on four base metals at room temperature, as a function of time

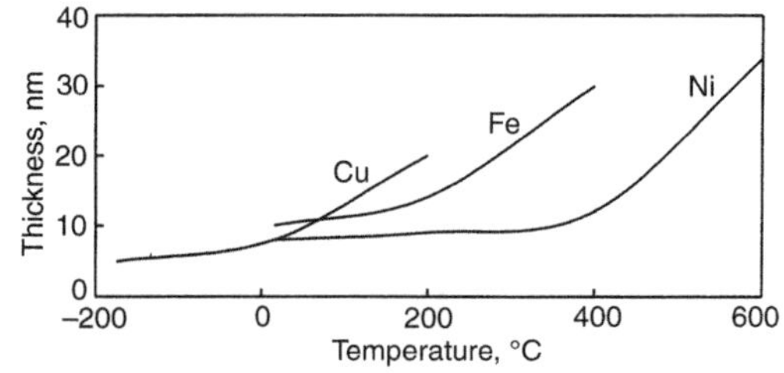

Fig. 3.16 Oxide growth on three base metals as a function of temperature. The time at temperature is of the order of a few minutes and has some correspondence with typical brazing cycle times once component mass and heating rates are taken into consideration.

to air or, indeed, left in an inert atmosphere of even the highest quality for more than a few minutes. Therefore, a fluxless brazing process, not involving a reducing atmosphere, must be able to cope with a thin layer of native oxide on the braze preform and component surfaces. Strategies for making a robust fluxless brazing process are described in the following sections.

A mechanical cleaning technique that has proved to be highly effective in removing even the most tenacious oxides of virtually all metals is ultrasonic tinning, and it is widely used with aluminum [Jones and Thomas 1956]. The use of ultrasonic cleaning has been practiced for many years, and there is a patent describing an ultrasonic soldering iron that predates World War II [Antonevich 1976]. Ultrasonic fluxing normally involves applying an ultrasonically activated soldering iron to "tin" the workpiece with a metal that will melt during the brazing operation [Scheffer et al. 1962]. For aluminum, the "tinned" coating is often zinc, mostly for reasons of electrochemical compatibility. In a variant arrangement, a heated bath is ultrasonically excited and the workpiece is dip coated with the braze. These ultrasonic systems typically operate at 20 to 80 kHz and 0 to 300 W of electrical power. Most of them operate at a single frequency, but some systems provide for application of high- and low-frequency excitation in the ultrasonic range simultaneously. Once the metal coating has been applied under ultrasonic agitation, it protects the workpiece surface against reoxidation, and, for that reason it is akin to chemical fluxing, as described in section 3.2 in this chapter. The coated part can be brazed subsequently using a conventional, fluxless, process because the oxide layer on the tinned surface will be disrupted when the coating melts.

3.4.1.4 Chemical Removal of Oxides

Oxide coatings on the components and braze can be removed using chemical agents. Acids, alkalis, pH-neutral organic chemicals, and salt bath cleaning methods have been developed for most common engineering metals and brazes. Recipes can be found in a number of reference works [Schwartz 2003, ASM Handbook 1993] (see also Chapter 1, section 1.3.2.7). Chemical processes have the merit that they can reach and clean the entire surface area of the parts, some of which may be physically inaccessible. They can also remove surface damage, if the chemical etch used directly attacks the material being cleaned. Multiphase materials are generally more difficult to clean than alloys of a simple constitution because of electrochemical differences that can exist between phases.

As most chemical agents are aqueous-based, it is good practice to precede chemical cleaning with a degreasing process. Low-vapor-pressure organic solvents (e.g., alcohol) or trisodium phosphate in water are commonly used for this process step. Alkaline salt baths are sufficiently aggressive not to require degreasing of components prior to use, and the salt chemical residues can be washed off with water prior to brazing.

3.4.2 Fluxless Brazing Processes

While the methods described in the preceding sections may be used to minimize the inhibiting effect of surface oxides on brazing, even when they are applied in combination, it is still normally not possible to make a fluxless brazed joint of the same quality as a fluxed joint. The reason is simply that these measures are not as effective at both cleaning critical interfaces in the assembly and then immediately protecting them against oxidation and other surface contamination up to the point where the brazing alloy flows over them. From this point of view, molten salt dip brazing is a superior process. Additional means are therefore necessary to encourage wetting and spreading of molten filler metals during fluxless brazing. In general, such methods involve utilizing nonoxidizible metallizations on the parent materials, minimizing the surface-area-to-volume ratio of the filler metal, applying a compressive loading to enhance joint filling and introducing small quanitities of active additions to the brazing alloy to lower the contact angle and enhance wetting.

3.4.2.1 Application of Metallizations Giving Improved Wettability

Because oxide films grow so rapidly on most base metals, as shown in Fig. 3.15 and 3.16, perfectly oxide-free and brazeable component surfaces can be achieved only if these are of gold or another platinum group metal. However, the fairly low free energy of oxide formation for copper and nickel (see Fig. 3.3) means that surfaces of these metals can be kept largely oxide free in most inert atmospheres, and therefore coatings of these metals are often preferred to expensive precious metals. This approach is commonly used in the brazing of stainless steel components, with the applied coatings typically tens of microns thick. Because the braze will not wet any areas of the stainless steels that are not plated, selective plating also provides a means for controlling braze spreading to defined regions.

An alternative is to "tin" freshly cleaned component surfaces with the braze itself. This approach is considered in the next section. Also, see section 3.4.1.3 on ultrasonic tinning.

3.4.2.2 Selection of a Suitable Braze Geometry

The form in which braze is presented to the joint can make a profound difference to the success of fluxed and, more especially, fluxless brazing processes. It is a general rule that the greater is the braze volume in relation to its exposed surface area, the more readily the brazing operation will be successful if other conditions, including the cleanliness of the faying surface, remain the same. This is simply because there will be proportionally less oxide on the brazing alloy to impede wetting and spreading. It is therefore perhaps not surprising that manufacturers of braze pastes go to quite considerable lengths to ensure that the granules of braze used in its preparation are spherical and have exceptionally high surface smoothness (low R_a). For similar reasons, a preferred form of braze preforms used is round wire, rather than, say, foil.

The most appropriate geometrical form for the braze preform depends on the shape of the components being joined. Ideally, these should be designed so as to have not only a high volume-to-surface-area ratio but also be orientated so that the advancing front of molten braze will sweep trapped gas out of the joint gap (see Chapter 4, section 4.3.1.1). Where possible, the flow direction of the molten metal should also be from a concealed area toward an external edge that can be inspected, as shown in Fig. 3.17. This provides a fairly unambiguous indication that spreading by the braze and joint filling have occurred.

A thin-foil braze preform is not a particularly advantageous method of admitting braze into a joint because:

- The surface area-to-braze volume ratio is unfavorable, with regard to accumulation of oxide.
- A foil is difficult to handle and clean if thin.
- There is a high risk of trapping pockets of vapor when the parts and foil are jigged together (see Chapter 4.3.1.1).

Thin foils also command a cost premium, especially when they have to be purchased to custom dimensions.

In situations where it is desired to make joints with very low aspect ratio, i.e., thin in relation to the plan area, it is often not possible to achieve sufficient braze spread so as to reliably achieve complete joint filling. The thinnest braze preforms that can be purchased economically are 15 μm (0.6 mils) thick, and these are available for only a few brazing alloys. Handling such thin foils requires considerable dexterity and mechanically cleaning them is virtually impossible. The high surface-area-to-volume ratio of such preforms also runs counter to the need to minimize the relative quantity of native oxides present.

A growing number of companies are now offering a solution to this problem in the form of substrates that are sold with the braze composition of choice, preapplied to surfaces, as a coating. The braze is usually applied by roll cladding (see Chapter 1, section 1.3.2.2 and Chapter 4, section 4.3.1.1). For high value-added applications it can be worthwhile to apply brazing alloys by electroplating (if technically feasible) or by vapor-phase techniques, usually in conjunction with a mask, so that only the required areas of the components are coated. Silver-copper eutectic is among the brazes that can be electroplated as a homogeneous alloy. The braze thickness available in this form typically ranges from 2 to 100 μm (0.08–4 mils).

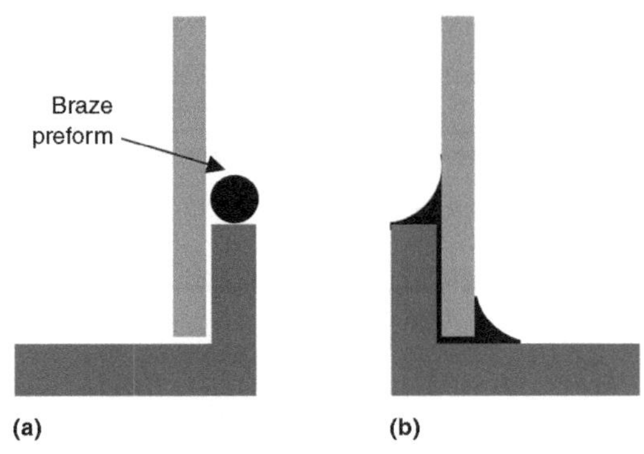

Fig. 3.17 Schematic cross section of a tube-to-plate joint designed such that braze flow will sweep gas and flux out of the joint gap. Formation of an external fillet provides evidence that some braze spreading and wetting has taken place. This design of joint also protects the braze preform from direct impingement by the flame of a brazing torch. (a) Braze preform pre-positioned. (b) After reflow

These braze-coated substrates offer a number of distinct advantages compared with a brazing foil or wire:

- Reducing piece-part inventory and number of suppliers by dispensing with preforms
- Simpler jigging requirements
- Decreasing the thickness of the braze joint because the braze layer can be substantially thinner than the minimum practicable thickness obtainable for a braze preform
- Improving brazing behavior by eliminating two joint surfaces with all the associated problems, including oxide layers, from the joint gap
- Spreading of the braze is promoted more strongly over the area defined by the coating

Because a fluxless brazing process is so critically dependent on there being virtually no oxide on the surface molten braze, it is good practice to test whether this condition is being achieved. An assessment of this kind may be accomplished readily if the furnace has a viewing port. On heating to the process temperature, a clean brazing alloy will melt, and its surface acquires a shiny, mirrorlike finish. This is often referred to as the "liquid lake" condition. Any defects in the liquid braze film, such as texture, a gray bloom, or brown spots indicate that some element of the process, usually either the braze itself or the furnace atmosphere, is in some way deficient and must be corrected.

3.4.2.3 Enhancement of Joint Filling through Compressive Loading

No matter what precautions are taken, in an industrial fluxless brazing process it will always be the case that by the time the components and preform have been set in jigs, loaded into the protective atmosphere, and heated to the process temperature, the braze, and quite possibly the components as well, will be covered by a substantial film of oxide. If it is possible to extrude virgin metal through fissures or other defects in the oxide layer on the braze preform, then there is an improved prospect of achieving a sound joint. One method of doing this is to apply a compressive force to a joint gap that contains a preplaced preform.

The loading is relatively easy to achieve with weights or spring-loaded jigs for all but the largest components. Precautions need to be taken to ensure that the load is applied uniformly and parallel to the joint gap and that the method of application does not impose a thermal sink on the assembly that would give rise to adverse temperature gradients. In some applications, the required load may be achieved simply by placing the largest or heaviest component uppermost in a stack of parts.

3.4.2.4 Improvement of Brazeability by Adding Activators to the Braze

Occasional reports appear in the published literature to the effect that low concentrations of other metals can make a significant difference by promoting wetting and spreading of molten brazes. Examples are discussed in Chapter 2, section 2.2.2 and Chapter 7, section 7.2. Lithium-containing self-fluxing brazes, discussed in section 3.3, offer this benefit and, given that they can be used without a chemical flux in a controlled atmosphere and that no significant quantity of slag is generated, the brazing process is essentially fluxless. Virtually all fluxless aluminum brazes take advantage of ppm additions of bismuth and beryllium to promote wetting and spreading of the Al-12Si eutectic alloy and of lower-melting-point aluminum brazes. This topic is discussed in the following section.

Small additions of rare earth elements are especially beneficial for improving the wetting characteristics of many brazing alloys. For example, cerium is known to produce a marked reduction in the contact angle of many brazes. Rare earth additions to established brazing alloys could be a fruitful area for further research. One example, cited previously, is the addition of rare earths to copper-phosphorous brazes (see section 3.3).

3.4.3 Fluxless Brazing of Aluminum

Despite the high oxygen affinity of aluminum and the stability of alumina, fluxless brazing of aluminum has been successfully developed for the fabrication of radiators and heat exchangers of this metal [Herr 1983]. It is routinely employed for mass production in the automotive industry. The process is performed in either high vacuum or in a high-quality nitrogen atmosphere and, for reasons of cost, uses wrought aluminum clad with aluminum-silicon eutectic filler alloy [Schultze and Schoer 1973]. To enable the process to operate successfully and consistently, four critical process requirements must be strictly met:

- *Surface preparation:* It is essential that the parts to be brazed are scrupulously cleaned and undergo a suitable surface treatment. This special procedure is an integral part of the fluxless process. First, the component surfaces must be thoroughly degreased because the subsequent process steps are aqueous based. The surfaces are then given a caustic etch for 20 seconds in a bath containing 5% (by weight) of sodium hydroxide in water at 45 °C (113 °F). After rinsing in deionized water, the parts are "desmutted" by immersion for 30 seconds in a bath of 7 vol% nitric acid at room temperature. The parts are then rinsed with dilute nitric acid, followed by deionized water, then alcohol, and finally, forced air dried. The alkali-acid pretreatment process removes about 10 μm (400 μin.) of material from the surface of the aluminum component and replaces it with a complex oxyhydride that is displaced readily during the brazing process. It is believed that, on heating above 400 °C (750 °F), the oxyhydride decomposes and the reaction products volatilize, exposing a clean aluminum surface [The Aluminum Association 1991]. An alternative cleaning treatment is an alkaline dip in a sodium hydroxide/bicarbonate solution; followed by a water rinse, an acid treatment, and accelerated drying (using a methanol or acetone rinse); followed by sealing of the cleaned surfaces with a film, of an organic material, such as polystyrene dissolved in a mixture of toluene and acetone. The sealed surfaces can be stored for several days before use without appreciably affecting the brazing process [Schwartz 2003, p 303–304].
- *Furnace atmosphere:* During the heating and brazing cycle, it is obviously necessary to minimize further oxidation of the aluminum. Even though this is intended as a large-scale industrial process, it is essential to achieve an atmosphere with a combined oxygen and water vapor content of 5 ppm or less. To accomplish this condition, considerable care needs to be taken with the furnace design and configuration, in terms of eliminating dead zones and leaks. During vacuum brazing, a dwell must be incorporated at around 200 °C (390 °F) so as to ensure that all of the evolved vapor can be pumped from the chamber. Heating of the chamber walls also helps to desorb moisture so that it can be pumped out before the parts are heated for any length of time. If a nitrogen flow furnace is used, its outlet does not need to be sealed, provided the flow rate of the nitrogen escaping to the external atmosphere is made to exceed a linear velocity of 0.5 m/s (1.6 ft/s). This gas velocity is greater than the diffusion rate of oxygen and moisture in air and, therefore, the furnace atmosphere will maintain the quality of the incoming nitrogen gas. All furnace furniture, pipe-work and seals must, of course, be in metal to achieve the atmosphere quality required.
- *Minor additions to the brazing alloy:* In addition to the process controls mentioned previously, it is necessary to include in the brazing alloy specified minor additions that promote spreading and fillet formation in order to induce aluminum-silicon eutectic braze to wet and spread on aluminum alloys. Additions of bismuth and beryllium at ppm concentrations have been found to be effective in this regard, and they have no adverse effects on the corrosion resistance of the joints. Exactly how these elements function in this context is not clear, but the beryllium appears to disrupt the surface oxides in a manner that permits wetting, while bismuth enhances the fluidity of the filler metal and promotes spreading (see Fig. 2.29, which shows the effect of certain elements on the surface tension of molten aluminum) [Schultze, and Schoer 1973, Singleton 1970, Terrill 1971].
- *Process temperature:* The prescribed process temperature for fluxless brazing of aluminum with the aluminum-silicon eutectic braze is 600 °C (1110 °F) and the time at peak temperature should not exceed three minutes. Overheating and extended processing times serve only to cause unfavorable metallurgical changes in many aluminum alloys that have an adverse effect on the mechanical properties of the assembly.

By using the same processing conditions (other than joining temperature) and utilizing the same minor elements in the filler metal, a majority of aluminum engineering alloys can be brazed using preforms of low-melting-point aluminum brazes (see Chapter 2, section 2.1.13.1) at temperatures as low as 525 °C (980 °F) [Jacobson, Humpston and Sangha 1996]. An example of a heat exchanger fabricated in this manner is shown in Fig. 3.18.

An aluminum brazing process that is frequently described as fluxless involves the use of magnesium, either in the brazing alloy or as an addition to the furnace atmosphere, in the form of chips loaded into a boat and placed adjacent to the workpiece in the hot zone of the furnace. The magnesium appears to act as a getter in the atmosphere and also as a flux that disrupts alumina films on the aluminum parts and the aluminum-silicon braze [McGurran and Nicholas 1984].

This process has been studied extensively [Ambrose and Nicholas 1986, Winterbottom and Gilmour 1976, Anderson 1977, Takemoto and Okamoto 1988, and Gempler 1983]. For the process to operate successfully, the furnace atmosphere must be a good vacuum (<10 mPa, or 10^{-4} mbar) and the heating rate must be controlled within defined limits. Also, the quantity of magnesium introduced to the furnace must be matched to the brazing job being undertaken. Too little of the metal will result in its ability to getter the furnace atmosphere being depleted prematurely, while excessive magnesium can depress the solidus temperature of the parent materials to the point where they sag or collapse. As mentioned previously, in addition to gettering the furnace atmosphere, the magnesium vapor aids wetting and spreading by disrupting the oxide films on the aluminum-base parent materials and the braze. The main drawback of this process is that magnesium vapor is corrosive and will attack furnace furniture. Also, the coating of magnesium deposited on the furnace walls and furniture will readily absorb moisture and water vapor when the furnace is open to air, which makes it difficult to obtain a good-quality atmosphere on the subsequent brazing cycles unless the chamber is cleaned regularly [Winterbottom 1984].

3.5 Stop-Off Compounds

Stop-off compounds may be considered as the antithesis of fluxes. The purpose of these materials is to prevent the braze from wetting and spreading onto areas of the assembly away from the joint region. Examples of their use include:

- Prevention of wetting of fixtures and furnace furniture
- Threaded parts, where the threads provide unwanted capillary attraction for the braze
- High-tolerance parts where braze spreading would render them out of tolerance and require post-brazing machining
- Components with fine features, for example, tubes that could become blocked or closed by the braze
- Situations where limiting the spread of the molten filler metal can help reduce the quantity of braze required to make a satisfactory joint, which could have a significant cost benefit for assembly using fillers containing precious metals
- Aesthetic considerations, where the braze has a different appearance or surface texture to the components and it is desirable to confine the former to the actual joint

Stop-off compounds are usually stable refractory oxides that obviously must not be reduced by the flux or brazing atmosphere. Titanium dioxide is a common constituent of such media. They take the form of as extremely fine powders suspended in alcohol, lacquer acryloid cement, water, or acetone. During the heating stage of the brazing cycle, the fine particle size of the oxides enables them to mechanically key strongly to the faying surfaces and, in consequence, they are not displaced easily by the advancing front of molten braze, although they can be undermined if the braze dissolves the component surface to a significant depth.

Stop-off materials can be removed from the component surfaces after brazing by mechanical abrasion, by pickling in 5 to 10% nitric/hydrofluoric acid, or in solutions of sodium hydroxide or ammonium biflouride.

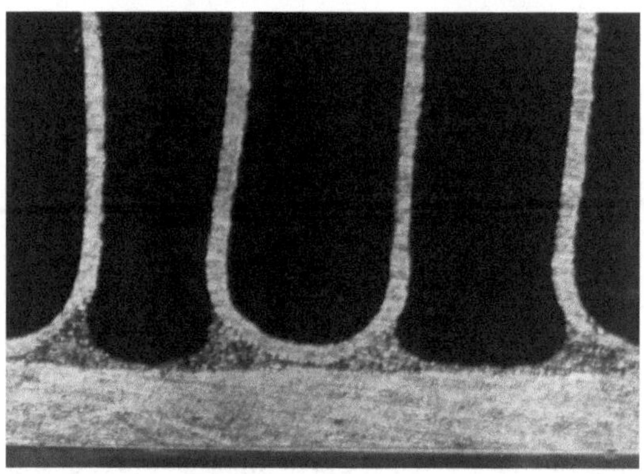

Fig. 3.18 Metallographic cross section through an aluminum heat exchanger fabricated using a foil preform in an entirely fluxless process. By using a low-melting-point braze, the mechanical properties of the heat exchanger face plate material are not degraded and there is negligible erosion of the thin fin members. The thickness of the braze preform is adjusted so that there is sufficient braze to form fillets that have adequate dimensions to effect good heat flow between the fins and the face-plates without unduly obstructing the passageways.

Appendix A3.1: Thermodynamic Equilibrium and the Boundary Conditions for Spontaneous Chemical Reaction

The thermodynamic function that provides a measure of the driving force of a chemical (including metallurgical) reaction is the Gibbs free energy, which is defined as:

$$G = E + PV - TS \qquad \text{(Eq A3.1)}$$

where E is the internal energy, S the entropy, T the absolute temperature, P the pressure, and V the volume of the materials system. The definition and physical meaning of internal energy and entropy are explained subsequently. As is shown, an important property of the Gibbs free energy function is that it is always a minimum at equilibrium, and the extent of its departure from the minimum value provides a measure of the tendency of a reaction to proceed spontaneously—that is, of the driving force for the reaction.

The First Law of Thermodynamics and Internal Energy

The subject of thermodynamics addresses energy changes in systems. In thermodynamics, the term *system* is used to describe a set of materials that are capable of undergoing a change, as, for example, through a chemical reaction.

The First Law of Thermodynamics is a statement of the Principle of Conservation of Energy. The various statements of this Law law are bound up with the differentiation of various types of energy and, in particular, with the concept of internal energy. The internal energy of a system may be considered the aggregate of the kinetic energies and energies of interaction (i.e., potential energies) of the atoms and molecules of which the constituent materials are composed. When the system is isolated from its surroundings, so that no exchange of energy can take place, then its internal energy remains fixed. However, if mechanical work can be done on the system, but no heat is exchanged with its surroundings (i.e., the system is adiabatically isolated)—for example, by an impellor stirring a liquid or gas in an insulated container—its internal energy will change by an incremental amount equal to the work performed:

$$dE = -dW \text{ (adiabatic)}$$

The minus sign denotes that work is done on the system to raise its internal energy. This expression provides a thermodynamic definition of internal energy.

The internal energy, E, of a system depends only on the state of the system (defined in terms of macroscopic or thermodynamic properties, such as the pressure and temperature of the system). For this reason, internal energy is termed a function of state.

Where work is done in changing the volume of a chemical system by an increment dV through the application of external pressure P, then:

$$dW = PdV$$
$$dE = -PdV$$

In practice, most systems are not totally insulated from their surroundings so that thermal energy may be exchanged between them.

If an increment of work dW is done on the system and an increment of heat dQ is exchanged with the surroundings, then the internal energy dE will change by the amount:

$$dE = dQ - dW$$

This equation is a mathematical expression of the First Law of Thermodynamics, which, for a chemical system, may be written:

$$dE = dQ - PdV \qquad \text{(Eq A3.2)}$$

Entropy and the Second Law of Thermodynamics

Internal energy alone cannot determine the equilibrium state of a system. Although when a system reaches a state of equilibrium, the internal energy achieves a fixed value, this may not be a minimum. For example, the internal energy will increase when a solid melts at constant temperature and pressure through the absorption of latent heat. For this reason, in addition to the internal energy, it is necessary to stipulate the value of another state function of the system—namely, entropy—which, together with the internal energy, measures the extent to which the system is removed from equilibrium.

The concept of entropy arises in connection with the conversion of heat into mechanical work and vice versa. The Second Law of Thermodynamics defines the conditions under which this conversion from one form of energy to the other can occur. The Kelvin-Planck statement of this law relates to a device that can perform work by extracting heat from a particular source and performing an equivalent amount of work without any other energy exchange with the surroundings. It follows that a reciprocating engine that operates by extracting heat from one source must reject some of this heat to a sink at a lower temperature. If the operating cycle of the engine is reversible, such that work can be performed to pump heat from the sink back to the source, it is possible to show that in accordance with the Second Law, the integrated ratio dQ/T over one complete cycle is zero:

$$\oint_R \frac{dQ}{T} = 0$$

The circle through the integral sign denotes that the integration is to be carried out over the complete cycle, and the letter R is a reminder that the equation applies only if the cycle is reversible. This result is known as Clausius' theorem.

If the integration is carried out over only part of the cycle, say between two states 1 and 2, then the integrated ratio dQ/T is not zero but equals the difference between the values of a thermodynamic function at the two states:

$$_R\int_1^2 \frac{dQ}{T} = S_1 - S_2$$

This thermodynamic function of state is called entropy. If the two states are infinitesimally near, then the relationship can be written:

$$\left(\frac{dQ_R}{T}\right)_R = dS \qquad \text{(Eq A3.3)}$$

Subscript R indicates that this equation holds only if the heat increment dQ is transferred reversibly. This equation provides a mathematical expression of the Second Law of Thermodynamics.

A consequence of the fact that entropy is a function only of state, a system that has changed from state 1 to state 2, always has entropy S_2, which differs from that of the initial state S_1 by $S_{1,2} = S_2 - S_1$, irrespective of the means used to drive the system. Thus, for example, the system may have been set in motion and some of the kinetic energy converted into heat in overcoming frictional forces, thereby raising its temperature to a value that takes the system from state 1 to state 2. In this irreversible process, the energy was not supplied to the system as heat so that:

$$_I\int_1^2 \frac{dQ}{T} = 0$$

The letter I denotes that the process is irreversible.

However, the entropy change $S_{1,2}$ is still the same as that obtained by a reversible change between states 1 and 2 because it depends only on these states and not on the process connecting them; that is, here too $S_{1,2} = S_2 - S_1$. Thus, in all irreversible processes, the entropy change is greater than ($\int dQ/T$, where dQ is the heat absorbed at each incremental step in the irreversible change. This result can be generalized to the statement that in a spontaneous irreversible change, the entropy of an isolated system will increase, and when in equilibrium, it will remain constant.

Considering the system and its surroundings together (i.e., the universe), any kind of process may be represented in entropy terms by:

$$dS\ \text{(universe)} > 0$$

Therefore, from a thermodynamic viewpoint, which is macroscopic, entropy can be understood as the propensity of a system to undergo a change, such as a chemical reaction. A clearer physical picture of entropy can be obtained at the microscopic level, where a system may be regarded as an ensemble of atoms or molecules. On this basis, it can be shown that entropy provides a measure of the degree of atomic or molecular disorder that exists in the system, and this will always tend to increase. This concept is consistent with the observations that all metals are intersoluble, albeit in some cases only to a small extent, and that all liquid metals will wet the clean surfaces of solid metals.

The Dependence of Gibbs Free Energy on Pressure

Having defined the thermodynamic functions internal energy, E, and entropy, S, and explained their physical significance, it is possible to demonstrate the significance of the Gibbs free energy function, G, to determine the temperatures and pressures under which chemical reactions are thermodynamically favorable, as well as and also the direction of the reactions.

In incremental form, Eq A3.1 can be written:

$$dG = dE + PdV + VdP - TdS - SdT$$

Substituting for dE and TdS from Eq A3.2 and Eq A3.3 (all chemical/metallurgical processes being reversible) gives:

$$\begin{aligned} dG &= dQ - PdV + PdV + VdP - dQ - SdT \\ &= VdP - SdT \end{aligned}$$

For a reversible process at constant temperature (isothermal) and constant pressure (isobaric), that is, when the system is in equilibrium:

$$dG = 0$$

and G is constant and has a minimum value. This is an important result for metallurgical reactions because these can be considered as taking place usually at constant temperature and pressure.

More generally, at constant pressure, $dP = 0$, and then:

$$dG = -SdT$$

and at constant temperature, $dT = 0$, so that:

$$dG = VdP$$

If the system is an ideal gas, the Gas Law:

$$PV = nRT$$

applies, where n is the number of moles of gas and R is the gas constant. Then:

$$dG = nRTdP/P \text{ at constant temperature}$$

so that the Gibbs free energy change resulting from a change from state 1 to state 2 at constant temperature is:

$$G_2 - G_1 = nRT \ln (P_2/P_1)$$

The Gibbs free energy, like any other measure of energy, must have some reference point. By convention, a zero value of G is assigned to the stable form of elements at 25 °C (77 °K) and 1 atm of pressure. Then the Gibbs free energy change of a gas at constant temperature from its value G^o at atmospheric pressure, which is defined as its standard state value, is given by:

$$G - G^o = nRT \ln P \qquad \text{(Eq A3.4)}$$

where P is the pressure corresponding to the free energy state G, expressed in atmospheres.

Although Eq A3.4 is strictly valid for ideal gases, it is also approximately applicable to real gases and can be used for them at pressures close to normal atmospheric pressure (1 atm).

In the case of solids, the molar volumes are small compared with those of gases so that the change in the Gibbs free energy of solids resulting from small pressure excursions ΔP, such that $\Delta P \ll 1$ atm (~100 kPa), at constant temperature is small and, to a first approximation, may be neglected in reactions involving solids and gases. It is also assumed that the solubility of the gaseous species in the solid phases is negligible at the temperatures of interest, as is largely the case in practice.

It is now possible to determine the pressure dependence of the Gibbs free energy of the reagents that participate in a chemical reaction. Consider a reaction involving four gases, A, B, C, and D and two solids, X and Y, all at constant temperature T, as follows:

$$x\text{X} + a\text{A} + b\text{B} \leftrightarrow y\text{Y} + c\text{C} + d\text{D}$$

where a, b, c, d, x, and y are the number of moles of each of the reagents. The gaseous reagents are assumed to behave as though they are ideal gases.

The Gibbs free energies G(X) and G(Y) of the solid constituents at moderate pressures are approximately equal to their values at atmospheric pressure, as explained previously. Therefore:

$$\begin{aligned} \text{Free energy of } x \text{ moles of solid X} &= xG(\text{X}) \\ &= xG^o(\text{X}) \end{aligned}$$

$$\text{Free energy of } y \text{ moles of solid Y} = yG(\text{Y}) = yG^o(\text{Y})$$

The Gibbs free energies of the gaseous constituents are:

$$\text{For } a \text{ moles of gas A: } aG(\text{A}) = aG^o(\text{A}) + aRT \ln P(\text{A})$$

$$\text{For b moles of gas B: } bG(\text{B}) = bG^o(\text{B}) + bRT \ln P(\text{B})$$

$$\text{For } c \text{ moles of gas C: } cG(\text{C}) = cG^o(\text{C}) + cRT \ln P(\text{C})$$

$$\text{For } d \text{ moles of gas D: } dG(\text{D}) = dG^o(\text{D}) + dRT \ln P(\text{D})$$

where G(A), G(B), etc. are the Gibbs free energies of one mole of the reagents A, B, etc. at pressures P(A), P(B), etc., and G^o(A), G^o(B), etc. are the corresponding values at 1 atm.

The free-energy change for the reaction is, from Eq A3.4:

$$\begin{aligned}\Delta G &= G(\text{products}) - G(\text{reactants})\\ &= cG^o(\text{C}) + dG^o(\text{D}) - aG^o(\text{A}) - bG^o(\text{B}) + RT \ln \frac{[P(\text{C})]^c[P(\text{D})]^d}{[P(\text{A})]^a[P(\text{B})]^b}\\ &= \Delta G^o + RT \ln \frac{[P(\text{C})]^c[P(\text{D})]^d}{[P(\text{A})]^a[P(\text{B})]^b}\end{aligned}$$

The Gibbs free energy changes of the solid reagents can be neglected, for the reasons given previously.

Under equilibrium conditions, temperature and the respective pressures P(A), P(B), and so forth are constant, and:

$$\Delta G = 0$$

Hence, the Gibbs free-energy change when the gaseous reactants A and B in their standard states are transformed to the products C and D in their standard states may be expressed in terms of the partial pressures of the respective reactants in equilibrium, thus:

$$\Delta G^o = -RT \ln \frac{[P(\text{C})]^c[P(\text{D})]^d}{[P(\text{A})]^a[P(\text{B})]^b} \qquad \text{(Eq A3.5)}$$

Since the Gibbs free energy change ΔG^o, for a particular reaction at a fixed temperature and at atmospheric pressure has a fixed value, then so too does the argument of the logarithm. This constant is called the equilibrium constant K_P because it can be used to determine the equilibrium state that a reacting system will attain.

$$K_P = \frac{[P(\text{C})]^c[P(\text{D})]^d}{[P(\text{A})]^a[P(\text{B})]^b}$$

The subscript P denotes that the equilibrium constant is specified in terms of pressures. Substituting, Equation A3.5 becomes:

$$\Delta G^o = -RT \ln. K_P$$

For an oxidizing reaction described by the equation:

$$x\text{M} + [y/2]\text{O}_2 \leftrightarrow \text{M}_x\text{O}_y$$

there is one gaseous constituent and two solids so that the equilibrium constant is simply:

$$K_P = \frac{1}{(P^{\text{M}}_{\text{O}_2})^{y/2}}$$

where $P^{\text{M}}_{\text{O}_2}$ is the partial pressure of oxygen required to effect the oxidation reaction, or the dissociation pressure of the oxide, and

$$\Delta G^o = RT \ln P^{\text{M}}_{\text{O}_2} \qquad \text{(Eq A3.6)}$$

per mole of oxygen participating in the reaction. That is, the driving force needed to oxidize a metal, as expressed by the Gibbs free energy change, is directly related to the oxygen partial pressure of the atmosphere according to Eq A3.6.

REFERENCES

- *Aluminum Brazing Handbook,* 1990. The Aluminum Association, Inc.
- Ambrose, J.C. and Nicholas, M.G., 1986. Alloys for Vacuum Brazing Aluminium, *Brazing Soldering,* Vol 11 (No. 3), p 374–379
- Anderson, W.A., 1977. The Effects of Metallurgical Structure on the Brazing of Aluminum, *Proc. Symp. Physical Metallurgy of Joining Metals,* Warrendale, Pennsylvania, p 222–243
- Antonevich, J.N., 1976. Fundamentals of Ultrasonic Soldering, *Weld. J. Res. Suppl.,* Vol 55 (No. 7), p 200s–207s

- Ashby, A.J., 1993. Equipment Used in the Controlled Atmosphere Brazing of Aluminium Heat Exchangers, *Proc. Conf. Advances in Brazing and Soldering Technology,* Solihull, October 20, 1993
- Bangwei, Z. et al., 1993. Fundamental Research for Cu-P Based Amorphous Filler Alloys, *China Weld.,* Vol 2 (No. 2), p 95–103
- Bannos, T.S., 1984. The Effect of Atmosphere Composition on Braze Flow, *Heat Treatment,* Vol 16, (No. 4), p 26–31
- Boughton, J.B. and Sloboda, M.H., 1970. Embrittling Effects of Trace Quantities of Aluminium and Phosphorus on Brazed Joints in Steel, *Weld. Met. Fabr.,* Vol 8, p 335–339
- Chatterjee, S.K. and Mingxi, Z., 1990. Tin-Containing Brazing Alloys, *Weld. J. Res. Suppl.,* Vol 69 (No. 10), p 37s–42s
- Cooke, W.E., Wright, T.E., and Hirschfield, J.A., 1978. Furnace Brazing of Aluminum with a Non-Corrosive Flux, *SAE Technical Paper Series No. 780300,* Society of Automotive Engineers
- Datta, A., Rabinkin, A., and Bose, D., 1984. Rapidly Solidified Copper-Phosphorus Base Brazing Foils, *Weld. J. Res. Suppl.,* Vol 63 (No. 10), p 14s–21s
- Dorofeeva, E.N., 1993. Modifying Copper-Phosphorus Alloys for Brazing Alloys, *Paton Weld. J.,* Vol 46 (No. 2), p 101–105 (Translated from *Avtomaticheskaya Svarka,* 1993, Vol 46 (No. 2), p 24–27)
- Esquivel, A.E. and Chavez, E., 1992. Benefits of Using Carbonic Gas in the Soldering Process and Curing Oven for Electronic Assemblies, *Proc. Conf., NEPCON West '92,* February 23–27, (Anaheim, CA), p 219–227
- Eustathopoulos, N., Nicholas, M.G., and Drevet, B., 1999. *Wettability at High Temperatures,* Pergamon Press
- Gempler, E.B., 1983. Parameters Evaluated in Long Cycle Aluminum Vacuum Brazing, *Weld. J. Res. Suppl.,* Vol 49 (No. 11), p 843s–849s
- Herr, H.K., 1983. *Aluminum Fluxless Vacuum Brazing,* The Aluminum Association Inc.
- Jacobson, D.M., Humpston, G., and Sangha, S.P.S. 1996. A New Low-Melting-Point Aluminum Braze, *Weld. J. Res. Suppl.,* Vol 75 (No. 8), p 243s–250s
- Jones, J.B. and Thomas, J.G., 1956. Ultrasonic Soldering of Aluminum, *Proc. Symp. Symposium on Solder,* American Society for Testing and Materials, June 19–20, Atlantic City, p 15–29
- Jordan, M.F. and Milner, D.R., 1951. The Removal of Oxide from Aluminium by Brazing Fluxes, *J. Inst. Met.,* Vol 85, p 33–40
- Martin, M. and Fromm, E., 1977. Low-Temperature Oxidation of Metal Surfaces, *J. Alloy. Compd.,* Vol 258, p 7–16
- McGurran, B. and Nicholas, M.G., 1984. A Study of Aluminum Brazing Filler Metals Using Hot Stage Scanning Electron Microscopy, *Weld. J. Res. Suppl.,* Vol 64 (No. 10), p 295s–299s.
- Milner, D.R., 1958. A Survey of the Scientific Principles Related to Wetting and Spreading, *Br. Weld. J.,* Vol 6, p 90–105
- Mottram, R.D., Wronski, A.S., and Chilton, A.C., 1986. Brazing Copper to Mild Steel and Stainless Steels Using Copper-Phosphorus-Tin Pastes, *Weld. J. Res. Suppl.,* Vol 65 (No. 4), p 43s–46s
- Scheffer, H. et al., 1962. How to Ultrasonically Seal Hermetic Ceramic Transistor Packages, *Ceram. Ind.,* Vol 79 (No. 6), 50–64
- Schultze, W. and Schoer, H., 1973. Fluxless Brazing of Aluminium Using Protective Gas, *Weld. J. Res. Suppl.,* Vol 52 (No. 9), p 644s–651s
- Schwartz, M.M., 2003. *Brazing,* ASM International
- Singleton, O.R., 1970. A Look at the Brazing of Aluminum—Particularly Fluxless Brazing, *Weld. J. Res. Suppl.,* Vol 49 (No. 11), p 843s–849s
- Solomon, H.D., Delair, R.E., and Thyssen, J., 2003. The High Temperature Wetting Balance and the Influence of Grit Blasting on Brazing of IN718, *Weld. J. Res. Suppl.,* Vol 72 (No. 10), p 278s–287s
- Stubbington, C.A., 1988. Materials Trends in Military Airframes, *Metals and Mater.,* Vol 4 (No. 7), p 424–431
- Takemoto, T. and Okamoto, I., 1988. Effects of Magnesium Content in Brazing Sheet Claddings on the Vacuum Brazeability of Aluminum in Relatively Enclosed Volumes, *Brazing Soldering,* Vol 15 (No. 3), p 32–36
- Terrill, J.T. et al., 1971. Understanding the Mechanisms of Aluminum Brazing. *Weld. J. Res. Suppl.,* Vol 50 (No. 12), p 833s–839s
- Timsit, R.S. and Janeway, B.J., 1994. A Novel Brazing Technique for Aluminum,

Weld. J. Res. Suppl., Vol 73 (No. 6), p 119s–128s
- *Welding, Brazing and Soldering,* 1993. *ASM Handbook,* Vol 6, ASM International
- Wicks, C.E. and Block, F.E., 1963. "Thermodynamic Properties of 65 Elements – Their Oxides, Halides, Carbides and Nitrides," U.S. Bureau of Mines, Bull. 605, U.S. Government Printing Office
- Wigley, D.A., Sandefur, P.G., and Lawing, P.L., 1981. Preliminary Results on the Development of Vacuum Brazed Joints for Cryogenic Wind Tunnel Aerofoil Models, *Proc. Conf. International Cryogenic Materials Conference,* August 10–14, (San Diego, CA)
- Winterbottom, W.L., 1984. Process Control for Brazing Aluminum Under Vacuum, *Weld. J. Res. Suppl.,* Vol 63, p 33s–39s
- Winterbottom, W.L. and Gilmour, G.A., 1976. Vacuum Brazing of Aluminum: Auger Studies of Wetting and Flow Characteristics, *J. Vac. Sci. Technol.,* Vol 13 (No. 2), p 634–643

CHAPTER 4

The Role of Materials in Defining Process Constraints

THIS CHAPTER CONSIDERS the role of materials in brazing operations and the manner in which they impact on the choice of processing conditions and their optimization.

Processing options are dictated by the set of components that have to be joined together to produce a unitary assembly and the service requirements that this product is expected to satisfy. The choice of the components is normally determined by functional requirements and considerations of cost. Often, therefore, the components are of different materials in order to maximize the performance of the product for a given price. Performance demands may be wide ranging and often encompass resilience to thermomechanical stresses and chemical attack. In meeting these requirements, the characteristics of the joint are often critically important.

Consideration of the functional issues, both separately and in combination, will reveal an array of constraints with which the brazing process will have to be compatible, some of which are not necessarily obvious. A flow chart illustrating the sequence of decision-making steps and the constraints through the stages of process design to implementation of assembly, involving brazing operations, is set out in Table 4.1. Constraints that are specifically related to the product include the cost tolerance of the product to the joining process, the scale and throughput of production that need to be satisfied, and also any statutory regulations that apply. The operating environment must then be considered. Here, the peak temperature, static and dynamic stress conditions, and the corrosion environment tend to be the critical parameters, although thermal conductance may also be an important factor. Finally, the materials used in the components, when taken individually, will impose an upper limit on the maximum joining temperature that can be tolerated, so as to avoid thermal degradation or, in the worst case, outright failure. This consideration may restrict the atmosphere in which the joining process can be carried out. When the overall assembly is considered, any mismatch in thermal expansivity of the abutting components can force compromises with regard to the choice of materials and processes, or even require redesign of the components and assembly.

Having taken account of the more obvious constraints in the design of an assembly, a selection of filler alloys can be made, each of which will impose its own set of limiting conditions. Among the most important of these conditions are the minimum practicable joining temperature (i.e., liquidus temperature of the filler alloy, with the addition of a margin to allow for possible temperature gradients across the joint), the geometries that the joint can assume and into which the filler can be fabricated, and the permissible joining atmosphere. At this point it will become clear what type of joining process can be used—in particular, whether soldering, brazing, welding, or adhesive bonding. Assuming that brazing is appropriate, a short list of brazing alloys which have a window of application conditions that is compatible with the combination of materials making up the assembly and also with the other steps of manufacture and the service environment, can be drawn up. Up to this point, the selection procedure is largely a paper exercise, but the viability of the proposed joining solution generally needs to be established by practical trials because a multiplicity of other factors also enters the equation. Among these are wetting behavior, erosion of the parent materials, and intermetallic phase formation within the joint. While any of these issues can radically affect the integrity of the product in service, with the exception of common combinations of parent materials and brazes, much of this type of information is unavailable from the published literature and cannot be surmised reliably.

If problems are identified at this stage of establishing the joining process, a number of avenues are available for seeking possible remedies. In certain situations, catastrophic mismatch stresses may be overcome by modifying the stress distribution in the vicinity of the joint. Problems associated with the formation of deleterious intermetallic phases in the joint, lack of wetting, and, at the other extreme, excessive erosion of one or more of the parent materials may often be circumvented by interposing a layer of a different metal between the braze and the parent material, thereby altering the metallurgical

Table 4.1 Materials systems approach to joining process development

Phase of development	Decision making	Boundary conditions
	Nature of the product and functional requirements ↓	Scale of production Cost constraints Size and weight limits Statutory requirements
Definition of the product	Service conditions ↓	Thermal environment Stress environment Chemical environment
	Parent materials ↓	Joining temperature Joining atmosphere Mismatch stress
Materials and process selection	Filler alloys ↓	Joining temperature Joining atmosphere Joint geometry Filler geometry and form
Process assessment: identification of critical materials problems	Determination of process viability ↓	Metallurgical constraints
	Mismatch ↓ — Alloying ↓ (Wetting, Erosion, Phases ↓)	
Identification and achievement of solutions	(Mechanical solutions) ↓ Interlayers Multilayers Graded structures Compliant structures Diffusion soldering/brazing (Metallurgical solutions) ↓ Surface processing Wettable coatings Barrier coatings Active filler alloys Fluxes Diffusion soldering/brazing ↓	
Prototyping and production	Process specified	Tolerance of process established

Adapted from *Principles of Soldering and Brazing*, 1993, p 112

constitution of the joints. This changes the constituents that will alloy in the course of the joining operation and also their relative proportions. Introducing shims (0.2–0.5 mm, or 8–20 mils, thick) of soft metal into the joint gap can preferentially absorb mismatch stresses in the joint. Fluxes and active filler alloys can be used to improve wetting and reduce void levels. These and other possible remedies to problems associated with the formation of brazed joints are detailed in the following sections.

If it transpires that there is no tractable joining solution with the choices made hitherto, or if the solutions cannot be economically justified, then changes earlier in the decision-making chain are required. In extreme situations, a drastic revision may be required, perhaps even to the extent that other means of assembly have to be considered or, alternatively, the functional requirements of the product may need to be relaxed.

4.1 Metallurgical Constraints and Solutions

In principle, most metals can be joined using filler alloys. However, when there is a requirement to braze two different parent materials together, the available choice of fillers that are compatible with both is narrowed somewhat, especially when the constraints on the braze, processing conditions, and properties of the joints mentioned previously are imposed. The problem is often made more acute by the fact that the joining processes tend to be left to a late stage of product design, which generally reduces the available options further. The joint should always be considered as an integral part of the overall design and dealt with from the outset.

Metallurgical incompatibility of materials and processing conditions will manifest itself through poor wetting, excessive erosion of the parent materials, the formation of undesirable phases, and, more alarmingly, mechanical failure. Means for preventing or suppressing these deleterious characteristics are described as follows.

4.1.1 Wetting of Metals by Brazes

Poor wetting of the component surfaces by the molten braze is a common cause of poor joints. Inadequate wetting may be the result of restrictions applied to the choice of joining temperature and atmosphere, including the use of fluxes, which leave the joint surfaces insufficiently clean. Active filler metals can often help to overcome this problem (see section 4.1.2.2 of this chapter), but it is not always possible to use brazes containing active wetting agents and alternative remedies are then necessary.

Many grades of stainless steels are quite difficult to braze satisfactorily. The problem may be generic to the grade or be inconsistent in its manifestation. As an example, the degree of surface oxidation of stainless steels at the instant that the filler metal melts can vary from one brazing operation to another, owing to small differences in component preparation, heating rate, and furnace atmosphere. The problem may be avoided in the case of grade 409 stainless steel, for example, by application of electrolytic nickel to the faying surfaces or a change to a niobium-stabilized grade of stainless steel. Niobium is less susceptible to oxidation than titanium, which is a minor constituent of 409 stainless steel (typically 0.5%), and the substitution greatly improves brazing consistency. Note that nickel platings are normally applied by electrolytic techniques. Electroless platings contain phosphorus and can melt on heating above 871 °C (1600 °F), making them unsuitable for brazing stainless steel assemblies required for elevated temperature service, not to mention the undesirable effect that the phosphorus may have on the joint metallurgy.

As in the case of grade 409 stainless steel, minor constituents in parent materials can be responsible for poor joint formation. Some grades of brasses and bronzes contain deliberate additions of aluminum, as mentioned in Chapter 2, section 2.2.1, and also, occasionally, of magnesium and zirconium in significant proportions. If the combined percentage of these elements is less than 10%, then using a suitable proprietary flux (which contains both chlorides and fluorides and designated as an aluminum bronze grade flux) can produce satisfactory brazing results.

Poor wetting is a particularly acute problem when the parent materials are refractory metals. Their reactivity with oxygen and, in some cases, with other elements in the atmosphere and the stability of the reaction products on the surfaces of refractory metals can impede wetting. Nonmetallic phases present at the surface of materials, such as graphite inclusions in cast iron and nonmetallic components in metal-matrix composites, represent other barriers to wetting. Special cleaning procedures may be effective in removing these inclusions, as in the case of cast iron [Schwartz 2003, 100–102].

Poor wetting is not restricted to the surface condition of parent metals but can also directly involve the filler metal. A prime example is provided by aluminum-base brazes. When these are used without an appropriate fluxing agent, foils and other preforms of these alloys tend to produce poor wetting and spreading over the joint surfaces. However, by making small additions of elements that lower the surface tension of the molten filler and also help to destabilize the native surface oxide layer, wetting can be improved considerably. This is illustrated for aluminum in Fig. 2.29. Other examples are mentioned in Chapter 2, section 2.2.2. The concentration of the individual additions is restricted to a maximum of about 1% in order to avoid perceptibly altering the bulk metallurgical characteristics of the filler alloy.

If the joining environment, which includes fluxes when these are permitted, cannot adequately clean and protect the components from oxidation during the process cycle, a favored solution is to apply a coating of a more noble metal to the precleaned joint surfaces. This coating may be sacrificial in that it is subsequently dissolved by the braze, which then proceeds to wet the oxide-free surfaces of the parent material. At the same time, the composition and thickness of the coating should be chosen such that when this dissolves it does not generate brittle phases by reaction with the filler and parent materials. As explained in Chapter 3, section 3.4.2.1, copper and nickel are the principal elements used as wettable metallizations for brazes because of the relative instability of their oxides; metallurgical compatibility with silver-and nickel-base brazes, respectively; ease of deposition, and low cost.

It was also pointed out that in many situations, metallizations that promote wetting can also be used to advantage to confine the molten filler to a specific area on the surfaces of components. By selectively applying the coating to prescribed areas, the spread of the braze can be restricted accordingly.

If the wettable coating is not adherent as applied, it can often be secured on the surface of a metal component by a subsequent heat treatment in a nonoxidizing atmosphere. This process can promote interdiffusion between the metal and the coating, resulting in a graded interface akin to that obtained using a carburizing (carbon-enriched) or sheradizing (zinc-enriched) surface treatment. For example, titanium is rendered brazeable with low-melting-point brazes, without flux, if it is electroplated with gold. As applied, the gold layer is not adherent and will readily flake off. However, after heat treatment at 750 °C (1380 °F) for 30 minutes in an inert atmosphere, some interdiffusion of titanium and gold occurs, but the free surface is left rich in gold and readily brazeable. This surface enrichment with gold does not alter the solubility of the parent metal in the filler alloy, as compared with untreated titanium.

In joints between dissimilar metals, coatings may be necessary on the components that are readily brazeable without them, due to problems arising elsewhere in the joint. For example, the brazing of molybdenum and tungsten to steel is often unsuccessful, owing to the formation of intermetallic compounds such as Fe_2Mo_6 and Fe_2W_6 adjacent to the surface of the refractory metal component in the assembly. Shielding the steel surface from the braze by applying a nickel or copper electroplate circumvents the problem.

4.1.2 Wetting of Nonmetals by Brazes

Nonmetals, namely ceramics, glasses, and plastics, are not wetted by most filler metals, even when their surfaces are scrupulously clean. This is because they are chemically very stable, with their atoms bound strongly to one another by ionic or covalent bonds, leaving no free charge to attract the free electrons associated with metal fillers or, in other words, no free bonds to form with the latter. Therefore, nonmetallic materials will not react with and be wetted by molten brazes unless the latter contains an active element that can attach itself to the anionic species of the nonmetallic material and thereby become bound up with the latter. In a compound or complex, the relevant anion is normally either oxygen, carbon, nitrogen, or a halide element.

There are two solutions to this problem. One is to use an active filler metal, as discussed in section 4.1.2.2 of this chapter. Active metal joining is effective only if sufficiently high temperatures, typically above 800 °C (1470 °F), can be used for the joining operation so that the active ingredient is able to react with the nonmetal. Where activated brazes produce successful joints, it is observed that the active element concentrates at the interface with the nonmetallic base material and imparts metallic characteristics to its surface region.

Where this approach is incompatible with the joining process, a very similar result can be achieved via a different route. This involves coating the joint surfaces with a metal that will bond strongly to the underlying nonmetal and that, once wetted by the filler, is then not entirely dissolved while the braze is molten. Any erosion through to the original component surface would result in dewetting. Full details of brazeable coatings for nonmetals are given in the next section.

4.1.2.1 Brazeable Coatings on Nonmetals

Wettable coatings can be applied to nonmetallic components by similar methods that are used on metals, namely, physical vapor deposition, chemical vapor deposition, and wet plating. Also widely used are fired-on glass frits loaded with particles of metal powder or flake, often referred to in the literature as thick-film metallization techniques. The key characteristics of these coating processes and the resulting metallizations are summarized in Tables 4.2 to 4.5.

Physical and chemical vapor deposition are suitable only for cost-tolerant products. Metallizations required for brazing are typically 5 μm (200 μin.) or more in thickness. This is because the high process temperature means that significant volume fractions of the metallization react with the substrate and are dissolved by the molten braze. The preferred physical vapor deposition method is sputtering because it offers the capability of reverse-biased sputter cleaning in the inert gas atmosphere immediately prior to the deposition process. This cleaning step greatly promotes adhesion of the deposited coating to nonmetal surfaces. Chromium and titanium are used as the active metals to bond to the nonmetals. It is usual to apply a layer of the active metal to a thickness of less than 1 μm (40 μin.), and then to immediately overcoat it with a layer of a more noble metal such as gold, silver, copper, or nickel with a similar or greater thickness. This outer layer serves both to protect the active layer from contact with oxygen from the air-ambient atmosphere and, at the same time, to present a surface that is compatible with a range of brazes. For some brazing applications, it has been found adequate to simply apply the active layer by exposing the component to vapor of the metal and relying on the thermal activation provided to achieve adhesion [Moorhead, Elliott and Kim 1993]. An example is the production of an adherent and mechanically robust coating of chromium carbide, Cr_3C_2, on graph-

Table 4.2 Techniques for applying metallizations: characteristic features

Process	Metallic materials capable of deposition	Suitable substrates	Throwing power	Film thickness achievable	Film thickness control	Throughput of process
Vacuum evaporation	Elemental metlas and some alloys	Most nonvolatile materials	Line-of-sight process	nm-μm	Good	Low, batch
Sputtering	Wide range of elemental metals and alloys	Most nonvolatile materials	Moderate (function of target size, gas pressure, and target-substrate distance)	nm-μm	Excellent	Low, batch
Chemical vapor deposition	Elemental metals	Materials that can withstand the high temperatures required	Good	μm-mm	Good, but need to stringently control several process variables simultaneously	High, many items at a time; batch or continuous
Electroplating	Elemental metals and some binary and ternary alloys	Electrical conductors	Moderate	μm-mm	Generally less precise than for vapor deposition	Very high, can be continuous
Electroless plating	Elemental metals and a few binary alloys	Wide range of materials	Good	μm	Generally poor	High, can be continuous
Thick film	Wide range of elemental metals and alloys	Materials that can withstand the firing temperatures	Physical access to surfaces is required	μm-mm	Moderate	High, batch or conveyor belt

ite. The process involves heating the graphite to 1400 °C (2550 °F) in the presence of chromium vapor, under a low partial pressure of oxygen. The outer surface of the resulting carbide layer formed is sufficiently metallic in character to enable it to be wet by common brazes [Hammond and Slaughter 1971]. An example of metallization by chemical vapor deposition (CVD) is the production of tungsten coatings on ceramic components by heating above 1000 °C (1830 °F) in a helium atmosphere containing WCl_6.

An alternative method of coating a nonmetallic material with an adherent metal coating is to use a two-stage joining process in which the first

Table 4.3 Metallization techniques: relative merits

Process	Advantages	Disadvantages
Vacuum evaporation	Relatively simple equipment required for resistance heating evaporation, which is suitable for coatings of most elemental metals	Not suitable for alloys that have constituents with greatly differeing vapor pressures. Meticulous substrate cleaning prior to deposition is required.
Sputtering	Possible to coat a wide range of compositions. Dense coatings and good adhesion obtainable	Requires sophisticated equipment. Low throughput. Heating of substrates and low deposition rates in conventional diode or triode sputtering
Chemical vapor deposition	High-quality coatings are obtainable. Output is generally high.	Equipment is sophisticated and is usually specific to particular coatings.
Electroplating	High throughput. Large areas can be coated with uniform thickness. Limited only by the size of the plating bath. Relatively easy to control	Film impurities and imperfections can also present problems. Can only apply coatings to electrically conductive materials. Thorough cleaning and chemical activation of substrates are required prior to plating.
Electroless plating	Large areas can be coated with uniform thickness. Good throwing power. Only very basic equipment is required.	As above, except that nonconducting materials can be plated. Range of available coatings is restricted.
Thick film	Requires simple equipment. Lends itself to high-volume production using screen printing and firing in belt furnaces	Relies on manufacturers' proprietary formulations. Relatively high process temperatures are used. Only thick films can be applied by this technique.

Table 4.4 Metallization techniques: important process parameters

Process	Rate of deposition, μm/min	Pressure in deposition chamber, mPa	Substrate temperature during coating process
Vacuum evaporation	0.001–5	0.01–10	Substrate is often heated to 200 °C (390 °F) to promote adhesion.
Sputtering	0.005–1	100–10,000	Mostly below 100 °C (212 °F)
Chemical vapor deposition	5–100	10,000–100,000	200–2000 °C (390–3630 °F), but usually 400–800 °C (750–1470 °F)
Electroplating	0.1–100	Ambient	10–100 °C (50–212 °F)
Electroless plating	0.1–1	Ambient	10–100 °C (50–212 °F)
Thick film	1000–10,000 (does not include firing times)	Ambient	400–1800 °C (750–3270 °F)

Table 4.5 Metallization techniques: coating quality

Process	Coating thickness uniformity	Coating continuity	Coating purity	Coating adhesion to substrate
Vacuum evaporation	Variable; determined by source-substrate geometry	Moderate to low porosity	Purity limited by source materials and deposition atmosphere	Fair
Sputtering	Higher uniformity possible than for vacuum evaporation	Low porosity	Purity limited by source materials and deposition atmosphere	Generally excellent
Chemical vapor deposition	Good uniformity possible; depends on design of the deposition chamber	Dense and essentially pore free	Purity is that of the starting materials	Variable; dependent on materials and processing conditions
Electroplating	Good uniformity on flats, nonuniform at edges	Susceptible to porosity and blistering	May incorporate salts and gaseous inclusions	Variable; often excellent
Electroless plating	Fair uniformity	Susceptible to porosity and blistering	May incorporate salts and gaseous inclusions	Variable; often excellent
Thick film	Variable	Dense coatings are achievable	Often contain glass and possibly organic residues	Variable; dependent on materials and processing conditions

step is to allow an activated braze to wet and spread over the component surfaces. The "tinned" surface is then usually mechanically dressed to present a flat metal plane for the actual joining step, which is performed at a temperature below the solidus point of the activated filler metal. Owing to the number of elements that are likely to be present in the joint, the resulting alloy constitution can be somewhat complex unless the process is designed such that the activated filler metal used for the coating and the braze have several constituents in common.

A somewhat different approach, used often in the electronics industry in joining to ceramic substrates, is to coat the latter with a paste containing fine powders of a suitable metal and a glass or low-melting-point ceramic. On firing, the glass or ceramic phase wets the component surface, much like an enamel, sometimes partly dissolving the substrate, while the metal particles percolate to the surface. Such a coating is described as a thick film because the thickness of the fired-on layer tends to be in the range 10 to 20 μm (400–800 μin.), to distinguish such layers from vapor-deposited metal coatings, which are generally much thinner [Bever 1986]. Adhesion between the solidified glass or ceramic phase and the metal particles is achieved by a combination of chemical and physical means.

Thick-film formulations usually comprise a slurry, containing the metals or metal compounds and sometimes a glass or ceramic as well, in an organic carrier, which is intended to be applied by painting or screen printing onto the desired areas. Subsequent firing drives off the organic fraction and stabilizes the metallization by producing a diffused interface with the nonmetal substrate. It is usual practice to apply and fire each thick-film metallization layer separately, although processes have been developed whereby at least two thick-film layers are fired together. Common thick-film metallizations are as follows:

- Systems based on reactive metals (zirconium, tungsten, titanium, manganese, molybdenum). These formulations are fired at about 1600 °C (2900 °F) in a reducing atmosphere. They are used because reactive metals are able to form adherent bonds directly to nonmetals. The relatively refractory nature of the resulting metal surface requires that either a strongly reducing environment is required to effect subsequent wetting by a braze, or a wettable metallization should be applied as an overcoat. It is often possible to apply the wettable surface layer over the reactive metal layer and fire the two layers together. An example is a tungsten-loaded frit overcoated with a nickel paste, which is cofired at 1300 °C (2370 °F). Even at this temperature, the interdiffusion between the two metals is slight so that a discrete layer of nickel forms on the surface after the heating cycle [Kon-ya et al. 1990].
- Systems based on noble metals (copper, silver, gold, palladium, platinum). These materials are fired between 850 and 950 °C (1560 and 1740 °F). The silver, gold, and platinum metallizations can be fired in air while a reducing atmosphere is generally required for the less-noble metals, copper and palladium. The more noble metals have the advantage that the resulting surface is guaranteed to be free of oxide films that could interfere with the wetting and spreading of a braze.
- Metal-loaded glass frits are fired on the surfaces of components above 400 °C (750 °F) to form a glaze that is strongly adherent to the nonmetal substrate. The glass fraction is typically 2 to 10% (by weight) relative to that of the metal. However, the glass tends to concentrate close to the ceramic so that after firing, the outer layer of the coating is sufficiently metallic in character, enabling it to be electroplated or brazed directly. The glasses are often high-lead borosilicates, typical compositions being B_2O_3-63PbO-12SiO_2 or lithium-borates (a mixture of Li_2O and B_2O_3).

Thick-film metallizations are supplied as complex proprietary formulations and are available in different physical forms, each tailored for a limited range of substrate materials. For example, thick film metallizations with a borosilicate glass base are generally used on oxide ceramics, while lithium borate and oxynitride glasses are preferred with nitride and carbide ceramics. It is advisable to consult the supplier on their conditions of use and likely properties.

Direct, selective metallization by a different route has been developed for alumina and aluminum nitride. This has a particular attraction for the electronics industry where there is a need to form wiring traces in addition to pad areas for component attachment. This selective process involves using an excimer laser to write a pattern directly onto the ceramic. The laser appears to locally decompose the surface, leaving a track that is sufficiently aluminum-rich so that it can be directly plated with copper, nickel, or gold by an electroless technique [Norton 1993].

Reference may sometimes be found in older technical publications to the active hydride process as a metallization method for ceramics. It involves applying a metal hydride, usually of titanium or zirconium, in the form of a paste to the component surface. On heating to about 900 °C (1650 °F) the hydride thermally dissociates, liberating hydrogen, to leave a metal film on the component surface. The metal then reacts with the ceramic to form a metallic surface coating [Pershall 1949]. This method has lost favor in recent years to alternatives because the quality of the resulting metallization tends to be variable, as reflected in the mechanical properties of joints made using this approach [Mizuhara and Mally 1985]. The unpredictable nature of the metallization quality is due to its high sensitivity to variations in the atmosphere during the firing stage. Oxygen and water vapor contents, in particular, affect the extent to which the highly reactive metallic constituent oxidizes after decomposition of the hydride and hence the ease of wetting by the molten filler.

A widely used method for metallizing oxide ceramics, in particular, alumina, is the socalled "moly-manganese" process ("moly" being a colloquial name for molybdenum) [Nolte 1954a,b]. This process has the advantages of being relatively straightforward to perform and is highly reproducible, and therefore amenable for processing large numbers of parts. In this process, a slurry comprising powders of molybdenum, manganese, and various glass-forming compounds is applied to alumina components as a paint. Tungsten is sometimes substituted for some of the molybdenum. The ratio of molybdenum to manganese is usually 90:10. The coated ceramic is fired in a wet hydrogen atmosphere (with a well-controlled dew point of around −30 °C, or −20 °F) at a temperature of 1450 to 1500 °C (2640–2730 °F). The firing operation in the reducing atmosphere enables the manganese to react with the alumina to form manganese aluminate spinel ($MnAl_2O_4$). This spinel is molten at the process temperature. It permeates the molybdenum and also penetrates grain boundaries in the ceramic, reacting with the glass phase in the ceramic in the process. The molybdenum remains reduced and becomes established as a metallic surface layer on the spinel. The spinel also bridges the thermal expansion mismatch between the molybdenum layer and the alumina, while, at the same time, it alters the glass transition temperature. Both of these characteristics favor formation of a strong interface [Pincus 1953]. The final metallization thickness after firing is 5 to 25 μm (0.2–1 mils) thick. Then, a regular (nonactive) braze can be used, with or without an overplating of nickel or copper. The minimum coating thickness of the moly-manganese paint needs to be about five times the mean diameter of the metal particles to ensure a continuous, metallic surface layer that can withstand some dissolution by a molten braze. If the coating is too thick, there is risk of its spalling off due to internal stress.

The bonding mechanism achieved by the moly-manganese process, between the interfacial spinel and the molybdenum, is essentially one of mechanical keying, which relies on the lower layers of the molybdenum metallization being porous. This aspect of the process was reviewed comprehensively in three papers by Twentyman and his coauthor [Twentyman 1975, Twentyman and Popper 1975a,b]. With regard to alumina, the moly-manganese process relies substantially on the presence of a glass fraction in the oxide ceramic and the latter must possess a fairly coarse microstructure (grain size, typically >5 μm, or 0.2 mils) with large pores. It is not suitable for application to ultrapure alumina and other nonoxide ceramics, including nitrides and carbides, not least because these tend not to contain glassy phases. This limitation may be overcome by adding manganese oxide to the paint so that the glass phase is formed in situ on the melting of the paint and therefore does not rely on extensive chemical reaction with the substrate.

Molten filler alloys designed for general use do not always readily wet metallic materials that have nonmetallic phases present at the surface. Two well-known examples are the graphitic phases in cast irons and the lead globules in free-machining steels. As an alternative to the application of barrier metallizations or the use of activated filler metals, it is often possible to use a mechanical or chemical process to remove the offending minority phase from the surface of the component, as mentioned briefly in section 4.1.1 of this chapter. A pretreatment of this type is often used when gray cast iron is to be brazed [Schwartz 2003, 100–102]. Components of this metal are "cleaned" by immersing them for several minutes in a salt bath at 400 °C (750 °F) that contains a mixture of sodium and potassium nitrides. The chemical treatment completely removes any exposed graphitic phase from the surface and leaves an iron-rich surface that is readily wetted by many filler alloys [Totty 1979]. The new generation of metal-matrix composites, which rely on nonmetallic reinforcement phases, can have their surface chemistry similarly modified, provided the reinforcement phase comprises discrete particles.

4.1.2.2 Activation of Joint Surfaces by Molten Brazes (Active Brazing Alloys)

Reactive elements added to brazing alloys can promote wetting of nonmetals and the formation of sound joints, if at least one of the products of reaction with the parent material is metallic and remains as a layer on the surface of the nonmetal. These brazing alloys are colloquially known as *active brazes*. When selecting an active braze, the choice of the reactive constituent, its optimum concentration, and the appropriate processing conditions need to be considered in relation to the nonmetallic parent material, or component. One of the most commonly used active constituents is titanium, as mentioned in Chapter 2, section 2.2.2. Less-reactive elements, such as chromium, and more reactive elements, such as hafnium, are also used. The subject of active brazes is considered in detail in Chapter 7, section 7.2.

4.1.3 Erosion of Parent Materials

When a molten filler wets the surface of a parent material, alloying occurs, leading to a degree of dissolution of the parent material (or metallization), commencing at the joint interface. Dissolution of the parent materials occurs because the materials system encompassing the joint is not in thermodynamic equilibrium. This situation provides the driving force for wetting and spreading. The maximum solubility of the parent materials in the molten filler can be predicted by reference to the appropriate phase diagram, as pointed out in Chapter 2, section 2.3. In summary, extensive erosion is likely where the liquidus surface on the phase diagram between the filler metal composition and that of the parent material has a shallow slope and where the alloying depresses the melting point of the filler in the joint. If the phase diagram exhibits either of these features, it is possible to limit erosion only by lowering the process temperature, shortening the heating cycle, and/or restricting the volume of molten filler metal. Such changes must obviously not compromise the integrity of the joints by, for example, reducing the effective fluidity of the molten filler alloy, which in turn will impede spreading and joint filling.

Erosion can be reduced somewhat if intermetallic phases form at the joint interface so as to attenuate the rate of dissolution of the parent materials into the molten filler. This approach is successful only if the intermetallic phases formed are ductile, otherwise, joint embrittlement results. Alternatively, it is often possible to protect the components against erosion by interposing a metallization that will act as a barrier to dissolution. An example is the application of a layer of copper, typically 10 μm (400 μin.) thick, on mild and austenitic stainless steels for brazing with self-fluxing copper-phosphorous filler metal. Although the copper is thin and almost completely dissolves during the brazing cycle, it moderates the reaction sufficiently so that phosphides do not form at the joint interface, to the extent that they compromise the mechanical properties of the joints. Thereby, joint strengths may be improved by a factor of ten for mild steel and a factor of two for the stainless steel [Eagles, Mitchell, and Wronski 1995].

4.1.4 Phase Formation

Alloying between a molten braze and parent materials can often result in the formation of intermetallic compounds. Because intermetallic compounds generally possess a higher elastic modulus than do many filler alloys, well-dispersed intermetallic phases are often beneficial to the stress-bearing capability of the filler metal. As a general rule, intermetallic compounds that exist over a range of composition have a tendency to be more compliant mechanically than others that have minimal phase width.

By and large, agglomerations of intermetallic compounds are deleterious to the mechanical properties of joints. This is particularly true where the compound has low fracture toughness and forms as a continuous interfacial layer between the component surfaces and the filler alloy. It is sometimes possible to restrict the coarsening of these phases by carrying out the joining operation under conditions that are unfavorable for their growth—namely, by restricting the duration of the heating, lowering the peak temperature, and employing strategies for rapid cooling. Sometimes, minor additions can be made to the filler alloy that will break up agglomerations of intermetallic phases present in the joint. Barrier metallizations applied to joint surfaces can be used to prevent alloying with the parent materials and the consequential formation of undesirable phases.

Some filler alloys are themselves hard because they contain intermetallic phases but yet form joints of high strength with certain parent materials, provided the intermetallic phases are finely divided within the joint microstructure. Many nickel-bearing brazes fall into this category, with rapid solidification of the braze being a prerequisite for forming robust joints. Further details are given in Chapter 2, section 2.1.10.

4.1.5 Filler Metal Partitioning

Partitioning of filler metals is a fabrication method that is becoming more common. It is an approach that can be beneficial if an alloy composition is relatively brittle when prepared by conventional ingot processing but where selected combinations of the constituents are ductile. Active brazes are often prepared in this manner where a foil of the active metal, often titanium, is sandwiched between sheets of the parent alloy, such as silver-copper eutectic. There are partitioned versions of the active brazing alloys (ABAs) referred to in Chapter 2, section 2.2.2, and listed in Table 2.17. As both silver-copper eutectic and titanium are ductile, so is the resulting composite structure. Other partitioned brazes of this type include gold/niobium/gold and other brazes with a vanadium foil in the center.

Another example of partitioning of braze constituents is a low-melting-point quaternary braze for aluminum, which contains aluminum, copper, silicon, and nickel as the main ingredients. Ingots of this alloy have only moderate ductility at room temperature and, although it can be hot rolled to foil (see Fig. 2.25), it is an energy-intensive and time-consuming process, which makes it economically unattractive. However, because both aluminum-silicon and copper-nickel alloys are ductile and commercially available, brazing foil can also be prepared as ductile composite from a sandwich of these two binary alloys. A metallographic section though such a trifoil is shown in Fig. 4.1. [Jacobson, Humpston and Sangha 1996]. A rolling mill with three feeder coils is required to make preforms in this geometry, as illustrated schematically in Fig. 4.2.

A number of brazing alloys are offered co mmercially as partitioned preforms. The objective is either one of attempting to improve the mechanical properties of the filler metal prior to melting, or one of cost economy through reducing the number of stock alloys required. It is usual to have the lower-melting-point combination in the cladding with the higher-melting-point element or alloy in the core. The low-melting-point fraction then more readily melts and wets the joint interfaces.

Sometimes a filler metal may appear to be partitioned, but the compositional separation may have been brought about for a purpose other than overcoming brittleness of a homogeneous alloy. A case in point is copper foil, clad on both sides with a similar thickness of silver-copper-zinc braze. Here, the cladding on its own is of the correct composition for the braze. The application of the braze as a cladding on a copper foil is intended to provide a compliant core that will be highly ductile and deform plastically as the brazed joint cools so that it is capable of relieving thermal expansion mismatch stress (see section 4.2). This behavior accounts for the use of braze-clad copper for attaching carbide cutting tool tips to steel shanks. Other benefits of admitting a braze into a joint gap in this form are explained in Chapter 1, section 1.3.2.2.

A soft iron core is sometimes used when brazing stainless steels with copper-base brazes. Under normal circumstances, chromium carbides $Cr_{23}C_6$ and Cr_7C_3 form at the braze/steel interface, and the fracture toughness of the resulting

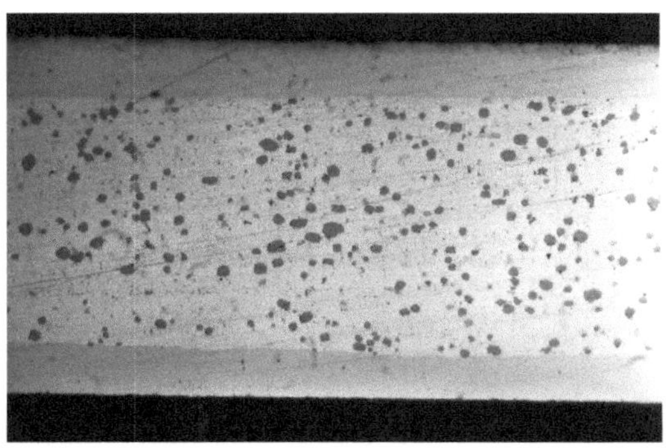

Fig. 4.1 Metallographic cross section through a 100 μm (4 mils) tri-foil braze preform. It consists of thin foils of a copper-nickel alloy applied by roll cladding to a core of aluminum-silicon alloy. The ratio of the thickness of the foils is adjusted to provide the correct aggregate composition when the preform melts.

assembly is poor. However, admitting a thin shim of iron into the joint gap, which partly dissolves in the copper brazing alloy cladding, permits the latter to dissolve chromium and carbon, which can result in a ninefold improvement in the impact toughness of the joint at room temperature. The iron core effectively provides an iron-modified variant of standard brazing alloy that would be expensive if it had to be produced in small quantities in the form of a homogeneous brazing foil.

4.2 Mechanical Constraints and Solutions

In an assembly composed of heterogeneous materials, there is usually a mismatch between the coefficients of thermal expansion (CTEs) of the abutting components. This manifests itself as stress on cooling from the solidus temperature of the filler metal and is normally a maximum at the lowest temperature that the assembly experiences. For aerospace and some terrestrial situations this temperature can be below −50 °C (−58 °F). For assemblies that are brazed, that is, joined using filler metals that melt at temperatures above 450 °C (840 °C), even modest differences in CTE between the abutting components can generate substantial mismatch stresses at ambient temperatures. Materials with a relatively low elastic modulus can accommodate strain and will tend to deform under the influence of the mismatch stress, while brittle materials, notably glasses and ceramics, have a tendency to fracture, particularly if the stress distribution places the component in tension. Even if a heterogeneous assembly survives the joining operation, the stresses arising from the thermal expansion mismatch can cause it to fail by fatigue or creep during subsequent thermal cycling in the field.

The stress σ_{12} in the region of the joint between two isotropic materials 1 and 2, with differing moduli and thermal expansivities, which develop on cooling from the freezing temperature of the filler metal, can be approximated by the following equation [Timoshenko 1925]:

$$\sigma_{12} = \left(\frac{E_1E_2}{E_1 + E_2}\right)(X_1 - X_2)(T_f - T_s)$$

where E is the modulus of elasticity of materials 1 and 2, X is the coefficient of thermal expansion of 1 and 2, T_f is the freezing point (solidus temperature) of the filler alloy, and T_s is the temperature of the assembly corresponding to the stress. In the derivation of this equation, it is assumed that the materials are deformed only within their elastic limits and that the joint is infinitely thin. Despite these simplifying assumptions and the inaccuracies that they introduce, this expression is useful in providing an indication of whether the stress due to thermal expansion mismatch is close to, or exceeds, the failure stress of either of the abutting materials, that is, whether failure of the assembly is likely to occur; this assumes that the joints do not fail before the components do. Some worked examples are described by Haug, Schaefer and Schamm [1989].

In an assembly with a planar joint between two elastic but different materials 1 and 2, the

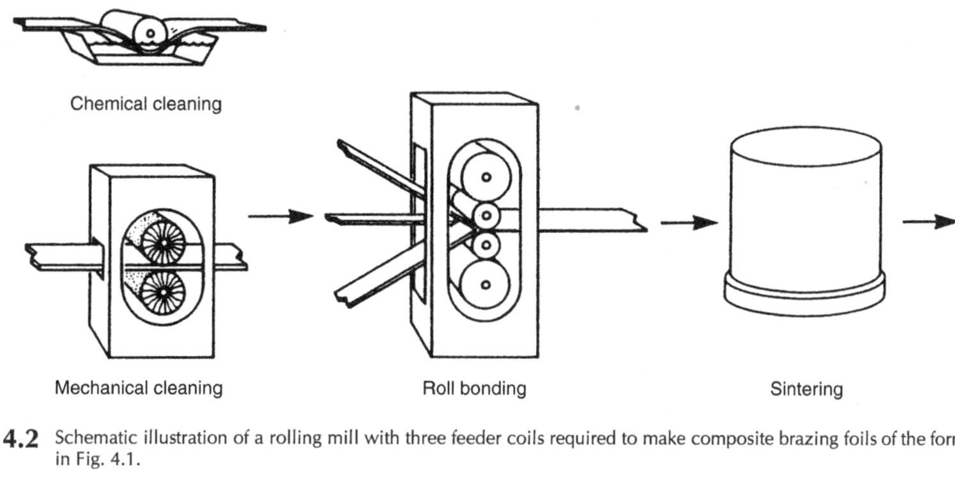

Fig. 4.2 Schematic illustration of a rolling mill with three feeder coils required to make composite brazing foils of the form shown in Fig. 4.1.

magnitude of the bow distortion in one dimension can be estimated from the physical properties of the materials using a simplified model [Timoshenko 1925]. With reference to Fig. 4.3, and assuming that the bow distortion, B, is small, this approximates to:

$$B = \left(\frac{L^2}{8}\right)\Big/R$$

and the radius of curvature, R, is

$$R = \frac{(A_1 + A_2)\left[3(1 + M)^2 + (1 + MN)\left(M^2 + \frac{1}{MN}\right)\right]}{6(X_1 - X_2)(T_f - T_s)(1 + M)^2} \quad \text{(Eq 4.1)}$$

where

$M = A_1/A_2$ and $N = E_1/E_2$,
L is the length of the joint
A_1 and A_2 are the thicknesses of materials 1 and 2
X_1 and X_2 are the coefficients of thermal expansion of materials 1 and 2
E_1 and E_2 are the elastic moduli of materials 1 and 2
T_f is the freezing point (solidus temperature) of the filler alloy
T_s is the temperature of the assembly corresponding to the bow distortion.

Equation 4.1 assumes that the joint in the heterogeneous assembly is infinitely thin and totally inelastic, which is not entirely correct. However, it does represent a worst case scenario and, in that sense, is useful.

From Eq 4.1, it can be seen that it is possible to effect some reduction in expansion mismatch stress, that is, minimize R, by applying one or more of the following measures:

- Decrease the solidus temperature of the filler alloy, T_f
- Increase the minimum service temperature of the assembly, T_s
- Reduce the dimensions of the joint area, L
- Change one or both materials to minimize the mismatch in thermal expansivity (CTE) $|X_1 - X_2|$

Occasionally it may be possible to implement one or more of these changes of this nature, but they are likely to conflict with processing constraints, if not with the intended functional requirements of the assembly. Alternative solutions must therefore be sought.

In practice, it is usually possible to obtain a small reduction in the distortion of a bowed heterogenous assembly by heat treating it at a temperature below the solidus temperature of the braze to enable stress relaxation and creep to occur in the filler metal. However, there is a limit to the reduction in distortion that can be obtained by this means, which is typically 10% or less. This stems from the fact that stress-reducing mechanisms are diffusion related and become more effective as the temperature is raised toward the melting point of the joint, while mismatch stress increases as the temperature of the assembly is reduced below the solidus temperature of the braze. Therefore, these two tendencies act in opposition, and the optimum condition for reducing the distortion of bonded components is a compromise between the two. In the absence of other indications, a good starting point is to use a temperature about 75% of the melting point of the braze, expressed in degrees Kelvin.

Some further improvement in the residual stress level can be obtained by using a more compliant filler metal, especially when the joint is reasonably wide (>25 μm, or 1 mil). Silver-base brazes have lower elastic moduli than nickel-base brazes have, but their interchangeability may be limited by metallurgical factors and temperature or chemical constraints.

Wide joint gaps (>500 μm, or 20 mils) can sometimes be used to minimize the effects of expansion mismatch between two components. The braze must have high viscosity in order to fill such wide joints. This is achieved by either

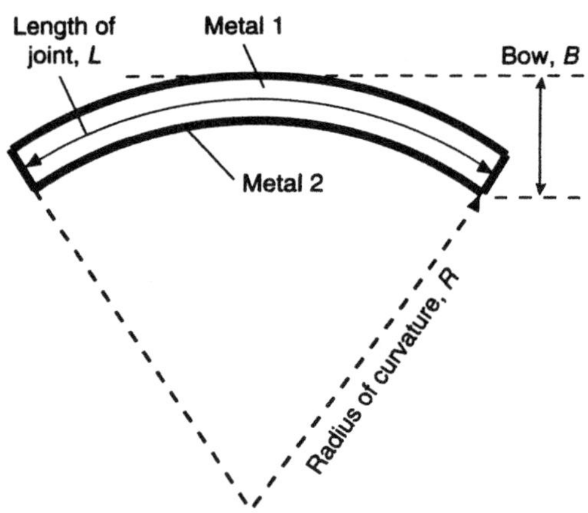

Fig. 4.3 Bow distortion of a bimetallic strip

using a filler metal with a wide melting range and performing the joining process at below the liquidus temperature, so that the alloy is not fully molten, or by mixing in an insoluble or partly soluble metal powder having a higher melting point and often referred to in the literature as the "additive." The joint gap may be controlled by spacers, by shims containing a mixture of the filler and "additive," or porous shims. Wide-gap brazing is described in section 4.3.4.2 of this chapter. One particular merit of wide joints to ceramic components is that they obviate the need to closely machine the mating surfaces of the components, which tends to be costly and can weaken the material by creating subsurface cracks. Where the joint gap is wide, the mechanical properties of the joint are essentially those of the bulk braze (see section 4.3.3.1 of this chapter).

A variety of mechanical schemes are available to assist in overcoming the problem of thermal expansion mismatch. Several approaches that have proved successful are described in section 4.2.3 of this chapter. However, there is an ongoing need for solutions that can achieve long service life (>30,000 hours) at high temperatures (>800 °C, or 1470 °F) in chemically aggressive environments, including air.

With the advent of low-cost computer software, it is relatively straightforward to model the stresses that arise when brazing together two different materials, with or without transition pieces. Thermomechanical finite element models are routinely used in industry. A description of the mathematical concepts underlying such models is provided by Messler, Jou and Orling [1995].

4.2.1 Controlled Expansion Materials

Fabrication of most products requires use of many different materials in their assembly, each selected for a particular property or compromise of properties that it offers. Ashby and co-workers have devised a scheme, in the form of materials selection charts that present a visual means of identifying materials with promising combinations of properties to consider [Ashby 1994]. Two examples are illustrated in Fig. 4.4 and 4.5. Because silicon and most other semiconductor and optical materials, as well as engineering ceramics, have low CTEs compared with metals, there is always interest in controlled expansion metals for use in assemblies that can bridge the CTE difference with metals over a specific temperature range.

Single-phase materials, which include many engineering ceramics such as alumina and aluminum nitride, change dimension in an approximately linear manner with temperature. This is certainly true over the temperature range of interest for consumer products (−50 to +150 °C, or −60 to +300 °F). For common metals, their CTE values at room temperature are directly proportional to their melting points. Thermal expansion is linked to the atomic vibrations of the crystal lattice. Raising the temperature increases the amplitude of vibration, which means that each atom occupies a greater volume and, hence, the material grows in size. Since the maximum amplitude of vibration is restricted by the melting point of the material, a low-melting-point material has a narrower temperature range over which the expansion can occur. In consequence, it will exhibit a higher CTE value than a metal with a relatively higher melting point, as demonstrated in Fig. 4.6, using the data in Table 4.6.

Thermal expansion curves for some common engineering materials are given in Fig. 4.7 and show linear and nonlinear behavior over this temperature range. It is evident that the principal low expansion metals (low slope and hence low CTE) are tungsten and molybdenum, which have CTE values at 20 °C (68 °F) of 4.5×10^{-6}/K and 5.1×10^{-6}/K, respectively. These metals, especially tungsten, are hard and stiff, offering little compliance, and they are also relatively difficult to machine. Both are dense materials so that components fabricated from them are heavy. Their high melting points mean that they contain significant energy content and, consequently, the net cost of parts fabricated from these metals can be high. Titanium has a low CTE for a metal (8.0×10^{-6}/K), and low density (4.5 g/cm^3), but has a poor thermal conductivity of only 15 W/m·K (8.7 Btu/h·ft·°F). This last property rules it out for any application where substantial heat transfer by conduction is an additional requirement. Alumina and aluminum nitride ceramics are also low-expansion materials. They have very high melting points, require energy-intensive methods of production, and are difficult to machine. These ceramics have to be fabricated in near-net shape forms by powder pressing and sintering, which makes the tooling costs of producing components in alumina and aluminum nitride somewhat high.

The shortage of single-phase materials that offer the combination of low CTE, high thermal conductivity, and, preferably, low density has led to the development of several families of multiphase materials with engineered properties. These include iron-nickel alloys, copper-tungsten and copper-molybdenum alloys, metal-metal laminates, metal-ceramic laminates, metal-matrix composites, and metal-metalloid composites. A comparison of some key physical properties of these materials is given in Table 4.7. Further information on each of these is given in the following sections.

4.2.1.1 Iron-Nickel Alloys

Most readers probably know iron-nickel alloys by trade names that include Invar, Kovar, Nilo, Alloy 42, and Alloy 45. Nilo K, for example, is an iron-nickel-cobalt alloy (practically identical to Kovar), of composition 54Fe-29Ni-17Co that has a CTE at 20 °C (68 °F) of 5.8 × 10^{-6}/K. These alloys are readily available from a number of manufacturers in many shapes and forms and are competitively priced. Iron-nickel alloys are widely used in electronic packaging. Being fairly soft and ductile in the annealed condition, these materials are frequently employed as mechanically compliant shims for joining low-expansion ceramics to higher-CTE metals. The low modulus and expansion characteristics of iron-nickel alloys enables them to distribute and buffer stresses arising from differential dimensional change between rigid ceramics and metals arising from variations in temperature.

Iron-nickel alloys offer abnormally low thermal expansion compared with their elemental constituents. Indeed, over a limited range of

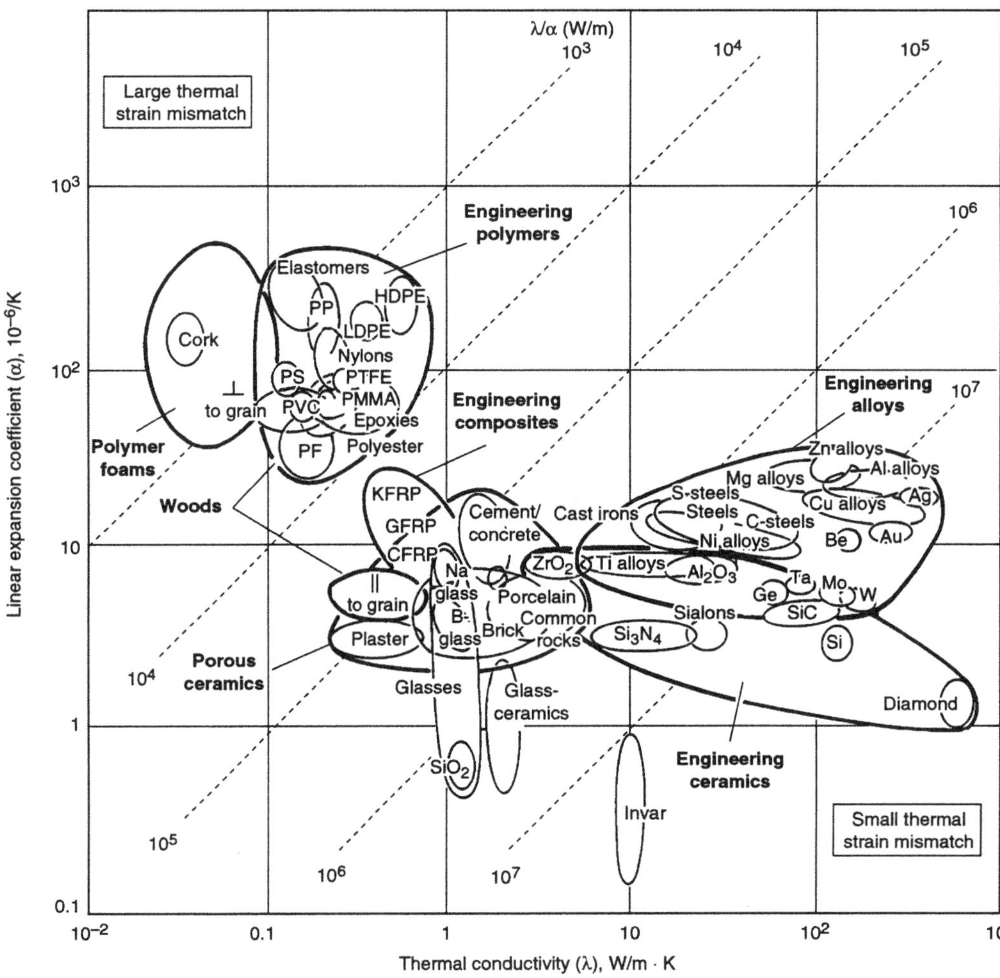

Fig. 4.4 An Ashby materials selection chart. The linear expansion coefficient, α, plotted against the thermal conductivity, λ. The contours show the thermal distortion parameter λ/α.

temperature it is possible to design an iron-nickel alloy that has zero expansivity. The unusual expansion characteristics of these alloys can be ascribed to the fact that they are ferromagnetic. At temperatures above the Curie point (approximately 450 °C, or 840 °F), these alloys become paramagnetic and then exhibit normal thermal expansion characteristics. Conse-

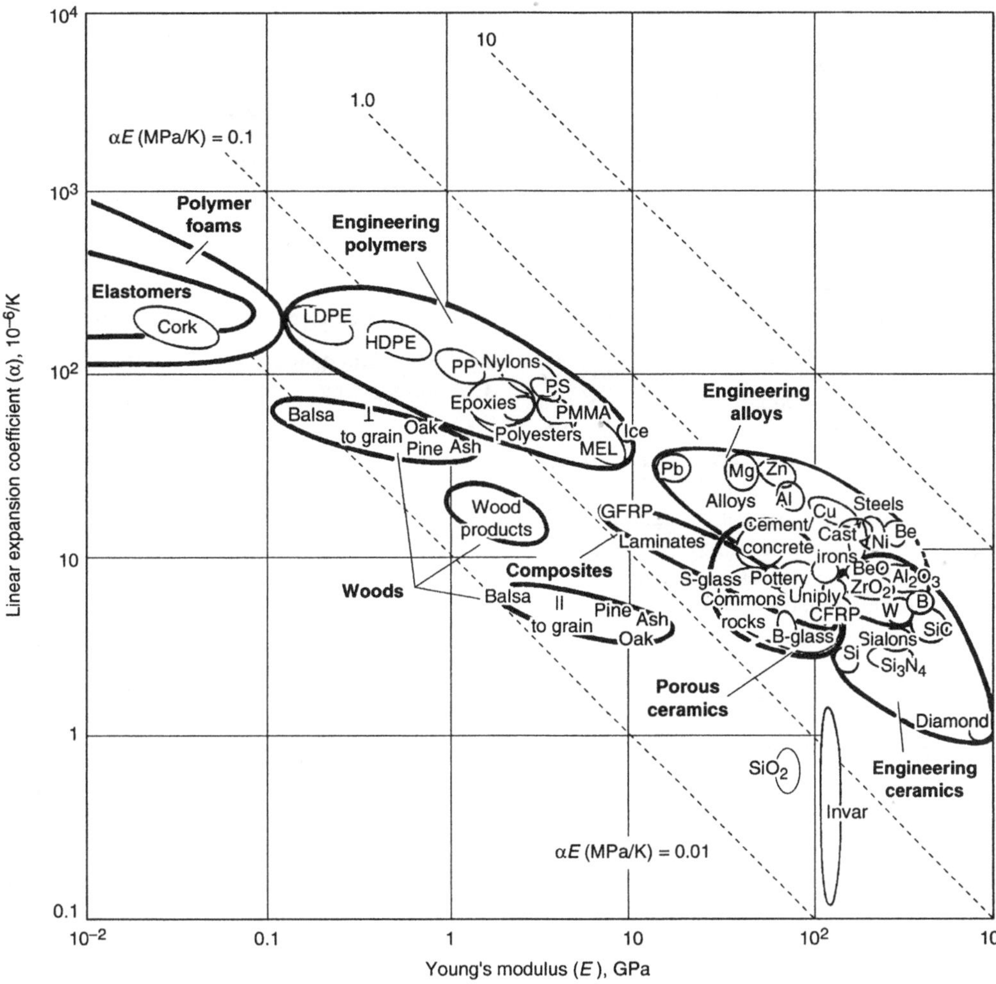

Fig. 4.5 An Ashby materials selection chart, showing groups of materials plotted in terms of their stiffness (modulus) and coefficient of thermal expansivity (CTE)

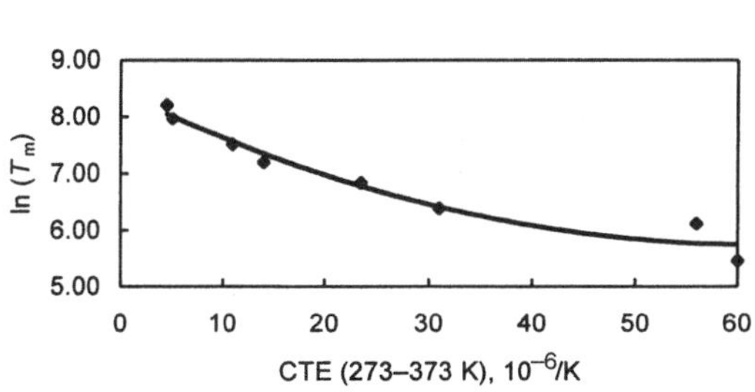

Fig. 4.6 General relationship between coefficient of thermal expansion, or CTE (between 273 and 373 K), and melting point for metals, T_m. Adapted from Li and Krsulich [1996]

Table 4.6 Metals and their properties used to prepare Fig. 4.6

Metal	Melting point °C	Melting point K	CTE at 300 K, 10^{-6}/K
Tungsten	3422	3695	4.5
Molybdenum	2623	2896	5.1
Palladium	1555	1828	11.0
Gold	1063	1336	14.1
Aluminum	660	933	23.5
Cadmium	321	594	31.0
Lithium	181	454	56.0
Mercury	−39	235	60.0

CTE, coefficient of thermal expansion

quently, their usefulness in brazed assemblies to modify thermal expansion mismatch tends to be limited, although they may still offer benefits in this regard, as explained in Chapter 7, section 7.4.2. The CTE curves for some iron-nickel alloys and low-carbon steel are shown in Fig. 4.8.

4.2.1.2 Copper-Molybdenum, Copper-Tungsten, and Tungsten-Nickel Alloys

Molten copper is virtually insoluble in both molybdenum and tungsten, but because it will wet refractory metals, various techniques have been devised to enable the manufacture of 100% dense alloys of these materials. Techniques that are used to produce copper alloys with molybdenum and tungsten include conventional powder metallurgy, as well as liquid infiltration casting. Controlled-expansion alloys of this type can be produced with a continuum of properties that range from essentially pure copper to pure refractory metal. By this means, families of materials that exhibit a range of controlled CTE values, as shown in Fig. 4.9, have been developed. A copper content in alloys of this type of 15% gives a reasonable boost to the thermal conductivity without greatly increasing the CTE above that of the refractory metals, which accounts for this being a much used composition. More importantly, perhaps, the addition of a soft copper phase in the otherwise refractory metal matrix greatly improves machinability and brazeability. These alloys still require considerable care to machine because the copper is much softer than the refractory metal matrix, and it is easy to cause damage in the surface region. Another drawback of these alloys, especially serious for aerospace applications, is their high density, although this is offset partly by their high modulus, which means that thinner sections can sometimes be used, depending on the functional requirements and design of the product. When used in precision assemblies, these alloys need to be in the annealed condition to ensure dimensional stability on thermal cycling.

Another family of material that can be fabricated with a wide range of properties tailored by composition is what are generally known as "heavy alloys." These are liquid-phase sintered materials comprising tungsten powder infiltrated with nickel-iron or nickel-copper braze. Representative compositions are W-7Ni-3Cu (which is included in Fig. 4.9), and W-3.5Ni-1.5Fe, although there are also "heavy alloys" containing molybdenum in place of either all or some of the tungsten, which possess higher CTE values. The ratios of constituents are frequently chosen to provide a good compromise between machinability and toughness. The presence of a continuous nickel-iron or nickel-copper phase means that these materials are wetted by most common brazing alloys. A remarkable example of a joint containing no less than seven transition pieces of different heavy alloy compositions to form a robust connection between tubular members of graphite and a nickel alloy is described by Hammond and Slaughter [1971]. The joints were all made simultaneously, with pure copper as the braze, by heating to 1130 °C (2065 °F) in argon.

4.2.1.3 Copper-Surface Laminates

Copper can be attached directly to alumina via the copper-copper oxide eutectic reaction (see Chapter 7, section 7.2.6). These products are often marketed as direct-copper-bonded, or DCB. The copper can be patterned so that it can also fulfill the function of a low-density printed circuit board, which has made it a very popular substrate for electronic power modules. A more recent development, now commercially exploited, is a DCB substrate containing arrays of through-thickness vias, each filled completely with copper. This does not change the in-plane thermal expansivity of the alumina but significantly improves the average through-thickness thermal conductivity.

Copper-molybdenum-copper and copper-invar-copper laminates also exist, offering a subtly different balance of properties. Figure 4.10 is a graph showing the expansion of copper-molybdenum-copper laminate measured at the

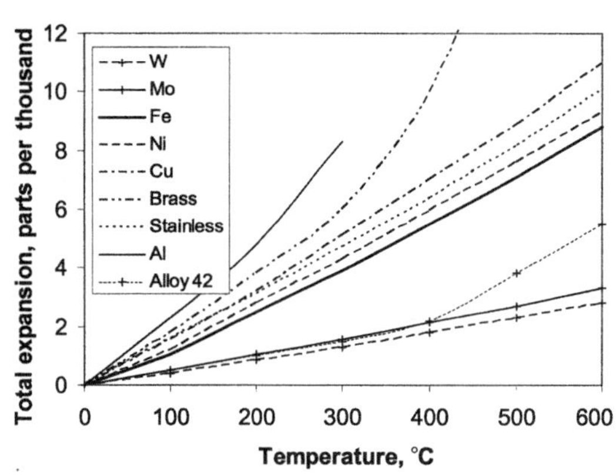

Fig. 4.7 Expansion of some common engineering materials as a function of temperature

surface as a function of the copper thickness relative to that of the molybdenum core. Because this type of laminate does not provide the same stiffness and patterning abilities as the ceramic-cored alternative, it is not as widely used. The invar-cored materials are obviously not dimensionally stable on heating to brazing temperatures.

4.2.1.4 Composite Materials

Several examples of composite materials are included in Table 4.7. Each is representative of a specific family: metal-metalloid, metal-ceramic, and metalloid-metalloid. These are all relatively new materials and many variations exist on the market. They have not been evaluated exhaustively under a wide range of application conditions and the sales brochures must be read with caution. Some of these materials will not withstand brazing temperatures without distorting or degrading.

One of the few types of composite that is unaffected by elevated temperatures is the family of metalloid-metalloid composites, represented by carbon-carbon fiber composites, which were originally developed for exceptionally high temperature applications such as aircraft brakes and

Table 4.7 Indicative physical properties for selected semiconductor and low-expansion materials at 20 °C (68 °F)

Exact values depend on the composition of the material, method of manufacture, test method, and test conditions. Reference should be made to suppliers' data sheets for precise values.

Material	Through-thickness thermal conductivity, W/m · K	In-plane thermal expansion coefficient, 10^{-6}/K	Density, g/cm³
Gallium arsenide	42	6.5	5.3
Silicon	84	2.5	2.3
Alumina	20	6.7	3.9
Aluminum nitride	165	4.5	3.3
Beryllia	260	7.2	2.9
Molybdenum	140	5.1	10.2
Titanium	15	8.0	4.5
Tungsten	174	4.5	19.3
Copper-alumina-copper	26	7.3	4.1
Copper-molybdenum-copper	166	5.5	10.0
Copper-85% tungsten alloy	180	7.2	16.1
Copper-85% molybdenum	160	6.7	10.0
Invar (Fe-36Ni alloy)	14	2.2	8.1
Kovar (Fe-29Ni-17Co alloy)	17	5.8	8.4
Aluminum-50% silicon alloy(a)	140	11.0	2.5
Aluminum-70% silicon alloy(a)	120	7.4	2.4
Aluminum-68% silicon carbide composite	150	7.2	3.0
Beryllium-30% beryllia composite (E20)(b)	210	8.7	2.1
Beryllium-51% beryllia composite (E40)(b)	220	7.5	2.3

(a) As supplied by Sandvik Osprey Ltd. (b) As supplied by Brush Wellman Inc.

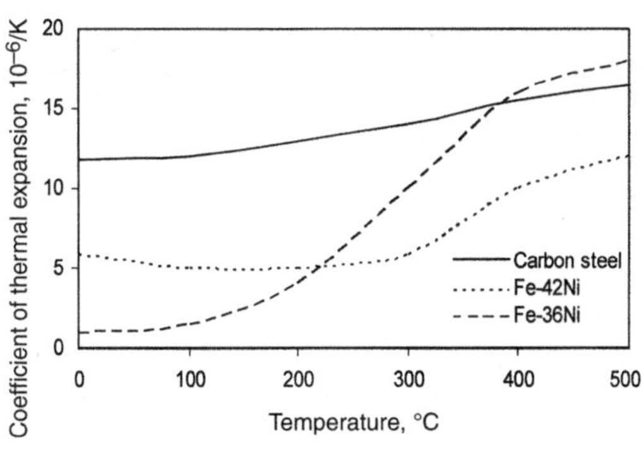

Fig. 4.8 Coefficient of thermal expansion (CTE) of low-carbon steel and iron-nickel alloys as a function of temperature. The low CTE of iron-nickel alloys exists only over a limited range of temperature. Normal expansion behavior is observed above about 400 °C (750 °F).

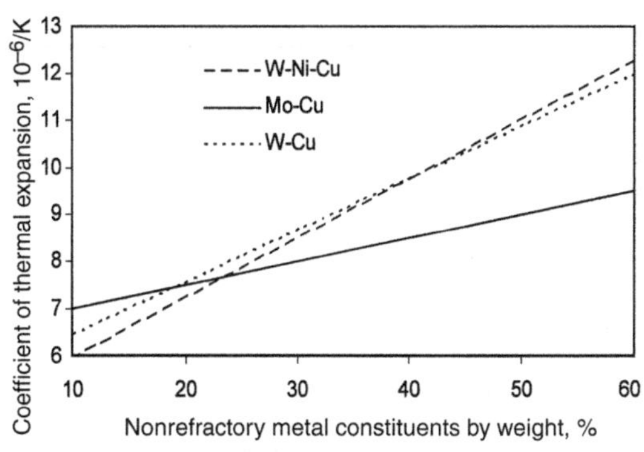

Fig. 4.9 Coefficient of thermal expansion of liquid-phase sintered tungsten and molybdenum materials as a function of the content of the main braze constituents, namely copper and nickel.

rocket motor nozzles. They can be metallized and also impregnated with copper. This allows the in-plane thermal expansivity to be increased to values that are representative of many metals and ceramics. Care needs to be taken when selecting the carbon fibers as well as the weave used for a particular component. Carbon fiber is a highly anisotropic material and hence the properties of parts fabricated using it can vary markedly with orientation. Likewise, there are many grades of carbon fiber, and, as one might expect, it is the more expensive ones that possess the most desirable and stable properties. Carbon-fiber reinforced aluminum and copper composites are also available, but these are highly anisotropic and tend to be dimensionally unstable. The aluminum-containing materials are unsuitable for use with brazes that melt much above 600 °C (1110 °F).

4.2.2 *Interlayers*

One route toward reducing the mismatch stress concentration that develops in brazed joints, on account of thermal expansion mismatch between the components, involves a redesign of the joint width to accommodate one or more interlayers. There are two basic configurations that can be employed.

In the first approach, a compliant interlayer is inserted into the joint gap that will yield when the joint is placed under stress and thereby reduce the forces acting on the components. High-purity copper is often used in brazed joints because it combines a low elastic modulus with good wetting characteristics, and it is also inexpensive to fabricate to the desired geometry. A shim of pure iron is an alternative material used for this application but only if its magnetic properties do not conflict with the application requirements. Nickel is relatively stiff, so it can be used only where one of the components is ceramic, while silver is too soluble in most brazes. Iron and copper interlayers are restricted in their application to temperatures below 400 °C (750 °F) in oxidizing environments.

Optimum stress reduction is normally achieved when the thickness-to-length ratio of the interlayer falls within a certain range, determined by the combination of materials used and the dimensions of the joint [Miyazawa *et al.* 1989; Xian and Si 1992]. This is illustrated for a joint between alumina and steel employing a silver-copper-titanium braze and a nickel interlayer in Fig. 4.11. For this joint, which has a length of 15 mm (0.6 in.), the interlayer should be between about 1.5 and 3 mm (0.06 and 0.12 in.) thick. If the interlayer is much thinner than the prescribed minimum thickness, it is unable to absorb a significant proportion of the applied stress, whereas if it falls outside the upper limit, the interlayer will not yield to the required extent. Numerical modeling is able to predict an optimum interlayer thickness for most situations. The fracture toughness of the ceramic is an additional and important determinant of the joint strength that may be achieved in metal-ceramic assemblies. Hence, brazed joints to silicon nitride are invariably "stronger" than SiAlON ceramic. A well-designed brazed joint will substantially enhance the fracture toughness of a ce-

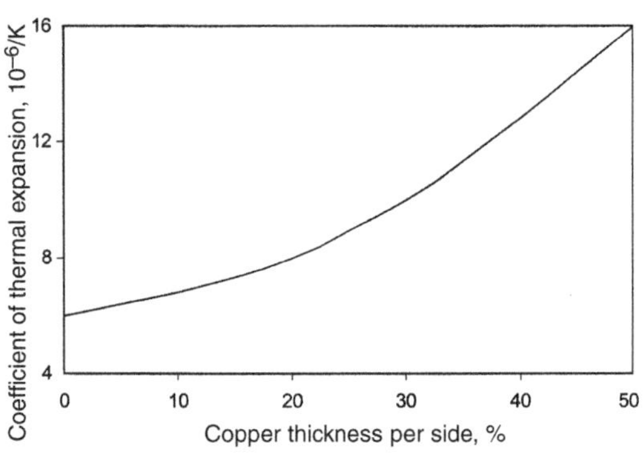

Fig. 4.10 Coefficient of thermal expansion of double-sided copper-clad molybdenum at room temperature as a function of the copper thickness

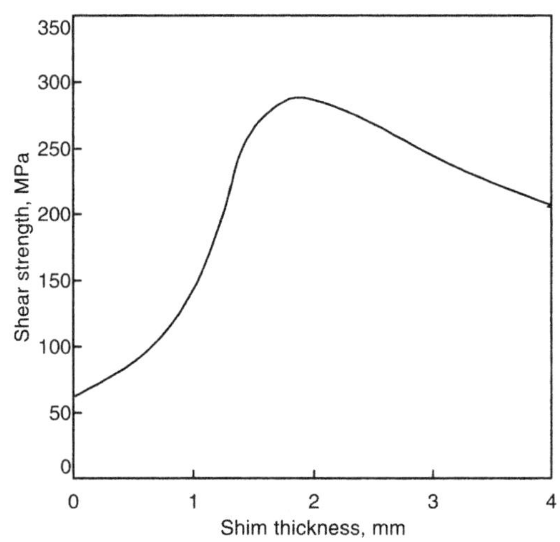

Fig. 4.11 Influence of shim thickness on the shear strength of an alumina-to-steel joint made using an Ag-Cu-Ti braze and a nickel interlayer. The joint length is 15 mm (0.6 in.). Source: After Miyazawa et al. [1989]

ramic member by filling surface defects and placing the wetted surface in a state of moderate compression.

An alternative approach for reducing mismatch stress concentrations is to redistribute the stresses across a much wider zone so that the stress is kept within tolerable levels everywhere in the assembly. A graduated redistribution of stress may be accomplished by inserting into the joint one or more thick shims or plates that have CTEs of intermediate magnitude between those of the abutting components. The plates must be sufficiently thick so that they are not significantly distorted by the imposed stresses and are therefore not usually less than 5 mm (200 mils) thick. An assembly containing a single plate with an intermediate thermal expansivity is shown in Fig. 4.12.

This approach is particularly suitable where there is a need to join metals to ceramics and other ceramiclike nonmetals where the component dimensions are large. If the intermediate plate is selected to have a CTE that is close to that of the nonmetal, then it is possible to transfer the major proportion of the stress to the more robust metallic part of the assembly. The joint should also be designed such that the part with the lower fracture toughness is always held in compression. This is illustrated in Fig. 4.13 for a tubular joint between a nickel-base alloy and a graphite member. The expansion mismatch (approximately 8×10^{-6}/K) is reduced by using a molybdenum interposer with scarf-profiled ends, which are suitably orientated to place the graphite member in compression [Werner and Slaughter 1968]. Where the two components have greatly different CTE values, it may be necessary to use a graded series of plates to reduce the mismatch stresses in each joint to an acceptable level. A schematic illustration of an assembly of this type is shown in Fig. 4.14.

A monolithic plate of graded composition and CTE can be used in place of a series of discrete homogeneous plates. Typically, these may be prepared from powder compacts. Copper-tungsten components of this type are made by infiltrating a loose compact of tungsten powder with molten copper. By adjusting the packing density of the powder through thickness, the relative proportions of copper and tungsten will vary and the properties of the component can vary from those of essentially pure copper to about 95% tungsten. This enables one side of the component to be made highly tungsten-rich, with a low expansion coefficient, and the other side highly copper-rich with a much higher expansion coefficient. Because there are no abrupt interfaces in such a component, it can survive thermal cycling over wide ranges of temperature almost indefinitely without suffering distortion through creep or fatigue fracture. A graded copper-tungsten plate is shown in Fig. 4.15. Brazing alloys

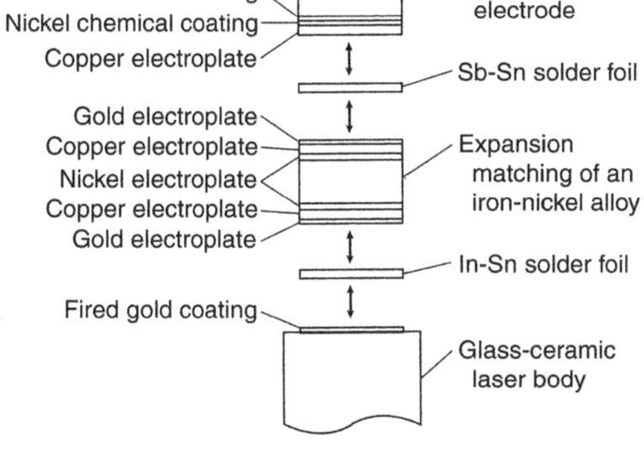

Fig. 4.12 Use of a plate of intermediate thermal expansivity to reduce the stress due to thermal expansion mismatch in an assembly between an aluminum alloy mount and the body of a solid-state laser

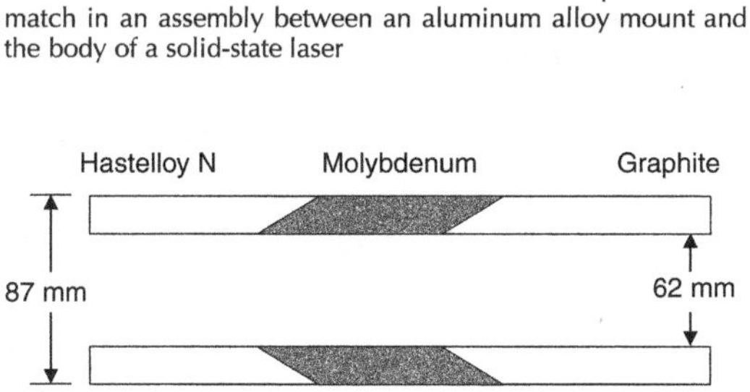

Fig. 4.13 Schematic cross section of a tubular joint between Hastelloy N (Ni-17Mo-7Cr-5Fe) and graphite, made with Pd-35Ni-5Cr braze. A transition piece, or interposer, of molybdenum decreases the expansion mismatch between the graphite and metal parts to an acceptable level. The scarf joint is orientated so that the graphite, which is the more brittle member, is always in compression.

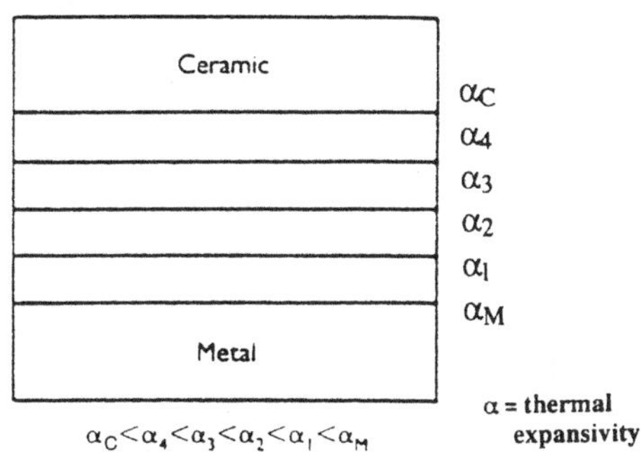

Fig. 4.14 A graded series of plates (1 to 4) designed to reduce the mismatch stress between ceramics and metals to an acceptable level. α, thermal expansivity.

with low and controlled thermal expansion coefficients are discussed in section 4.3.3.2 of this chapter.

The following disadvantages are associated with the use of graduated joined assemblies incorporating intermediate plates to accommodate mismatch stresses:

- An increase in the thickness and often in the weight of the assembly, which may be significant. This modification will also introduce additional materials and fabrication costs.
- At least two brazed joints are used in place of a single joint. Because further materials are introduced to the assembly, alternative filler alloys and joining processes may need to be used and qualified for the requisite application.
- The thermal and electrical conductance between the joined components is likely to be degraded. This is a consequence of the increase in the overall thickness of and number of interfaces in the assembly. Materials with low CTE values tend to be inferior conductors, especially of heat. A notable exception is diamond and a thin shim (300 μm, or 12 mils) of this material will function as both a buffer against strain and as an effective heat spreader for optical and electronic components.
- This type of configuration is difficult to apply to assemblies and joints that do not have simple planar geometries.

4.2.3 Compliant Structures

Equation 4.1 given for calculating the bow distortion of a bimetallic assembly implies that the mismatch stress is a sensitive function of joint dimension or, more specifically, joint area. Although the overall size of the assembly is likely to be fixed by the functional requirements of the product, it may be possible to replace one of the monolithic components with a filamentary, brushlike structure. Then, the dimensions of each individual bimetallic joint can be made as small as necessary, thereby effectively eliminating the mismatch stress from this source, while the high aspect ratio of the filaments confers a degree of lateral compliance that can accommodate the mismatch strain. Examples of these highly compliant structures are illustrated in Fig. 4.16 and 4.17, and others are described in the scientific and technical literature [Huchisuka 1986, Hanson and Fernie 1994].

The use of compliant structures of the forms shown obviously incurs a cost penalty due to the greater complexity of manufacture. The conductance between the components via the filamentary member will also be impaired. Even with filaments having a hexagonal cross section to produce a close-packed structure, it is difficult to obtain a compliant structure that will work effectively with a packing density of greater than about 85% [Glascock and Webster 1983]. More typically, the total joint area is only 10 to 30% of the entire plan area so that the mechanical properties are similarly compromised. Furthermore, the ability to simultaneously make large numbers of small area joints is by no means a trivial exercise but one that demands stringent control of tolerances and highly specified joining processes.

4.2.4 Dynamic Thermal Expansion Mismatch

A common error, which is implicit in the preceding discussion, is that materials expand and contract in a regular manner in response to thermal excursions. This is true only for simple single-phase materials. Many engineering materials have nonlinear thermal expansion characteristics, as can be seen in Fig. 4.7. One metallic example referred to in section 4.2.1.1 of this chapter is iron-nickel alloys that exhibit anomalous thermal expansion over a specific temperature range. A further complication arises when two joined materials possess different thermal emissivities, specific heat capacities, and thermal conductivities. The combination of these factors means that in a thermally dynamic

Fig. 4.15 Monolithic plates of graded composition, varying from essentially pure copper to approximately 95% W. The thermal expansivities of the two surfaces differ by approximately 12×10^{-6}/°C (22×10^{-6}/°F).

environment, such as exists on cooling from the peak temperature of a brazing process, transient differences in thermal expansion mismatch and accompanying stresses will be manifested. These can be many times greater in magnitude than the corresponding static coefficients of thermal expansion mismatch.

Dynamic thermal expansion mismatch is relatively simple to model and the necessary properties required for solution of the equations are known for many materials. Consider, for example, a 50 mm (2 in.) diameter rod of nickel-base alloy brazed to an equally sized disk of a machinable ceramic (Macor). The CTE of the nickel alloy is 14×10^{-6}/K and that of the ceramic is 9.5×10^{-6}/K. One might therefore assume that the thermal expansion mismatch between these two materials is, roughly, 5×10^{-6}/K. For simplicity, assume that the solidification temperature of the braze is 950 °C (1740 °F) and that the nickel alloy and Macor parts simultaneously and uniformly attain the peak process temperature. During ensuing air cooling, the nickel alloy component will cool faster, largely

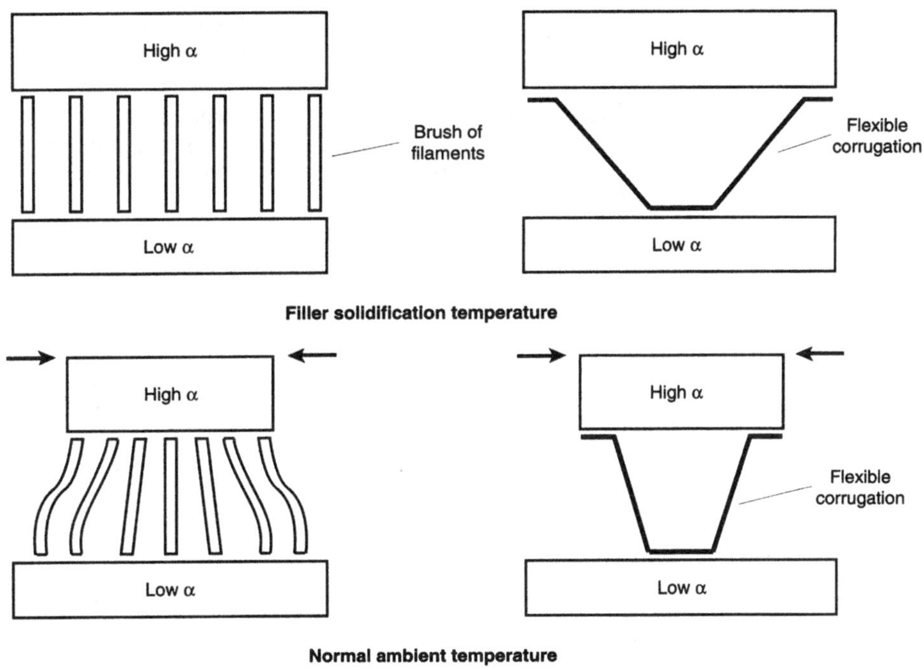

Fig. 4.16 Examples of compliant structures for mitigating mismatch expansivity (α) of the abutting components

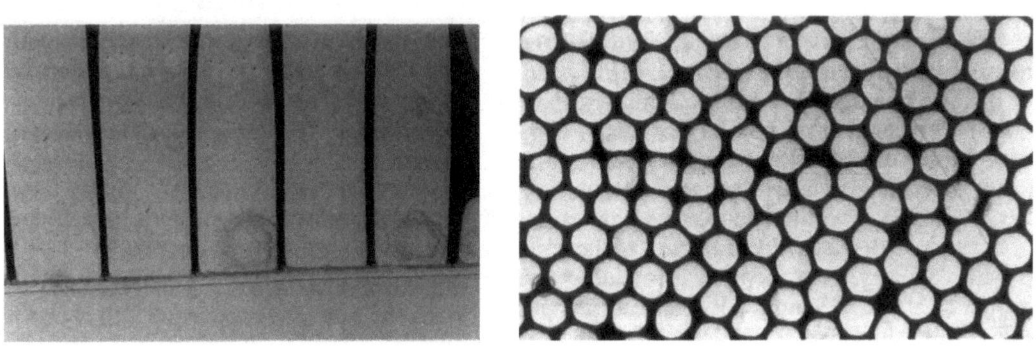

(a) (b)

Fig. 4.17 (a) Longitudinal and (b) transverse sections through a compliant structure that is capable of accommodating a thermal expansivity difference between joined components. (a) 99×. (b) 450×

by virtue of its thermal diffusivity being 20 times higher than that of the Macor ceramic. Therefore, the temperature differential between the metal and ceramic parts reaches a maximum and then declines as both parts approach room temperature. This process is shown in Fig. 4.18.

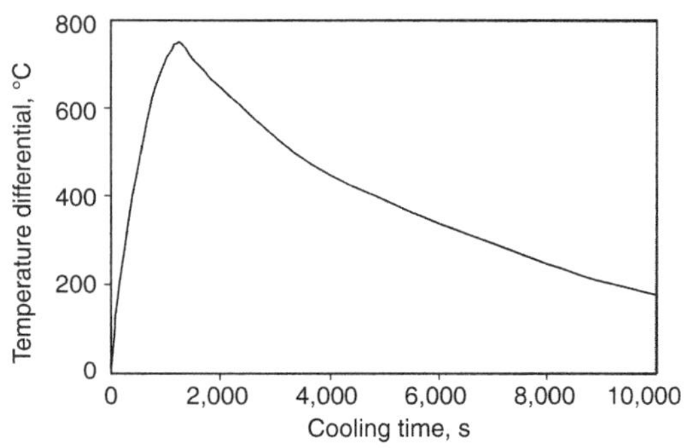

Fig. 4.18 Variation of temperature mismatch between ceramic and nickel-base components, 50 mm (2 in.) diameter on air cooling from the solidus temperature of the braze (950 °C, or 1740 °F) with time.

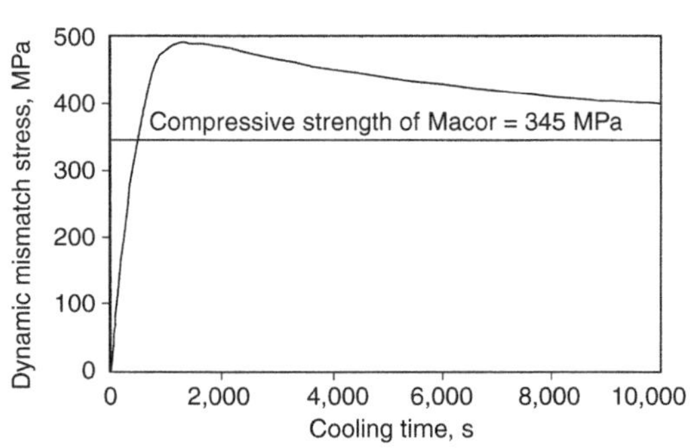

Fig. 4.19 Variation of mismatch stress between Macor ceramic and nickel-base components, 50 mm (2 in.) diameter on air cooling from the solidus temperature of the braze (950 °C, or 1740 °F), with time.

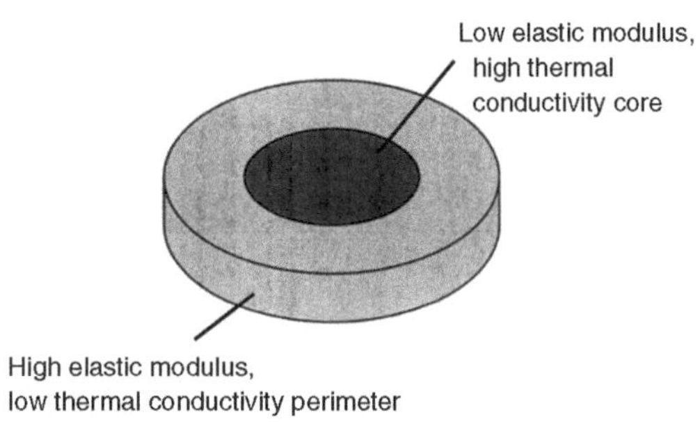

Fig. 4.20 Schematic illustration of a radially graduated interposer designed to minimize dynamic thermal expansion mismatch between brazed components of different thermal attributes

The temperature differential gives rise to strain across the brazed joint, and the manner in which this is accommodated, in the form of stress, depends on the elastic moduli of the two materials at the temperature that each has reached. This parameter can also be calculated as shown in Fig. 4.19, which shows that the peak stress in the assembly in question exceeds the compressive strength of the Macor ceramic component (345 MPa, or 50 ksi). Hence, joint failure is likely in this case [Li 1993].

The key to minimizing dynamic thermal expansion (assuming that the materials and braze cannot be changed for functional reasons) is to design joints that are compliant in the plane of the joint. Two solutions have been considered already. One of these is to use compliant interlayers, and some mechanically compliant structures will also fulfill this requirement (see sections 4.2.2 and 4.2.3 of this chapter). Radially graded intermediate materials can achieve the same result. For example, the interposer might comprise a brass washer with a copper core, as shown schematically in Fig. 4.20. Relevant thermal and physical properties for these materials are listed in Table 4.8. This combination possesses a soft, compliant, and high-conductivity member in the center of the joint, capable of absorbing stress and, being highly conductive, it also helps to reduce the temperature differential between the ceramic and metal components. At the periphery of the joint, the temperature differential is smaller so that close to the perimeter, a material with a lower thermal conductivity, such as brass, can be tolerated. The fact that brass is less compliant than copper is an advantage because this enhances the overall joint rigidity. Using this interlayer structure, specimen parts of a nickel alloy brazed to machineable Macor ceramic were able to withstand quenching from the braze solidus temperature into iced water [Li 1993].

Table 4.8 Selected mechanical and thermomechanical properties of annealed copper and a brass that can be combined to form radially graded shims in joints

Property	Annealed copper	Annealed Cu-30Zn brass
Coefficicient of thermal expansion, 10^{-6}/K	16.5	19.9
Thermal conductivity, W/m · K	400	120
Young's modulus, GPa	85	100

4.2.5 *The Role of Fillets*

Wherever possible, it is good practice to design joints so as to encourage the formation of rounded fillets. Then, even if the joint contains voids, for whatever reason, the fillet will serve to seal the joint. This is because there is a higher probability that where the formation of fillets is promoted, these tend to be continuous and void-free. Fillets also have a beneficial effect on the mechanical properties of joints. Well-formed fillets of filler metal can enhance the tensile, shear, and peel strengths by as much as an order of magnitude, the value depending on the geometry of the joint and the mode of stressing. This is illustrated by Charpy impact test results shown in Fig. 4.21. The improvement in mechanical properties may be attributed to the gradual transition in edge profile that the fillet provides, which distributes stress concentration evenly over the entire joint periphery. By contrast, sharp joint corners tend to focus stresses and thereby encourage joint failure through crack and peel initiation. The stress concentration can be calculated and is presented as a function of contact angle in Fig. 4.22. [Eley 1961]. Provided the braze wets to form a fillet with a contact angle below 30°, there is no appreciable stress concentration at the gradual change in edge profile. The subject of stress concentration is discussed in section 4.3.3 of this chapter.

Because the formation of fillets of reproducible geometry and constant profile around the entire periphery is often difficult to achieve, test pieces for mechanical testing are often designed to exclude fillets, either by preventing their formation through the use of nonwettable surfaces outside the edge of the joint (e.g., through the application of stop-offs) or by removing any fillets that happen to form. Although this practice makes measurements more readily reproducible, it modifies joint strengths to an extent whereby they may not be representative of most practical situations, where fillets normally form. At least, the mechanical properties measured on assemblies where the brazed joints have sharp corners are likely to be inferior to those achieved normally.

Examination of edge fillets can provide an indication of the filler/substrate contact angle at the onset of solidification. Low-contact angles are usually held to be a mark of the overall quality of the joint. While such a judgment is generally justified, the reader is cautioned that visual inspection of edge fillets can be misleading regarding the quality of the interior of the joint. Information on the overall soundness of the joint can only be obtained by x-radiography, scanning acoustic microscopy, transient themography, or destructive testing to ascertain its mechanical integrity.

4.3 Constraints Imposed by the Components and Solutions

A large area brazed joint may be defined as one where the total joint area exceeds about 20 mm^2 (0.03 in.^{2}) and the length in any direction is greater than about 5 mm (0.2 in.). This definition is based on the following practical criteria:

- The significant distortion of assemblies that have joints between materials with CTEs differing by as little as 2×10^{-6}/°C (1×10^{-6}/°F). Distortion is related to the joint dimensions and solidus temperature of the filler alloy, and this problem is therefore most acute for large area joints made with high-melting-point brazes.

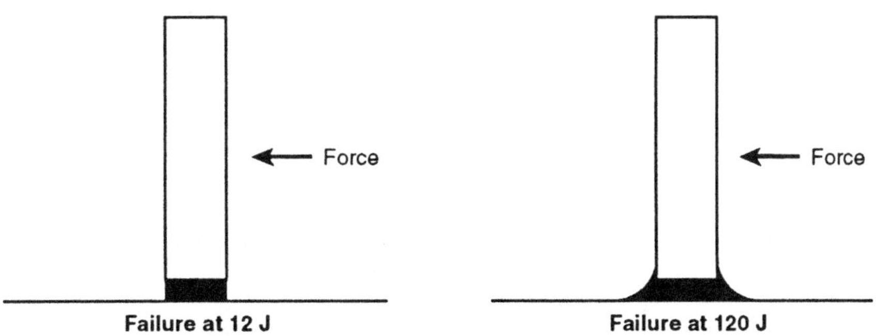

Fig. 4.21 Impact test on brazed T-joints, clearly demonstrating the role of fillets in enhancing joint strength. Substrate: mild steel. Braze: Ag-Cu-Cd-Zn

- The incorporation of significant void levels (above 10%) in joints. This problem tends to be more pronounced with brazing processes conducted using foil preforms at lower temperatures and without a significant compressive load applied to the joint. The incidence of voids declines with increasing temperature and may be significantly reduced by applying a compressive load to the joint through the brazing operation (see Chapter 1, section 1.3.2.1). It should be added that there are better methods of introducing brazing alloys into a joint gap than using foil preforms, as mentioned in Chapter 3, section 3.4.2.2 and discussed further below.

Distortion of assemblies can arise from a number of causes, some of which are discussed in Chapter 1, section 1.3.2. Apart from CTE mismatch, the most common sources of warping or bowing are uneven heating, which leads to temperature gradients in the components, and residual stress from earlier stages of fabrication that is relieved in the heating cycle. Distortion from these causes can be avoided by performing the joining operations under carefully controlled conditions. Notably, heating should be carried out only in furnaces that provide highly uniform temperature zones, with the rate of heating tailored to the thermal mass of the components and after appropriate stress-relief routines have been performed. Joints that are largely free of voids can be produced by careful attention to the cleanliness of the components and the directionality of flow by the molten braze.

4.3.1 Joint Area

The strength per unit area of a joint between components of the same material tends to reduce in proportion to the area above some lower threshold, often around 50 mm^2 (0.08 in.^{2}) for a brazed joint. The reason for this is that large area brazed joints generally contain more voids and other defects. Figure 4.23(a) shows a wedge-shaped test piece brazed to a flat plate using a foil preform and Fig. 4.23(b) shows an x-radiograph of the same component in plan view. Although continuous and well-rounded fillets are formed along the entire periphery of the joint, implying good wetting by the braze, the radiograph reveals that there is a propensity for void formation that increases with the size of the joint.

The voids have two causes:

- Gas trapped or generated within the joint
- Solidification shrinkage of the molten filler metal

Each of these causes is considered in further detail in the following sections.

4.3.1.1 Trapped Gas

By far the largest source of voids in brazed joints made using foil preforms is trapped gas,

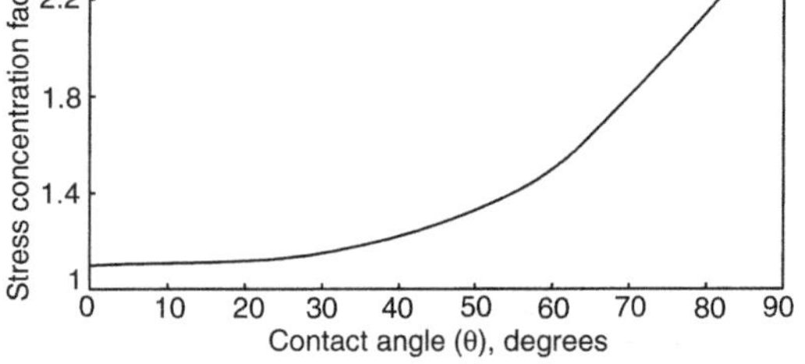

Fig. 4.22 Role of fillets in reducing stress concentration at the changes in section between abutting components

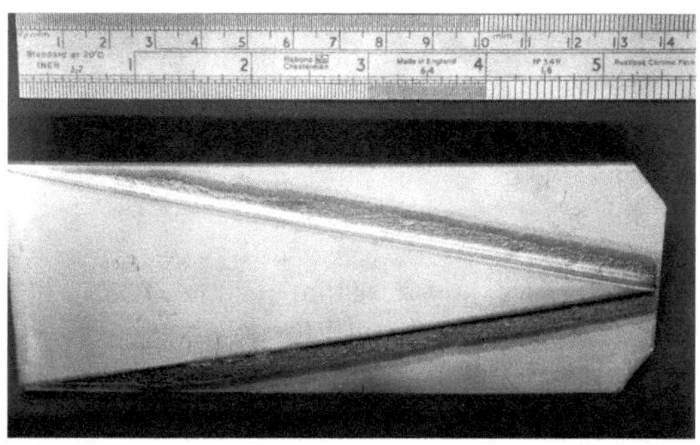

(a)

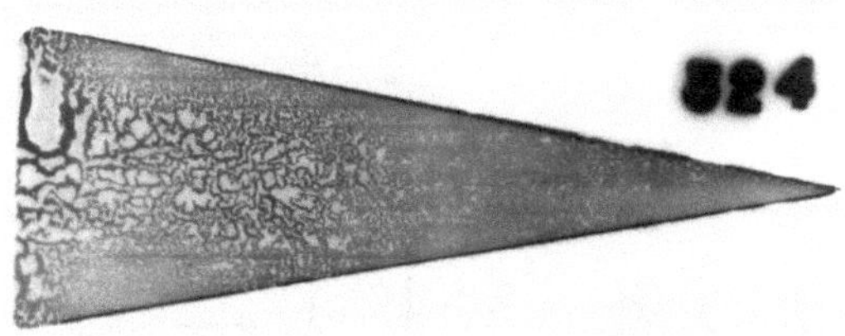

(b)

Fig. 4.23 (a) Test piece used to determine the propensity for void formation as a function of joint dimensions. In this case, the components are an aluminum engineering alloy (AlMn1), joined by fluxless, vacuum brazing, using the Al-12Si braze, which is admitted to the joint gap in the form of a foil preform of matching dimensions to the upper, wedge-shaped component. (b) Radiograph of the brazed assembly in plan. There is evidence of voids as the joint dimensions increase.

even for joints made in high vacuum where the components and filler alloys have been given a vacuum bakeout prior to the joining cycle. Void levels in brazed joints larger than about 100 mm^2 (0.16 $in.^2$) in area and that have a width greater than 5 mm (0.2 in.) can reach 30% of the joint volume, which is consistent with the entrapment of air or an evolved gas such as water vapor. Figure 4.24 shows a scanning acoustic microscope image of a 50 mm (2 in.) diameter assembly formed by vacuum brazing. Three voids are evident in the joint gap, including a large edge-opening void at the right.

Air becomes trapped when the components are assembled, the mating surfaces forming an effective seal. As the temperature of the assembly is raised to the peak process temperature, the trapped air is augmented by gas (usually moisture) evolved from the joint surfaces. The total gas volume will increase as the temperature is raised and the ambient pressure is reduced, in accordance with the gas law (pressure × volume = constant × absolute temperature).

Evolved gas can originate from several sources, in particular:

- Organic residues and adsorbed water vapor on the surfaces to be joined. These species volatilize as the temperature of the components is raised. Residues can be minimized by carefully precleaning the surfaces. A bakeout in vacuum immediately prior to joining is usually effective in removing water vapor and organic residues, provided the temperature used exceeds about 150 °C (300 °F). Reactive ion-etching, using a hydrogen or halogen plasma, and oxygen plasma-ashing are cleaning methods that have recently gained in popularity for cleaning components, owing to their effectiveness at dealing with organic contamination and also due to the wider availability of off-the-shelf equipment. These processes usually involve a combination of elevated temperature and reduced pressure, coupled with a chemically active ingredient that helps remove organic species.
- Volatile materials within the bulk of the components. Problems with volatile materials are most pronounced when the components are porous or include polymeric materials and when the components (including any metallizations present), the braze, or the flux contain constituents that volatilize during the heating cycle. Elements such as zinc; magnesium; and cadmium; and, to a lesser extent, manganese; and also fluxes and pastes, which generally contain volatile constituents, can be the cause of voids in joints.
- Water introduced as an essential part of the joining process. Water is a common carrier for brazing fluxes, used in the form of pastes. It is particularly detrimental in this regard because water vapor expands rapidly with temperature above its boiling point. Thus, at 500 °C (930 °C), water will expand to 1,000 times its volume at 20 °C (68 °F).

If the path length between a gas bubble and the joint periphery is small, the gas pressure can normally exceed the hydrostatic force exerted by the molten filler metal, allowing the gas to escape to the surrounding atmosphere. However, the limit to the path length for this to occur is of the order of 5 mm (0.2 in.) for joints brazed below 700 °C (1290 °F) but longer at higher joining temperatures. For some large components it may be possible to incorporate vents through the components to provide a passage for trapped gas to escape from within the joint. This approach is commonly employed in brazing complex assemblies to alleviate blind spaces.

When undertaking brazing operations, a common mistake is to use a foil preform of similar dimensions to the plan area of the joint. This approach typically results in a high level of voids owing to the large surface area exposed, i.e., an unfavorable volume-to-surface area ratio, as can be seen from Fig. 4.23.

An effective method of removing trapped air is to design the joint in such a manner that the molten metal is made to flow from the center of the joint out toward the periphery or through the joint from one edge. Both of these tend to occur naturally when the braze is introduced into the joint in the form of a rod or a wire preform. Suitable arrangements for achieving this type of flow are illustrated schematically in Fig. 4.25. The advancing front of molten metal is then able to displace the gas and air ahead of it as it flows into the joint gap. However, neither approach is entirely satisfactory. A preform of increased thickness and reduced area can make jigging of the components difficult. Moreover, the expedient of introducing the braze from one side of the joint is effective only with filler alloys that, when molten, do not react strongly with the substrate materials to stifle flow of the braze in its path through the joint (see section 4.1 of this chapter). Nevertheless, admitting the braze in the form of round wire preform overcomes the two fundamental deficiencies associated with using foil preforms:

- The surface area-to-volume ratio is reduced considerably. Correspondingly, the detrimental effects of oxide on the surface of the braze are diminished greatly. Furthermore, the braze preform can be easily cleaned and stripped of thick oxide by wiping the wire several times with an abrasive cloth containing an organic solvent. If this operation is conducted immediately prior to the heating cycle, then voids stemming from the oxides on the surfaces of braze preforms can be effectively eliminated.
- A round wire has only one small area of contact with each of the faying surfaces. This precludes gas pockets being trapped during jigging, and the advancing braze front is always in the optimal location to sweep out air from the joint gap. As a practical side benefit, for most brazing alloys, wire is readily available at a relatively low price premium over the metal content, compared with foil performs, and a few stock sizes can be used for a wide variety of joining applications.

An alternative approach to reducing the volume of trapped gas is to lessen the number of surfaces in the joint. Brazes can be applied as roll claddings, or as vapor-deposited or electroplated coatings to the components, thereby eliminating two free surfaces from the joint. But, considerable care must be taken to ensure that the coated layers do not themselves contain significant volumes of gases or other volatile constituents. Electroplatings often contain small quantities of organic materials that are deliberate additions of the bath chemistry to function as "brighteners" (surface leveling agents) and grain refiners. Furthermore, the aggregate composition of the coating must be checked for conformance to that of the requisite brazing alloy.

Substrates with the braze preapplied, either in the form of roll claddings or in defined areas using one of the more advanced methods outlined in the preceding paragraph, have the advantages of:

- Reducing piece-part inventory and number of suppliers

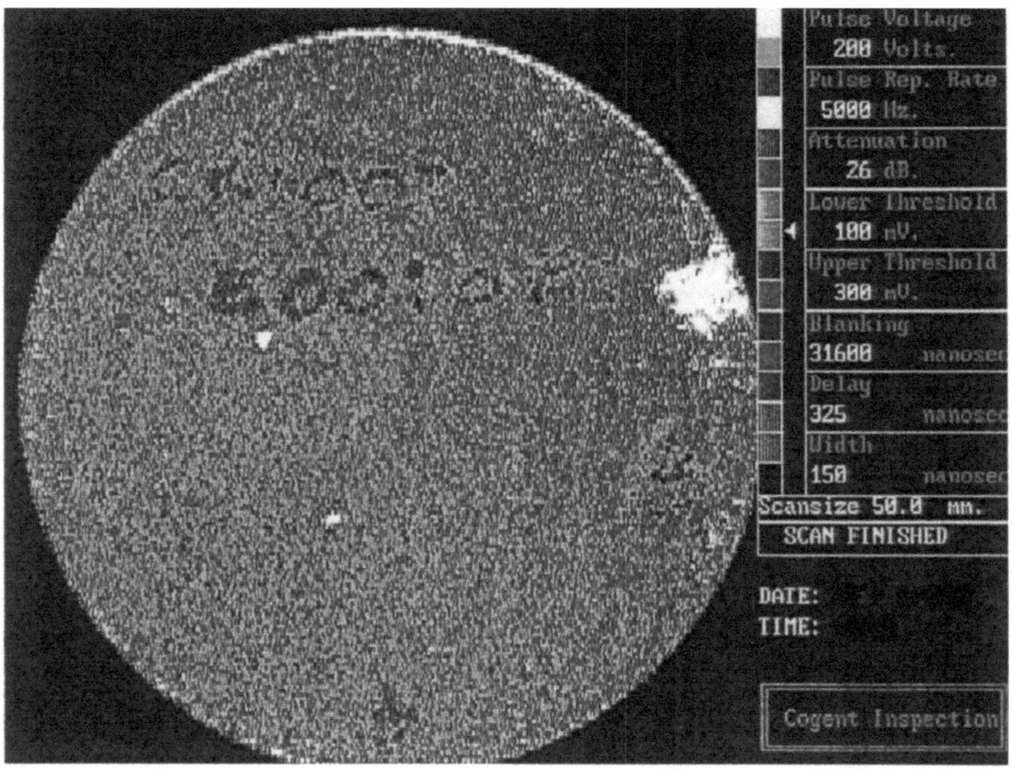

Fig. 4.24 Scanning acoustic microscope image of a 50 mm (2 in.) diameter brazed assembly showing a large edge-opening void at the right

- Simplifying the jigging
- Reducing the thickness of the braze joint because the braze layer can be substantially thinner than the minimum practicable thickness of circa 25 μm (1 mil) required for a foil braze preform
- Improved brazing behavior by eliminating two joint surfaces with all the attendant problems from the joint gap
- Automatically confining the spread of the braze to a precise area
- Helping to make the brazing operation reproducible

A common method of introducing a braze into joints is in the form of a fluxed paste, comprising the brazing alloy in the form of a powder of a controlled shape and particle size distribution and a suitable flux in the form of a powder, compounded in an organic medium. The aggregate paste must have satisfactory tackiness to ensure good adhesion to the faying surfaces of the components and should burn cleanly during the heating cycle so that carbonaceous residues, which might degrade wetting by the molten braze and its infiltration into the joints, are not left behind. Often, the metal powder is spheroidal in shape to provide maximum volume-to-surface area ratio (i.e., metal-to-oxide) and ensure good syringing properties in a commercial paste dispenser. The heating cycle should be designed to allow sufficient time for water vapor and organic carriers in the flux paste to vent completely from the joint gap before the braze melts.

Brazing materials suppliers offer a wide range of ready-made braze pastes correctly formulated to suit different parent materials, processes, and methods of application. The user must, however, take considerable care to store these pastes in cool and dry environments, as recommended by suppliers for specific products. Many commercially available brazing fluxes, especially those designed for use below about 800 °C (1470 °F) are to some extent hydroscopic and will hydrate over time if kept in a warm, humid environment. When moisture is present in the pastes, the flux in the mixture will progressively corrode the braze powder. In consequence, its brazing characteristics will become degraded, which will be manifested in poor wetting and joint infiltration, and also a significant increase in the incidence of voids. Often, but not always, discoloration of the paste will provide a tell-tale sign that the braze paste has expired.

4.3.1.2 Solidification Shrinkage

In any brazed joint, a fraction of the residual voiding does not derive from trapped air, moisture, or gas. These residual voids are extremely difficult, if not impossible, to remove because they are intrinsic to the filler metal, being caused by the shrinkage when it solidifies and further cools. Table 4.9 lists values for the thermal contraction of elements common to many filler metals. The reservoir of filler metal represented by the edge spillage fraction or fillet is seldom able to feed large-area joints and compensate, even in part, for the contraction because the outer extremities of joints usually solidify first through radiative heat losses to the surroundings.

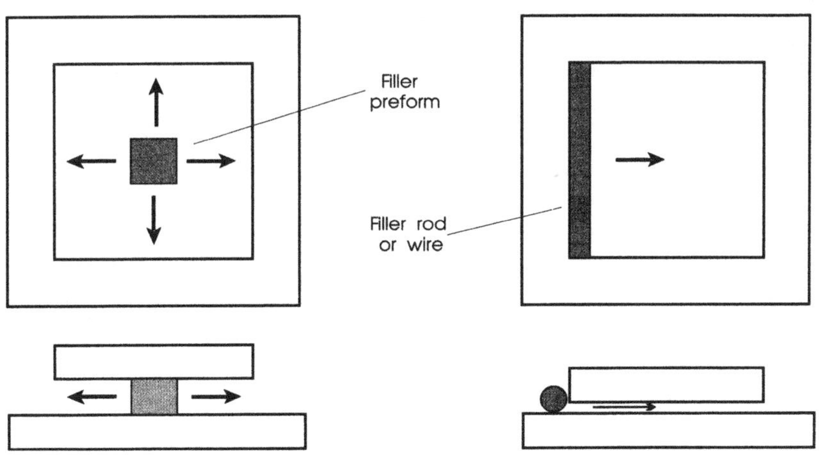

Fig. 4.25 Two configurations showing flow by a molten filler designed to sweep trapped gas out of a joint

The magnitude of solidification shrinkage, as given in Table 4.9, accounts for the fact that it is difficult to make joints of large area that contain less than about 3 to 5% voids by volume. Shrinkage voids tend not to occur in small or narrow joints (<5 mm, or 0.2 in., in one of the joint area dimensions). This is because the thermal gradients that usually develop along a joint when the assembly is cooled from the joining temperature are large in relation to the joint dimensions in these cases, and this causes the filler to directionally solidify from one edge to the other. Thereby, voids are prevented from forming within the joint. It is possible to achieve the same effect in large area and wide joints by imposing a temperature gradient on the assembly, from either center-to-edge or edge-to-edge such that some periphery of the joint is always the last portion to solidify. However, this is not always easy to achieve, especially when large numbers of components are involved. Furthermore, the imposition of a temperature gradient on a large assembly may produce stress gradients, leading to dimensional distortion of the components, which becomes fixed when the filler solidifies.

Solidification shrinkage voids are not usually a source of mechanical weakness. This is because these voids are usually fine and well distributed. Indeed, their presence is often difficult to discern in metallographically prepared samples and in examination by nondestructive techniques. Solidification shrinkage voids, however, can be normally differentiated from gas voids. The latter are generally spherical with smooth interior surfaces, while shrinkage voids are mostly irregular and often concentrate at boundaries between phases in the braze microstructure.

Where the parent materials and braze are of similar composition, maintaining the assembly at elevated temperature but below the solidus temperature of the filler can result in a gradual reduction in void levels arising from solidification shrinkage through vacancy diffusion. This is the mechanism by which dry joint interfaces are removed in diffusion bonding (see Chapter 1, section 1.1.7.3). The process times then become relatively long and not less than one hour, although these can be contracted by applying a compressive stress to the joint during the heat treatment. The magnitude of the stress that is effective in removing voids depends on the particular joining process and materials associated with the joint.

Void levels can actually increase during isothermal processing for certain combinations of materials if there is nonsymmetrical diffusion of different elements across the mutual interface (the Kirkendall effect). The application of a compressive force also helps to suppress voids arising from this cause. Although interdiffusion between tin and copper gives rise to high levels of Kirkendall porosity at temperatures above 400 °C (750 °F), a process has been developed for joining copper components using a thin layer of tin at 500 to 800 °C (930–1470 °F), which can see subsequent service at temperatures up to 900

Table 4.9 Volume contraction at the freezing point on solidification of selected major and minor constituents of brazes. Of these elements, only silicon expands on freezing

Element	Volume contraction, % of solid	Solid expansivity (CTE)(a), linear 10^{-6}/K	Liquid expansivity (CTE)(b), cubic 10^{-6}/K
Silver, Ag	5.0	19	97
Aluminum, Al	6.5	23	117
Gold, Au	5.2	14	86
Cadmium, Cd	4.0	31	144
Cerium, Ce	1.0	8	34
Cobalt, Co	3.5	12	127
Copper, Cu	4.8	17	100
Iron, Fe	3.5	12	126
Indium, In	2.5	25	96
Magnesium, Mg	4.2	26	166
Manganese, Mn	1.7	23	122
Nickel, Ni	4.5	13	147
Phosphorus, P	3.5	...	...
Silicon, Si	−10	7	127
Tin, Sn	2.6	23	87
Palladium, Pd	...	11	120
Platinum, Pt	...	9	152
Titanium, Ti	6.9	9	170
Zinc, Zn	6.9	31	167

(a) CTE, coefficient of thermal expansion; 273–373K. (b) Close to the melting point

°C (1650 °F). Provided a pressure of at least 5 MPa (700 psi) is applied throughout the brazing cycle, voiding in the joint can be eliminated. Although the filler metal is tin, which is conventionally classed as a solder because it melts at 232 °C (450 °F), the elevated process temperatures that are necessary for the removal of voids and brittle interfacial phases to be effective, and also the high joint remelt temperature, qualify this as a brazing process. This and other diffusion brazing processes are described in Chapter 6. Diffusion soldering processes are described in Chapter 5 of the companion volume *Principles of Soldering*.

4.3.2 Tests for Braze Wetting and Joint Filling

The ability of a braze to wet, spread, and also fill joints can be assessed by suitably designed tests [Eustathopoulos, Nicholas and Drevet 1999, 106–107].

The most commonly used means of assessing wetting by a braze is the sessile drop test. This test involves measuring the contact angle of a pellet of braze after being melted on a flat coupon of the substrate of interest, which is held at a fixed temperature for a specified length of time. A flux is generally added if the test is performed in air. A dynamic assessment of wetting may be made by measuring the contact angle as a function of time, temperature, and atmosphere quality (oxygen partial pressure) [Levi and Kaplan 2002].

A test that is widely used to quantitatively measure wetting of solders as a function of time, the wetting balance "solderability" test, has been adapted for higher temperatures to measure wettability of brazes. A wetting balance comprises:

- A load cell and signal processing system that furnishes a measurement of load versus time and provides automatic tareing of specimen weight
- A temperature-controlled bath for the brazing alloy
- A bath lift or specimen fall mechanism with speed and positional control
- A computer to display the force/time curve and derive key metrics from the data

A wetting balance specially designed for wetting measurements above 1000 °C (1830 °F) under controlled atmospheres is shown in Fig. 4.26 [Solomon, Delair, and Thyssen 2003]. The sample chamber is hermetically enclosed to enable measurements to be made in vacuum or in a controlled gas atmosphere. The specimen under test is held in a holder that is suspended from the load cell. The ceramic crucible containing the molten filler metal is supported on a ceramic rod connected to a steel bellows and a linear motor, enabling the crucible to be raised and lowered at a preselected speed. The raising of the crucible immerses the test piece to a prescribed depth in the molten braze (on other equipment the test piece is lowered into the braze reservoir). The bath is held in this position for a set dwell time and is then returned to its rest position.

The resolved vertical forces acting on the specimen are recorded as a function of time over the whole test cycle. Fig. 4.27 shows the typical form of the trace that is recorded, together with the corresponding position of the specimen relative to the bath of braze at each stage.

The wetting balance provides a measurement of the vertical component of the force exerted

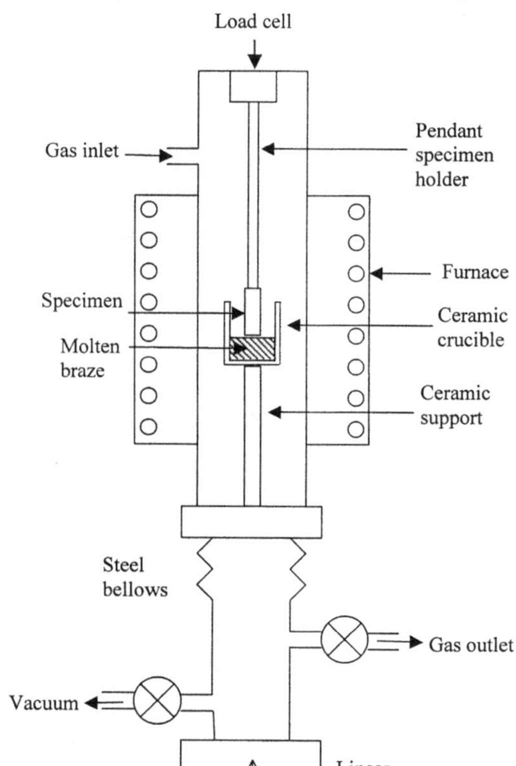

Fig. 4.26 Schematic representation of a wetting balance tailored for use above 1000 °C (1830 °F). Adapted from Solomon, Delair, and Thyssen [2003]

on the test piece as it is lowered into a reservoir of the molten braze, as a function of time. This force is theoretically equal to the sum of the vertical component of the surface tension force, F_γ, between the filler and the test piece and the buoyancy of the test piece, F_B. Figure 4.28 shows an equilibrium situation appropriate to partial wetting. The resolved force in the vertical direction, F_R, is the parameter measured in the test. The variation of this force as a function of time provides information on the dynamics of the wetting process. The time dependence of wetting is determined by a number of factors, including the rate that surface oxides are removed or penetrated, the kinetics of diffusion of interdiffusion of constituents of the braze and those of the parent material to create interfacial intermetallics.

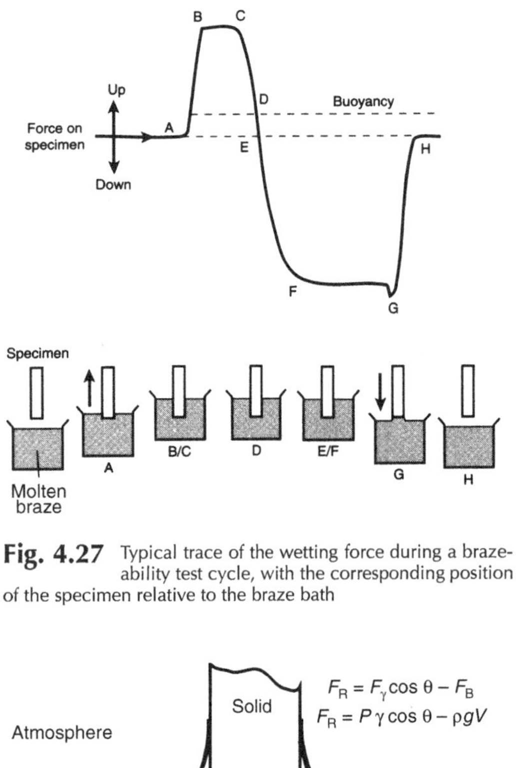

Fig. 4.27 Typical trace of the wetting force during a brazeability test cycle, with the corresponding position of the specimen relative to the braze bath

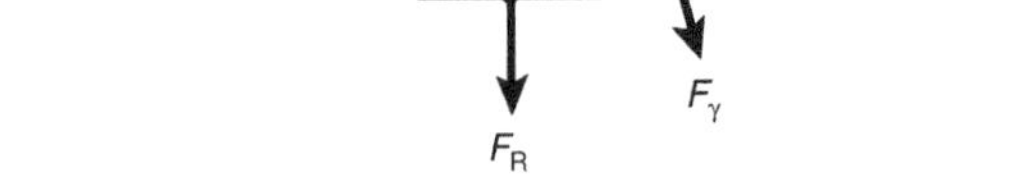

Fig. 4.28 Forces diagram for a solid plate partially immersed in a liquid. *P*, specimen periphery length; γ, liquid surface tension; θ, contact angle, ρ, liquid density; gravitational constant, *g*, 9.81 m²/s; *V*, immersed solder volume

A wetting balance test can be performed rapidly and the results are quantitative, inasmuch as reproducible numerical data can be obtained for a well-defined set of sample and instrumental parameters and operating procedures, as explained in Barranger [1989] and Lea [1991]. Furthermore, the change in wetting as a function of time can be monitored. The surface tension of the molten filler can be calculated from data obtained on the wetting balance using nonwetted ($\theta = 180°$) ceramic coupons, using the equation:

$$F_R = P\gamma \cos\theta - \rho g V$$

From this value and the measured wetting force, the angle of contact between the molten filler and the test piece can be calculated.

A high-temperature wetting balance of the type shown in Fig. 4.26 has been used to quantitatively establish the effect of grit blasting with different abrasive materials on the wetting of an IN718 nickel-base alloy (Ni-19Cr-18.5Fe-3Mo-0.9Ti-0.5Al) by the complementary AMS 4777 (=AWS BNi-2) brazing alloy nickel-base filler metal of composition Ni-7Cr-4.5Si-3.1B-3Fe.

For assessing spreading and propensity for joint filling, simple tests using suitably designed test pieces have been devised. Two variant test configurations that have been used are represented in Fig. 4.29 and 4.30. In the first configuration (Fig. 4.29), a small tube is fixed to the inside of a larger diameter tube by spot welds at the top and bottom. The joint width then varies from an interference fit to a wide gap between

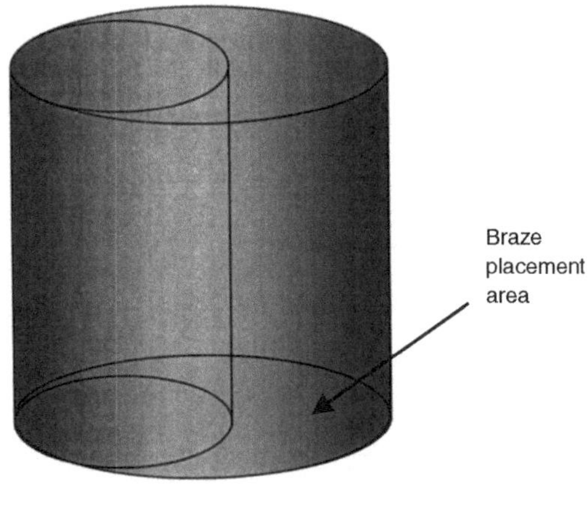

Fig. 4.29 Test piece comprising concentric tubes used to assess the ability of a brazing alloy to spread and fill a vertical joint gap

the two tubes. The tubes are placed on a nonwettable plate, and the lowest portion of the area between the inner and outer tubes is packed with braze (and, optionally, flux) as required. On melting, the braze will spread across the joint width and climb up the narrower portion of the joint gap by capillary action, in a manner depending on the materials combination and process parameters of the brazing cycle. Nondestructive and destructive assessment can then be made to establish the efficacy of braze spreading, joint filling, and their limits.

To eliminate the influence of gravity, the same test may also be conducted with the tubes laid horizontally. Then, an alternative test piece configuration may also be used. The brazing alloy (and flux) reservoir are contained in two parallel trenches machined in one component (Fig. 4.30). If these trenches are covered by an angled plate of similar area, the effect of joint width for a specified dimension of the joint size can be determined. The data provided by these tests are of great assistance in specifying joint tolerances that need to be achieved when jigging components prior to brazing. Machining mating surfaces to tight tolerances is expensive and should be avoided unless required for other reasons.

A general approach that has been found to be effective in producing well-filled and hermetic joints involves generating strong metallurgical reactions across the joint and the displacement of voids during the heating cycle while a compressive force is applied. The void-free joints obtained using the diffusion brazing processes are associated with such reactions. See Chapter 6 for further details on diffusion brazing processes.

4.3.3 Joints to Strong Materials

New materials with enhanced strengths are continually coming onto the market. Notable examples are composite materials, such as metal-matrix composites (MMCs) and precipitation-strengthened and dispersion-stabilized alloys. There is a desire to exploit these materials in a range of applications, but their widespread adoption is often contingent on being able to exploit their superior bulk strengths in joined assemblies. The strength of components fabricated by joining high-strength materials, even when welded, is generally inferior to that of the materials in monolithic form. Moreover, the heating cycle used in the joining process can itself degrade the properties of these materials. For example, aluminum/silicon-carbide MMCs are susceptible to degradation when heated above about 500 °C (930 °F) due to reaction between the constituents, which results in the formation of a brittle interfacial layer of Al_3C_4 [Iseki, Kameda, and Murayama 1984]. Usually this problem is more severe in welding, owing to the higher temperatures involved, and therefore efforts have focused on attempts to devise joining methods based on brazing and diffusion bonding, while keeping the process temperature as low as possible.

It is possible to produce brazed joints to enhanced-strength materials that are metallurgically sound, by which is meant that the joints have a low incidence of voids, are free from embrittling intermetallic phases, and achieve smooth fillets with low-contact angles. Such joints can be sufficiently strong for fracture to occur preferentially through the parent materials. However, the fracture stresses in joined assemblies involving these enhanced-strength materials are usually lower than those involving conventional engineering alloys! This paradox can be explained by the combination of relatively high elastic modulus coupled with the low ductility and fracture toughness possessed by these strong materials. Induced stresses cannot find relief through plastic deformation, owing to the high elastic modulus and therefore remain localized, their concentration depending on the joint geometry and mode of stressing, as described subsequently. The assemblies are therefore vulnerable to brittle fracture at relatively low levels of applied stress.

In most circumstances, application of stress to a joint does not result in all regions of the joint sharing an equal proportion of the load. The unevenness of the stress distribution is called the stress concentration, *K*. Mathematically, this is a dimensionless number that simply describes the

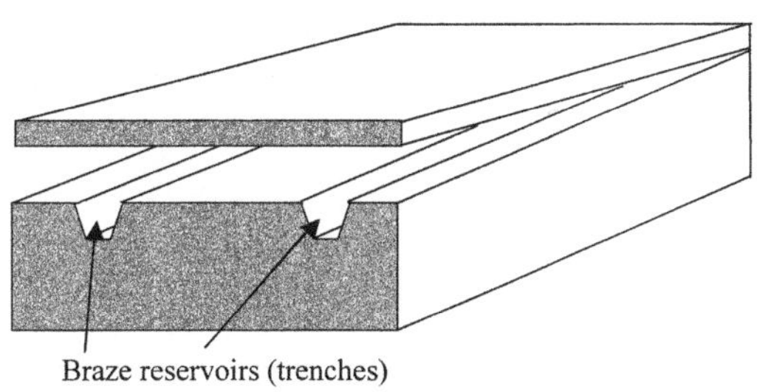

Fig. 4.30 Schematic illustration used to assess the ability of a braze to form void-free joints as a function of the joint thickness at constant joint width

magnification factor of the actual stress at one location compared with the uniform stress that would prevail in the absence of any stress concentrations. Expressed as an equation:

$$K \text{ (At a specific location)} = \frac{\text{Local stress at that location}}{\text{Average stress}} = \frac{\text{Local stress}}{\text{Applied force/Joint area}}$$

Thus, the key to producing high-strength joints is to prevent the development of stress concentrations and, at the same time, the strength of the braze must be optimized. These aspects are considered in turn next.

4.3.3.1 Joint Design to Minimize Concentration of Stress

Analysis of the stress distribution in joints is relatively straightforward provided the joints have simple geometry. Some common examples are considered later. The ideal joint is one in which, under all practical loading conditions, the filler metal is stressed in the orientation in which it best resists failure. The complexity of the joint design should also take into account the load intensity to be sustained and also any aesthetic considerations. In general, simple and hidden joints are suitable for low-level loads, while higher and more complex loading situations require more elaborate joints that are generally visible. In this context, a hidden joint is one contained wholly within the dimensions of the components, such as a scarf joint, while a visible joint requires additional members that protrude beyond the components, as in the case of a strap joint. Three common joint designs are analyzed in the following paragraphs.

Longitudinally Loaded Lap Joints and the Effect of Fillets. In order to understand the origin and magnitude of stress concentrations that can arise, reference will be made to a single lap joint loaded in tension. Stress concentrations arise from two sources, namely, differential straining of the components and the braze, and the eccentricity of the loading path. In lap joints, the strength of the joint per unit area (or length of overlap) actually decreases with increase in the joint length, as shown schematically in Fig. 4.31. This apparent anomaly can be explained by the fact that the shear stress is highest toward the ends of the joint. This buildup of stress concentration at either end of a lap joint is primarily a consequence of the difference in elastic moduli of the components and the braze. Solidified brazes will usually have a lower modulus than the components so that when stressed in shear, the joint will deform slightly more, per unit length, than the components. This means that at any given section of the joint not all of the strain is transferred from one component to the other. Therefore, if the length of the joint is increased beyond a certain limiting value, the filler in the central portion of the joint will carry little or no stress, with the applied stress concentrating at both ends, as depicted in Fig. 4.31. Hence, simply increasing the length of the overlap does not improve the strength of this type of joint beyond a certain level. Lap joints that are longer than the optimum, waste materials and result in unnecessarily bulky assemblies. A good rule of thumb for lap joints is that the overlap length should be three times the thickness of the thinner of the two piece parts being joined and that the

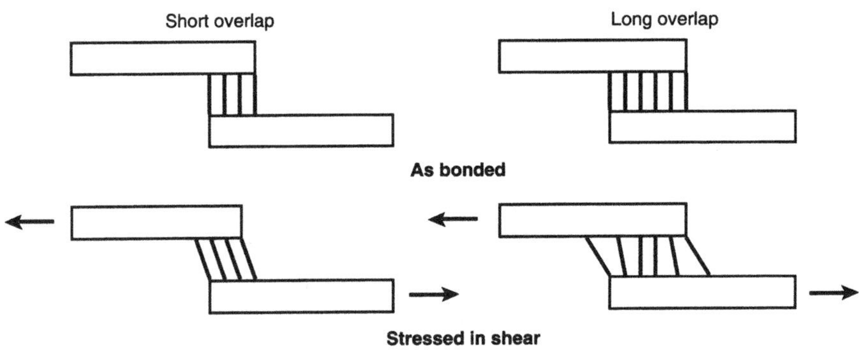

Fig. 4.31 Schematic illustration of the stress distribution in the filler metal of lap joints of short and long overlap. When stressed in shear, the central portion of a long lap joint carries little or no load.

joint gap at the brazing temperature is less than 75 μm (0.03 in.) [*Brazing Handbook* 1991].

The effect of materials with different elastic moduli on a simple (i.e., single) lap joint has been analyzed for the simple case of two rigid plates of equal thickness, adhesively joined with an overlap of $2c$ [Lancaster 1965]. The thickness of the adhesive layer is d. In this context, an adhesive may be considered an extreme case of a soft braze with a low elastic modulus. Here, the stress concentration factor at the ends of the lap joint was found to be approximately:

$$K = \left(\frac{2c^2G_a}{Etd}\right)^{1/2} \qquad \text{(Eq 4.2)}$$

where E is the Young's or elastic modulus of the plates, t is the thickness of the plates, and G_a is the shear modulus of the adhesive.

From Eq 4.2, it can be seen that the stress concentration factor, K, increases linearly with the overlap (although, in reality, only up to a certain limiting value, as noted previously). This relationship is to be expected in a situation where the stress concentration arises essentially from the difference in elastic modulus between the components and the filler. According to this equation, the stress concentration factor decreases with increasing thickness of the components, with a square root dependence.

The strength of single lap joints is even more sensitive to the tensile or "peeling" stresses that act in a normal direction to the ends of the joint and originate from the eccentricity of the loading of the assembly. The elastic analysis is relatively complex, but the result obtained is that longitudinal loading of a single lap joint effectively applies a perpendicular tensile stress of approximately four times the applied load, to the ends of the overlaps [Harris and Adams 1984]. These perpendicular tensile forces initiate failure of the joint by peel. With the continued application of stress, the joined assembly rotates in an attempt to correct for the axial misalignment and the fracture continues to propagate due to peel-type debonding. The stress concentrations in a simple lap joint and their influence on its resulting failure mode are illustrated in Fig. 4.32. The influence of overlap length on the failure mode of simple lap joints has been verified experimentally and is indicated in Fig. 4.33.

Fillets at the edges of a joint act to reduce the stress concentration in that region, as indicated in Fig. 4.34. They accomplish this by distribut-

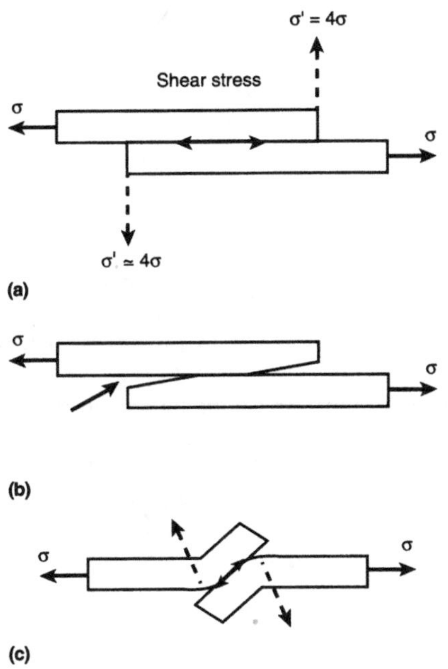

Fig. 4.32 Failure in a simple lap joint loaded in tension. (a) Stress concentrations. (b) Initiation of failure. Edge-opening crack (free arrow) formed and propagated by the high normal stress concentration. (c) Progression of joint rotation to fracture. Plastic bending of the joint region results in the majority of the failure being due to peel-type debonding. Adapted from Dunford and Partridge [1990]

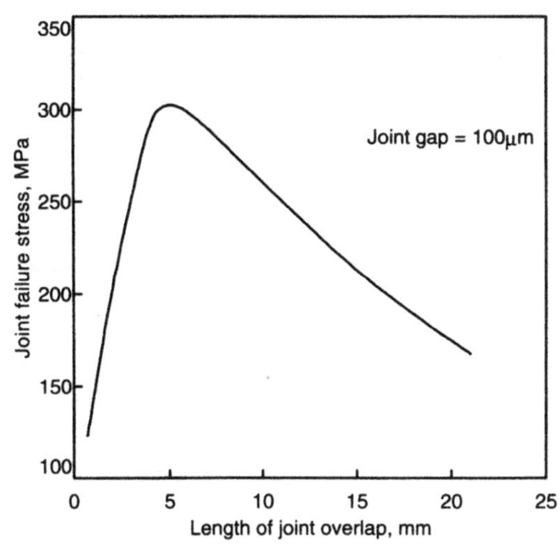

Fig. 4.33 Effect of overlap length on the failure stress in shear for a simple lap joint between mild steel components joined using a silver-base braze [Sloboda 1961]. For short overlaps, failure is by shear. As the overlap length increases, the forces in the joint change from shear to peel as the specimen rotates prior to failure, to accommodate the eccentricity of loading.

ing some of the applied stress around the ends of the laps, thereby reducing the differential straining between the components and the filler in the joint and also shifting the position of the maximum perpendicular tensile stress (originating from the eccentricity of loading) to outside the joint. The magnitude of these effects depends on the radius of the fillets, the step height, and the elastic properties of the filler. For the fillets to play an effective role in reducing the stress concentration, the radius of the fillets must exceed the step height, and hence it is desirable for brazed joints to have large and well-rounded fillets at their peripheries.

In a double lap joint, of the form shown in Fig. 4.35, some bending moment still exists, but the symmetry of the configuration results in the failure load of the joint being increased by a factor of three to four. This improvement has been verified by experiment [Kinloch 1982]. Lap joints are among the most common configurations used in brazing, with insertion lugs widely used for this purpose in tubular assemblies. Further improvements to the design of lap joints can be made by tapering the ends of the overlap, which is easily achieved if the components are thin, and ensuring a fillet is allowed to form in this region. Lap joint styles for different stress regimes are illustrated in Fig. 4.36(a).

Axially Loaded Butt Joints and the Effect of Voids and Cracks. An axially loaded butt joint in a cylindrical assembly is shown schematically in Fig. 4.37(a). The tensile stress σ_z produces an axial strain ε_z and lateral contractions $-\varepsilon_x$ and $-\varepsilon_y$, with accompanying shear stresses σ_x and σ_y. It is assumed that the material is randomly polycrystalline and therefore isotropic, in which case $\varepsilon_x = \varepsilon_y$ and $\sigma_x = \sigma_y$.

On the central axis of the assembly, the shear stresses σ_x and σ_y, which are physically manifested as necking when homogeneous materials are loaded in tension, act equally in the radial and circumferential directions. Figure 4.37(b) represents a solid element of the joint lying on the central axis of symmetry. Provided the joint is thin, then the filler is constrained by the higher Young's modulus of the components from deforming, and the shear and tensile strength of the joint will increase with reducing joint gap, up to a maximum [Sloboda 1961; Bredzs 1954]. This effect, which has been verified by experiment for silver-base brazes, as illustrated in Fig. 4.38, can be explained as follows. A very thin layer of braze in a joint to components that are considerably stronger than the filler metal itself will deform in a manner very different from that when it is in bulk form. When the stress applied to a joint with a narrow joint gap reaches the yield point of the filler, the latter cannot deform plastically, owing to the adjacent layers of the component materials that are still stressed within their elastic range. In consequence, the filler is subjected to a triaxial (hydrostatic) tension and does not fail until the applied stress has reached the brittle fracture strength of the filler. Thus, for example, silver, which does not alloy significantly with iron and whose ultimate tensile strength as measured in a standard tensile test is 150 MPa (25 ksi), will sustain a stress of 680 MPa (100 ksi) when in the form of a filler in a joint to high-strength steel. As the width of the joint is increased beyond the optimum value, the effect described diminishes and the strength declines toward that of the bulk filler. Recent advances in elucidating the mechanical behavior of constrained thin brazed joints are reviewed by Kassner et al. [1992].

Very narrow joints are prone to be compromised by incomplete filling if the joining operation relies mainly on the molten braze being drawn into the joint gap by capillary action, which is common industrial practice. Incomplete filling by the braze accounts for the decline in tensile strength seen in Fig. 4.38, when the joint thickness is reduced below ~50 μm (2 mils). Capillary action becomes progressively impeded by phenomena such as gas entrapment and constrained flow by the molten filler (due to its viscosity) as joints are made narrower. Some methods for producing very thin and well-filled brazed joints are described in section 4.3.4.1 of this chapter.

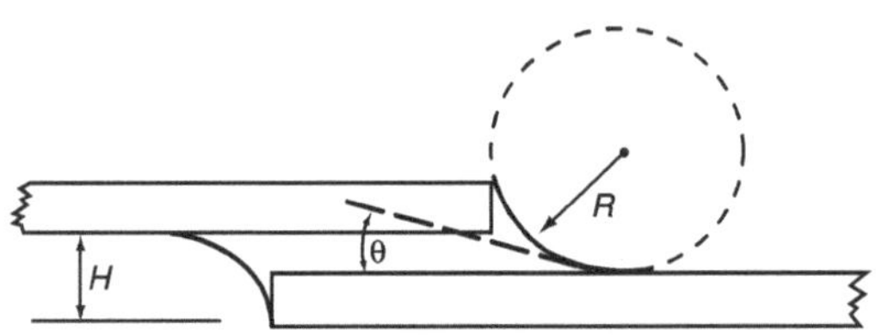

Fig. 4.34 A lap joint showing step height, H, fillet radius, R, and contact angle, θ

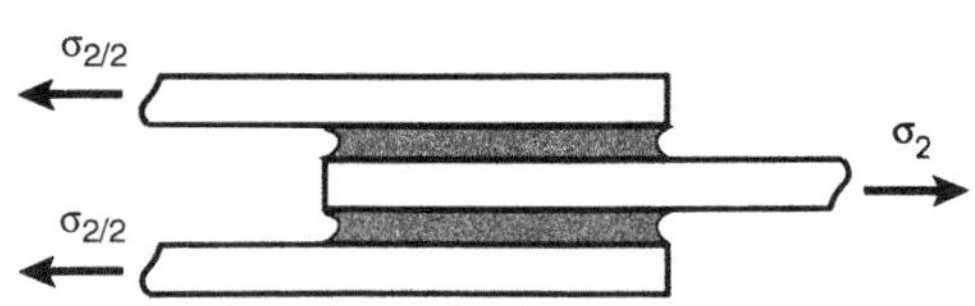

Fig. 4.35 Double lap joint

Although, as mentioned previously, the tensile strength of a brazed assembly with narrow joints may be high, an applied tensile stress as low as one-tenth of the failure load, sustained over a period of months or years can result in (delayed) ductile failure (Fig. 4.39) [Kassner et al. 1992]. Cavities nucleate and coalesce internally, driven by the potential energy provided by the unremitting triaxial stress experienced by the filler (this energy being available for vacancy diffusion), until rupture occurs.

It has so far been assumed that the strain ε_{Ez} parallel to the applied load σ_z is uniform throughout the assembly. In practice, however, this is not the case because the mechanical properties and, in particular, the Poisson's ratio, ν, of the filler alloy and the components differ. The Poisson's ratio is given by:

$$\nu = \frac{\varepsilon_x}{\varepsilon_z} = \frac{\varepsilon_y}{\varepsilon_z} \text{ for an isotropic material}$$

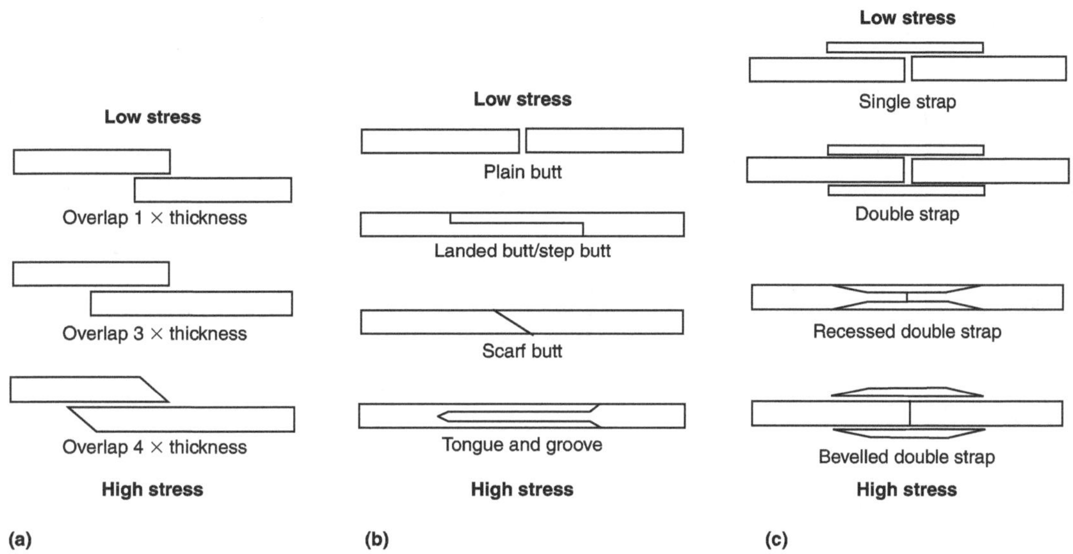

Fig. 4.36 Recommended designs of (a) lap, (b) butt, and (c) strap joints for different stress environments

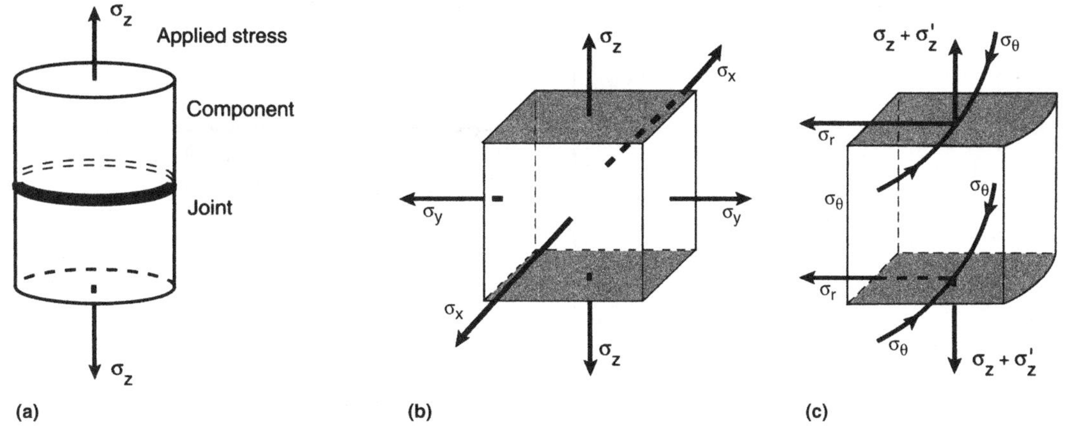

Fig. 4.37 Butt joint loaded axially in tension. (a) A stress concentration exists at the periphery of the joint due to a difference in the Poisson's ratios of the filler and of the components. (b) Stress distribution in the axial center of the joint. Deformation of joint surfaces is constrained by the components, and thus the filler is subject to triaxial tension. (c) Stress distribution at the periphery of the joint. The difference in Poisson's ratio between the components and the filler generates a shear stress at the joint interface and an additional normal tensile stress, σ_z'.

These differences in the mechanical properties mean that the tendency for lateral deformation in response to the normal strain, ε_z, will be unequal in the two materials and give rise to an additional stress at the filler/component interface in the vicinity of the joint periphery. In the cylindrical assembly under consideration, the magnitude of the resulting stress is dependent on the radius of the joint and its thickness and is least for thin joints in small-diameter components. The perpendicular component of the additional stress at the outer circumference of the joint, σ'_z, is aligned with the applied tensile stress, σ_z, and is depicted schematically in Fig. 4.37(c), where the curved surface of the element lies on the cylindrical face of the assembly.

Hence, the effective tensile stress acting at the periphery of the joint is higher than in the center by the ratio, K, where:

$$K = \frac{\sigma_z + \sigma'_z}{\sigma_z}$$

where the factor K is the stress concentration. In simple butt joints that are well filled, the stress concentration, K, is usually small and less than about 1.2.

Consider now a highly eccentric elliptical hole, that is, a crack in the joint gap (see Fig. 4.40).

The maximum stress at the end of a crack is given by the equation [Dieter 1976]:

$$\sigma'_z \approx 2\sigma_z\left(\frac{a}{b}\right)$$

and

$$K \approx 2\left(\frac{a}{b}\right) >> 1$$

where a is half the crack length, b is half the crack width, and $a >> b$.

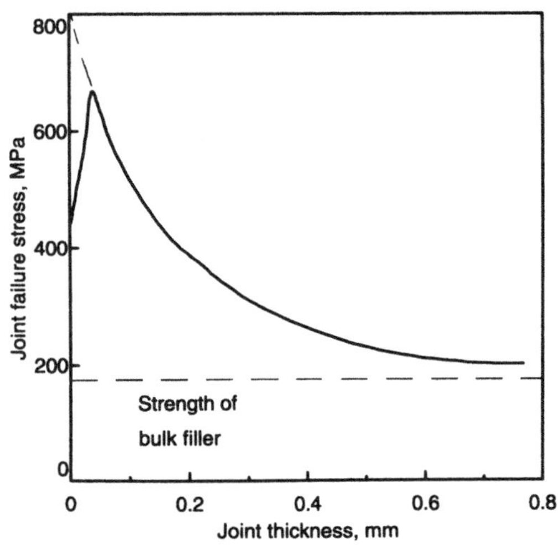

Fig. 4.38 Relationship between the fracture stress and joint thickness of butt joints in medium-carbon steel test pieces made with a silver-base braze. Very narrow joint gaps tend to be inadequately filled, thereby causing the measured joint strengths to decline. Adapted from Sloboda [1961]

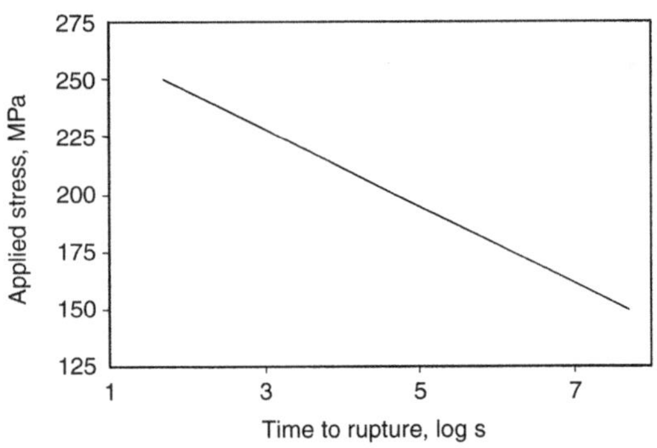

Fig. 4.39 Rupture time as a function of applied stress at room temperature for 0.15 mm (6 mils) thick joints made with silver-base filler metal between type 304 stainless steel components. Failure occurs through the thickness of the joint. The triaxial stress regime results in failure in the braze sooner than would be expected than for classical creep rupture of monolithic pieces of the filler metal.

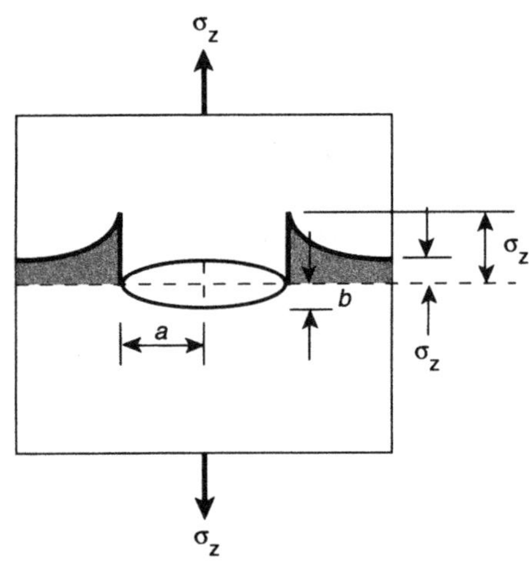

Fig. 4.40 Stress concentrations caused by an elliptical hole in a component. For a crack, $a >> b$, giving a high stress concentration at its ends. For a circular void, $a = b$, so that $\sigma'_z = 2\sigma_z$, and $K = 3$. Thus, voids are not as critical as cracks to the mechanical properties of joints.

Thus, at the end of a very narrow crack, the stress concentration can be extremely high so that a low applied stress can easily exceed the tensile stress of the filler in that region, enabling the crack to propagate and cause the joint to fail. Note that for a circular hole, $a = b$ and, hence, $K = 3$. Therefore, voids caused by trapped gas are not nearly as detrimental to the mechanical properties of the assembly as are cracks [Dieter 1976].

For a perfectly filled butt joint between two circular rods subject to tension, there is negligible stress concentration. The strength of such joints is therefore proportional to area. However, the plain butt joint is suitable only for the least demanding of applications. The main reason for this is that the joint has very low resistance to bending forces. The scarf butt joint has the merit of only requiring simple machining to prepare the faying surfaces, yet it is highly efficient at resisting deformation under load. Scarfing results in differential strains (because there is now a shear component of stress), but the stress concentrations at the ends of the joint are less than for lap joints (because the shear component of stress in a scarf joint is smaller than in a pure lap configuration). The landed or step joint relies on the step sizes being small to achieve the same effect. These two configurations are symmetrical and therefore axial stresses will be balanced over the area of the joint. They are illustrated in Fig. 4.36(b). From a theoretical perspective, a radially symmetric tongue and groove joint should be the best able to resist all loads, but achieving adequate filling of a joint having this type of geometry could present a major difficulty, and, in addition, the costs associated with the preparation of the components could be quite high.

Strap Joints and Other Configurations. Strap joints are often considered to be strong because, when loaded in tension, it is the parent material that fails, usually just outside of the joint region. However, the fracture stress is substantially lower than in a monolithic body, owing to the stresses being concentrated in the components close to the edges of the straps. The origin of this stress concentration is explained in Fig. 4.41. Strap joints, both single and double, are widely used in engineering structures because little or no machining of the components is required, although the thickness and weight of the assembly are increased, and its aero/fluid dynamic performance is often impaired. Some recommendations for different stress regimes are shown in Fig. 4.36(c). The main problem with this style of joint, as with lap joints, is that any asymmetry in thickness or material properties results in stress concentrations that cause the parent material to fail prematurely. This is often taken erroneously as an indication that the joint is stronger than the parent material.

For strap joints, as for lap joints, stress concentrations can be reduced by tapering the ends of the overlapping material. This modification is equivalent to having very large fillets in that region. A point to be aware of is that tapering of the strap and lap ends boosts strength only substantially when the tapering is taken right to the edge of the strap, or at least to a thickness where fillets can complete the graduation [Thamm 1976].

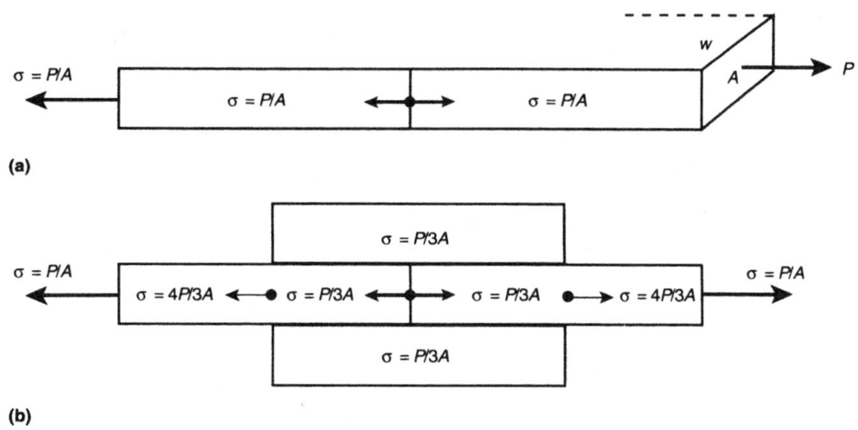

Fig. 4.41 Stress concentrations in butt and strap joint configurations. (a) Simple butt joint. (b) Butt joint with a double strap. The reinforcing straps transfer the stresses from the butt joint to the edges of the straps. This arises from the reduced load in the reinforcing strap(s) constraining Poisson contraction parallel to the edges of the straps. Adapted from Breinan and Kreider [1969]

In the preceding discussion, the assemblies were considered to be loaded solely in uniaxial tension. The location of stress concentration and its magnitude will change as the stressing mode is altered and hence the optimal style of joint varies depending on the stress environment in which the component is required to operate. Fortunately, numerical modeling techniques are able to accurately accommodate descriptions of virtually all joint styles, under complex loading scenarios. This enables right-first-time designs to be realized, obviating the need for traditional and expensive experimental programs to achieve optimization.

4.3.3.2 Strengthened Brazes to Enhance Joint Integrity

One of the limiting parameters of joint strength, especially of wide gap joints, is that of the filler metal itself. Brazing alloys can be strengthened by metallurgical mechanisms involving elements placed in solid solution, microscopic second phase particles (of either intermetallic precipitates or a dispersed refractory phase), and refinement of the grains of the filler. Brazes modified in this manner offer significantly improved mechanical properties, particularly over the range of normal ambient temperatures. These modifications often involve the incorporation of small additions of which the user may not be aware because these minor constituents do perceptibly alter the melting range or affect the wetting characteristics of the braze.

Brazing alloys can be augmented with refractory metals such as molybdenum or tungsten, dispersed as fine particles. The reinforcement is selected to be readily wetted by the braze but have low solubility in it. The presence of such hard particles is reported to improve the high-temperature properties of brazed joints, and the reduced fluidity makes them suitable for applications where the joint gap is wide (see section 4.3.4.2 in this chapter) [Chekunov 1996].

The incorporation of particles of ceramics into active brazing alloys, in particular those based on Ag-Cu-Ti, has been shown to both increase their strength and reduce their coefficient of thermal expansion (CTE). An addition of just 5 vol% of SiC to these brazes has been found to increase the bending strength of joints to ceramic components by 67% [Fernie and Ironside 1999]. The CTE reduction effected by the ceramic addition decreases the mismatch stresses.

Another approach has been to incorporate into the filler chopped carbon fibers electroplated with nickel or copper. This has been achieved with a braze (Ag-28Cu) and also with a solder (Pb-60Sn), and the results show a threefold enhancement of the shear and tensile strength of the joints with respect to the unmodified fillers and, more particularly, a significant reduction in the CTE of the filler [Ho and Chung 1990, Cao and Chung 1992]. The fraction of fibers was up to 55% of the volume of the composite filler. At a fiber content of 42 vol%, the CTE of the composite braze declined to zero over the temperature range 25 to 100 °C (77–212 °F). About 15% by volume of fibers is the maximum that can be incorporated while retaining acceptable workability of the braze in the molten state. The benefits are most pronounced when the composite braze is used to join metal to ceramic, where the reduced CTE of the filler metal is an important consequential benefit in reducing the overall stress in the assembly on cooling from the freezing point of the braze. The high CTE of a conventional filler alloy (typically $>18 \times 10^{-6}$/K) introduces a shear stress at the component/filler interface, which is overcome by using a carbon-fiber-loaded filler. The reduction in the CTE brought about by the fiber addition accounts for a threefold enhancement in fatigue life on thermal cycling that was observed in the bonded assemblies.

The carbon-fibers need not be coated if an active braze is used, which, in certain circumstances, can bring additional benefits. For example, when brazing alumina to stainless steel using a standard active silver-base brazing alloy 63Ag-34Cu-2Ti-1Sn, a problem with joint strength can arise because of excessive interfacial reaction between the titanium and alumina. When carbon fibers are introduced into the filler metal, some titanium is consumed in wetting the additional interfaces, which consequently reduces the thickness of the titanium-rich layer formed on the alumina component, as compared with the layer that forms in the absence of the carbon fibers. Adding chopped carbon fiber to the filler metal necessarily means that wider joint gaps must be used simply to admit a uniform distribution of composite filler metal. On account of the inverse relationship that exists between joint gap and strength, there is an optimum combination of joint gap and volume fraction of fiber to achieve maximum joint strength. For the materials combination cited, a volume loading of fiber of 12% and a joint gap of 150 μm (6 mils) resulted in a shear strength of brazed joints that was approximately 25% higher than for a joint made with the basic active braze (Fig. 4.42) [Zhu and Chung 1997].

4.3.4 *Wide and Narrow Gap Brazing*

Under normal circumstances, a brazed joint will naturally tend to be a few tens of microns (of the order of a mil) thick. Recommended joint clearances for brazing are generally in the region of 50 μm (2 mils) for reduced atmosphere brazing and closer to 100 μm (4 mils) for molten flux brazing (see Table 4.10). Sometimes it is necessary to create joints that are either significantly thinner or thicker. Thin joints benefit from capillary flow and may possess superior mechanical, physical, and aesthetic characteristics, although they are more vulnerable to brittle failure when stressed, as pointed out in section 4.3.3.1 of this chapter. Thick joints tend to be encountered where:

- The mechanical tolerance of the components does not allow for joints to be consistently made narrower.
- Machining tolerances are not maintained within tight limits, usually in order to reduce production costs.
- Mating parts have surfaces of different and possibly nonuniform curvature.
- The original joint surfaces have been dressed by the physical removal of material, as in repair work.

Although wide gap joints can be filled with preforms or paste of the filler metal at room temperature, if the joint gap is too wide, then the molten filler will simply run out of the joint. The term "wide gap" brazing normally refers to joints with a clearance of between 500 μm (20 mils) and 4 mm (160 mils), and "narrow gap" brazing where the clearance is less than 30 μm (1 mil).

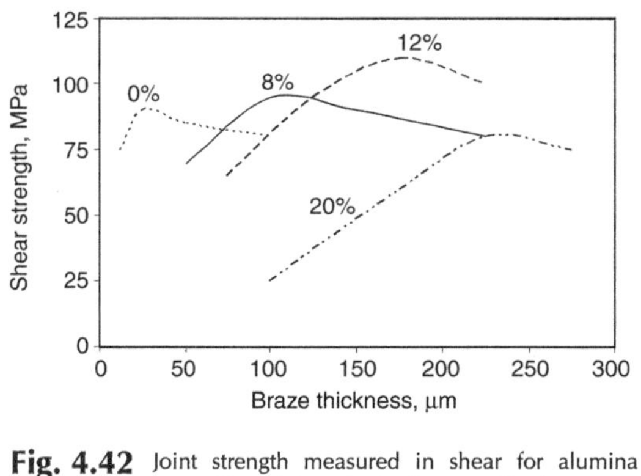

Fig. 4.42 Joint strength measured in shear for alumina brazed to stainless steel using an active braze as a function of the joint (braze) thickness and loading of the braze with chopped carbon fiber. Carbon fiber content in volume percent

4.3.4.1 Narrow Gap Brazing

It is practicable to make brazed joints that are as thin as 1 to 2 μm (40–80 μin.), even in volume manufacturing. Joints of this thinness require that the braze is preapplied to one of the joint surfaces as a high-quality film. As pointed out in section 4.3.3.1 of this chapter, it is generally not possible to make thin brazed joints by simply using a narrow joint gap and hoping that the braze from a reservoir will be drawn in by capillary action and fill it. Copper brazing of mild steel is one of the few exceptions. In thin brazed joints, molten fluxes interfere with wetting and spreading because the brazing fluxes tend to be fairly viscous. Similarly, volatile species that are evolved during the heating cycle have trouble escaping from a narrow gap. Narrow joints are therefore best made using a self-fluxing braze, a gaseous flux, or fluxless (see Chapter 3, sections 3.3, 3.1 and 3.4, respectively).

Another method of making a thin joint is to carry out a conventional brazing operation with a standard amount of braze, either drawn in from one or more edges of the joint or inserted as a preform. Once the joint surfaces have been wetted by the molten filler, sufficient compressive stress is applied to overcome the hydrostatic pressure of the braze, and the surplus material is simply extruded from the joint gap. Physical stops can be used to control the final joint gap. If the lower component has a larger area parallel to the joint than the upper one, lands can be provided to catch the overspill in a controlled manner. The stress required to reliably force a molten

Table 4.10 Recommended joint clearances, at the brazing temperature, for common brazes used in conjunction with engineering materials

Wide ranges imply that the joint gap requirements are sensitive to the joint dimensions, composition of the braze, and process superheat.

Braze alloy family	Joint clearance, mm (in.)
Silver-base	0.05–0.13 (0.002–0.005)
Copper-base	0–0.05 (0–0.002)
Gold-base	0.03–0.13 (0.001–0.005)
Nickel-base	0.03–0.61 (0.001–0.024)
Palladium-base	0.03–0.1 (0.001–0.004)
Aluminum-base	0.12–0.75 (0.005–0.03)
Self fluxing, Cu-base	0.03–0.13 (0.001–0.005)
Self-fluxing, Nickel-base	0–0.03 (0–0.001)

braze to flow out of a joint gap is of the order of 0.1 kgf/mm^2 (i.e., 1 MPa, or 145 psi). The principal drawbacks of this approach are the complexity of the jigging to apply the compressive stress and the possibility of damaging more delicate components.

A frequently overlooked consideration when attempting to make thin joints is the cleanliness of the components and particularly the environment in which the jigging prior to joining is conducted. When the desired joint gap is just a few microns (sub-mil) wide, there is no point jigging the components in a room where the airborne particles are larger! Therefore, the brazing process must be undertaken in a semiconductor-grade clean room, and great attention must be paid to the particulate content of all process gases, cleaning chemicals, and tools. Table 4.11 shows the correlation between the various classes of clean room and their particle size distributions. Clearly, if the requirement is for a joint gap less than 5 μm (200 μin.), then a class M4 (class 100) or better clean room is required.

4.3.4.2 Wide Gap Brazing

When endeavoring to make particularly thick brazed joints (>500 μm, or 20 mils), the problem encountered is how to retain the braze in the joint gap. The traditional method of solving this problem has been practiced for generations in the jewelery and plumbing industries and for the manufacture of musical instruments. The approach is to select a braze that has a wide melting range and to conduct the joining operation below the liquidus temperature when the filler alloy is in a pasty state. This process is carried out with commensurate skill by craftsmen in these trades. The presence of the solid phase drastically modifies the viscosity of the alloy and prevents it from flowing out of a wide joint gap. However, brazes with wide melting ranges are generally effective only for joint clearances less than 1 mm (40 mils), although applications such as lugless joining of tubular frames usually lie within this constraint.

A number of strategies have been developed to address wide gap joining. Most of these comprise mixtures of a conventional brazing alloy and a high-melting-point metal that provides "body" for bridging a gap. The high-melting-point material may simply be powder of the parent material. Joints of up to several millimeters in width in Inconel components can be bridged by this means using nickel-phosphorus as the braze, and in titanium aluminides using copper as the filler metal [Gale et al. 2002]. An advantage of using the parent material in the joint gap is that it helps speed removal of low-melting-point components by diffusion (see Chapter 6 on diffusion brazing).

Some methods of wide joint gap brazing involve inserting into the joint gap, at the time of jigging the assembly, a solid structure such as a plate, mesh, or honeycomb, which acts as a spacer that the filler alloy can wet. One of the more interesting types of such structures that is now available for this purpose is metal foam. Nickel foams, in particular, are now commercially available with a wide variety of pore dimensions and packing densities [Liu and Liang 2000]. Inserting thin parallel shims of copper,

Table 4.11 Relationship between clean room class designation and airborne particle size distribution

Federal standard 209F airborne particulate cleanliness classes

Class name		Class limits: 0.1 μm Volume units		0.2 μm Volume units		0.3 μm Volume units		0.5 μm Volume units		5 μm Volume units	
SI	English	m^3	ft^3	m^3	ft^3	m^3	ft^3	m^3	ft^3	m^3	ft^3
M1	...	350	9.91	75.7	2.14	30.9	0.875	10.0	0.283	...	...
M1.5	1	1,240	35.3	265	7.50	106	3.00	35.3	1.00	...	...
M2	...	3,500	99.1	757	21.4	309	8.75	100	2.83	...	...
M2.5	10	12,400	350	2,650	75.0	1,060	30.0	353	10.0	...	...
M3	...	35,000	991	7,570	214	3,090	87.5	1,000	28.3	...	...
M3.5	100	...	...	26,500	750	10,600	300	3,530	100	...	...
M4	...	...	...	75,700	2,140	30,900	875	10,000	283	...	...
M4.5	1000	...	...	...	...	...	...	35,300	1,000	247	7.00
M5	...	...	...	...	...	...	...	100,000	2,830	618	17.5
M5.5	10,000	...	...	...	...	...	...	353,000	10,000	2,470	70.0
M6	...	...	...	...	...	...	...	1,000,000	28,300	6,180	175
M6.5	100,000	...	...	...	...	...	...	3,530,000	100,000	24,700	700
M7	...	...	...	...	...	...	...	10,000,000	283,000	61,800	1750

for example, into the joint effectively partitions the joint gap into a series of much thinner joints and enables conventional joining methods to be employed. This approach is also used to reduce the effects of thermal expansion mismatch between abutting components (see section 4.2.2 in this chapter).

One of the more common approaches to wide joint gap brazing is to use a composite filler metal, comprising a regular brazing alloy, such as a copper- or nickel-base braze and spherical powder of a higher-melting-point component, usually called the "gap filler," or "additive," that does not substantially melt at the process temperature. This constituent can be appropriately selected to improve the mechanical properties of the braze [Chekunov 1996]. One example of the latter is provided by powder of Ni-20Cr alloys used in conjunction with nickel-base brazes. The two alloy fractions are introduced in the joint as a compact of mixed powders, held in a polymeric binder [Yu and Lai 1995; Lim, Lee, and Lai 1995; Tung and Lim 1994]. Provided the high-melting-point metal is largely insoluble in the braze, the apparent viscosity of the filler metal can be altered independently of temperature by varying the ratio of braze to gap filler material. Thus, by judicious selection of the gap filler proportion and size distribution of the powder, it is possible to fill joints over 4 mm (160 mils) in width [Radsijewski 1992]. Other combinations reported include iron powder in copper braze and iron-nickel powder in copper-zinc and copper-manganese brazes. Because the powder remains solid at the brazing temperature, the particle morphology, size, and size distribution all need to be controlled to get the desired balance of wetting and spreading characteristics for the composite filler. The optimum proportion of the high-melting-point fraction is typically 30 to 40 vol% of the mixture. Investigations have shown that this balance achieves sufficient stiffness in the mixture to bridge the gap, while ensuring that there is sufficient fluidity to fill interstices within the joint.

Besides incomplete infiltration of the braze into the "gap filler" when carrying out joining operations using powder mixtures, sponges, or foams, defects such as porosity arise from shrinkage, and this problem grows as the joint width, depth, and volume fraction of the gap filler material increases. For satisfactory results, pressures of 5 to 10 MPa (700–1,400 psi) need to be applied to assemblies during the joining operation, which must be carried out in a vacuum or in a reducing atmosphere because brazing flux residues only exacerbate the formation of porosity and other defects in the joint. Shrinkage can be reduced by hot isostatic pressing (HIPPing) the braze/"gap filler" compact into a dense preform prior to the brazing operation. This approach, combined with the application of 1 MPa (145 psi) pressure during the brazing cycle, has proved effective in achieving fully filled joints when using a mixture of nickel-base filler metal (AMS 4777 = AWS BNi-2) and up to 30 vol% nickel [Wu, et al. 2001]. A fast HIPPing cycle was used (10 min holding time). Also, the temperature in this pressing operation was kept below the solidus temperature of the filler 988 °C (1810 °F), in order to prevent the melting depressants in the alloy, namely boron and silicon, from diffusing into the nickel "gap filler," and so lose the melting point differential between the two constituents.

Infiltration of the molten braze into interstices is aided by high process temperatures provided that the fluidity per degree change in temperature increases faster than does consumption of the braze by reaction with the "gap filler" material. Obtaining void-free joints when a "gap filler" is present becomes more difficult the wider is the joint since empty spaces (regions of lower packing density) are always the last to fill, owing to the reduced capillarity in these air- and vapor-filled regions [Tung and Lim 1995]. Often metalloids such as boron and silicon are included in the braze formulation as both melting point depressants and wetting promoters, as in Nicrobraz LC (Wall Colmonoy, UK) (Ni-14Cr-4.5Fe-4.5Si-3B). Where these metalloids are used, a relatively high superheat is required to disperse them and prevent formation of brittle intermetallic inclusions. The process is then akin to diffusion brazing (see Chapter 6). This restricts the use of wide gap brazes based on metalloids to applications where high temperatures (>950 °C, or 1740 °F) and extended cycle times can be tolerated, which considerably limits their applicability.

The reason that much of the work on wide gap joining processes pertains to nickel-base alloys owes to the requirement to repair cracks that develop in aeroengine components. This has been satisfactorily addressed by the dual powder approach as a method for repairing defects that developed in service. The dressed crack will be of the order of 0.5 mm (20 mils) wide, which is too wide to be bridged by most conventional brazing alloys. Welding as a repair solution is also not usually a practical option because the nickel-base superalloys used in aeroengine hot section components are difficult to weld, and their complex geometries are not compatible with this method.

The methodology that is usually followed is to pack the dressed crack with a mixture of "gap filler" and braze powder, with an additional supply of the braze deposited outside of the joint gap. The gap filler in this case takes the form of spherical particulate with a mean diameter in the region of 75 μm (3 mils). Aeroengine parts are generally (fluxless) vacuum-brazed. The process requires some degree of skill to successfully implement, with minimum shrinkage voids and other joint defects, which would be highly deleterious to the fatigue life of the repaired component. The process window with regard to gap filler fraction and joining temperature can be represented in a braze quality map, an example of which is given in Table 4.12 [Tung and Lim 1995, Lim, Lee and Lai 1995].

When contemplating using fiber or particulate-reinforced brazes, one of the key targets is to obtain a void-free joint; otherwise, poor joint filling mitigates the strengthening effect. This end is greatly assisted when the infiltration of braze into the gap filler is promoted not only by metallurgical wetting of the braze, but surface tension forces are exploited to achieve spontaneous infiltration into the interstices. This situation has been studied from a theoretical standpoint, albeit simplified and some of the key results are presented in Table 4.13. In summary, provided the wetting angle of the lower-melting-point braze to the reinforcement material (or metallization applied to it) is below 45°, then spontaneous infiltration should take place irrespective of the aspect ratio of the reinforcement. If the reinforcement medium (gap filler) is not closely packed, then the critical wetting angle decreases accordingly. The corollary is that unless the minimum conditions given in Table 4.13 are achieved, the resulting joint will contain voids, unless external pressure is applied to force the molten metal into the interstices of the reinforcement material [Yang and Xi 1995].

The composite filler approach for wide-gap brazing has been successfully combined with diffusion brazing (otherwise referred to as transient liquid-phase bonding) to produce well-filled and high-strength joints. This topic is discussed in Chapter 6, section 6.4.

4.4 Service Environment Considerations

Brazed assemblies generally serve engineering applications, so the joints must be compati-

Table 4.12 Braze quality control map delineating regions of joint quality as a function of the "gap filler" content, brazing temperature, and gap width

For In-625 nickel-base superalloy brazed with Nicrobraz LC (74Ni-14Cr-4.5Fe-4.5Si-3B) braze and 80Ni-20Cr "gap filler" powder

	Process temperature											
	1125 °C (2057 °F) Gap width, mm			1150 °C (2102 °F) Gap width, mm			1750 °C (3182 °F) Gap width, mm			1200 °C (2192 °F) Gap width, mm		
Gap filler content, %	0.6	0.8	1.0	0.6	0.8	1.0	0.6	0.8	1.0	0.6	0.8	1.0
0	(a)	(a)	(a)	...	...	...	...	...	...	...	...	...
10	(a)	(a)	(a)	...	...	...	...	...	...	...	...	...
20	(a)	(a)	(a)	...	...	...	...	...	...	...	...	...
30	(a)	(a)	(a)	...	...	(a)	...	...	(a)	...	...	(a)
40	(a)	(a)	(a)	...	(a)	(a)	...	(a)	(a)	...	...	(a)
50	(a)	(a)	(a)	(a)	(a)	(a)	...	(a)	(a)	...	(a)	(a)

(a) Unsound joint containing microvoids

Table 4.13 Calculated critical angle for a liquid to spontaneously infiltrate the interstices in selected close-packed structures, and the minimum packing density necessary to achieve filling even with perfect wetting

Above the minimum packing density, spontaneous infiltration is relatively easy to achieve, even when the wetting is relatively poor.

Reinforcement type	Critical wetting angle for spontaneous infiltration, degrees	Minimum volume fraction of reinforcement for spontaneous infiltration at 0° contact angle, %
Unidirectional fibers	45	40
Body-centered cubic packed mono-sized spheres	65	20
Face-centered cubic/hexagonal close-packed mono-sized spheres	50	40

ble with the service environment. The suitability of a joint to a particular stress regime can be predicted with a fair degree of accuracy by numerical modeling. Provided this analysis is done and the joint is fit-for-purpose, the principal failure mechanism of brazed joints then tends to be by corrosion.

The corrosion of pure metals and common alloys such as steel is reasonably well understood. See, for example, the *ASM Handbook* [Vol 13A 2003]. However, brazed joints are often metallurgically complex, because they generally contain a plethora of phases, the relative proportions and distributions of which can alter over a wide range simply through variations in the process conditions. Compare, for example, the microstructure of a joint in a thin-walled pipe that has been made with the same materials combination, made by torch brazing in a short heating cycle with a brazing flux and by furnace brazing in a protective atmosphere, with a much longer heating cycle, lower peak temperature, and without a chemical flux. The different microstructures will result in different responses to a chemical environment. Disparate metallurgical microstructures can also arise merely when the brazed assemblies are of different size, as the result of different cooling rates in the two cases. Laboratory test pieces may therefore not always provide a reliable guide to durability in service.

A good illustration of the potential pitfalls in making intuitive judgments about corrosion resistance of brazed joints is provided by the corrosion that has been observed in joints made to copper and stainless steel pipe work for potable water. After all, pipe work for conveying domestic drinking water is, at first sight, a relatively benign environment because the temperature is maintained within a fairly narrow range (5–90 °C, or 40–195 °F); the mechanical stresses applied to joints are usually low, and the water is chemically neutral, having a pH value maintained close to 7. Nevertheless, tap water frequently contains significant volumes of dissolved gases and salts, which confer a degree of electrical conductivity, thereby enabling galvanic-type corrosion to occur between different metals in contact.

A study was made by Nielsen [1984] to establish the ranking order of two brazes (and two solders) with regard to their corrosion resistance. Lengths of copper and stainless steel pipe work were joined by standard methods using representative alloys of each of these systems and inserted into a hot water circuit. The variables of the test conditions included composition of the water (characteristic of three different geographical locations), its temperature (60 to 87 °C, or 140 to 190 °F), and flow rate (0.3–0.5 m/s, or 1–1.6 ft/s). After approximately a year of the test, the couplings were cut through and the joints examined in metallographic section. The results for the brazed assemblies were:

- Silver-copper-zinc brazing alloys were found to be extensively corroded, with an attrition of the filler of about 5 mm (0.2 in.) per year, due to the dezincification of the braze. This form of corrosion, which is illustrated schematically in Fig. 4.43(a), occurs because the zinc phase in the braze has a large and negative electrochemical potential (i.e., is anodic) with respect to the other phases in the filler and especially to the copper pipe. The leaching of the zinc-rich phase out of the joint microstructure loosens the silver- and copper-rich phases, which are then washed away by the flowing water.
- Silver-copper-phosphorous alloys: These brazes, which are self-fluxing (see Chapter 3, section 3.3), exhibited no corrosion at all. However, the copper pipe work immediately adjacent to the joints was moderately corroded, at 0.2 to 0.3 mm (8–12 mils) per year (see Fig. 4.43b), which suggests that the solidified filler is significantly cathodic with respect to copper. Although copper phosphide is strongly cathodic to copper, corrosion from this source is unlikely in view of the fact that the phosphorus gets incorporated in

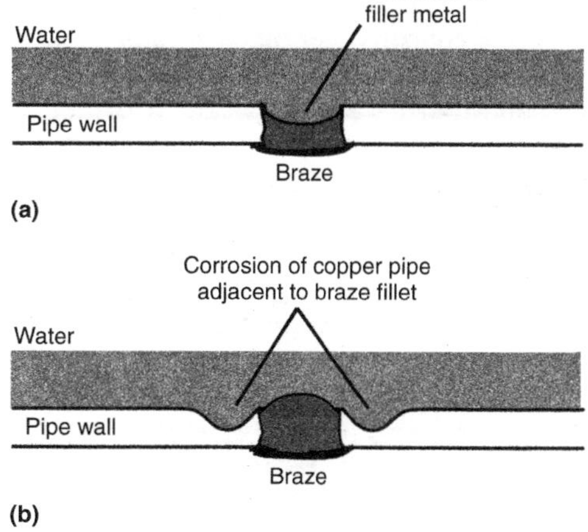

Fig. 4.43 Schematic illustrations showing the corrosion of butt joints between copper pipes conveying tap water. (a) Ag-Cu-Zn braze. (b) Ag-Cu-P braze

the oxide slag during the brazing cycle and the residual silver-copper alloy is only slightly cathodic to copper. A more likely cause of the observed corrosion is that the braze fillet is covered with a tenacious layer of the slag, which consists primarily of cuprous oxide and is electrically conductive and also strongly cathodic with respect to copper (i.e., the copper is anodic toward the slag).

- Water pipes of stainless steel brazed with silver-copper-zinc failed catastrophically within a few months at the braze/steel interface, as illustrated schematically in Fig. 4.44 [Jarman, Myles, and Booker 1973]. The mechanisms responsible have been established and are as follows. First, dezincification of the braze fillet occurs, as described previously, which enables water to reach the braze/steel interface, within the joint. At this interface there is a thin zone that consists essentially of copper, iron, and some nickel, which is generated in the brazing operation [Kuhn, Rawlings, and May 1984]. Being less noble than both the braze and the stainless steel, this zone corrodes rapidly, particularly as its width is small [Jarman, Linekar, and Booker 1973].

Joints in stainless steel, that are resistant to interfacial corrosion by tap water, can be made using silver-copper-zinc brazes, if these contain more than about 1% nickel [Jarman, Myles, and Booker 1973]. Then, as described in Chapter 2, section 2.1.5, this element concentrates at the joint periphery, where it attenuates the electrode potential difference between the braze and the stainless steel. This is also the preferred solution for joining copper water pipes because it is difficult to scrupulously remove the deleterious slag deposit from a joint made with copper-phosphorous braze when it is located on the inside of a long length of pipe.

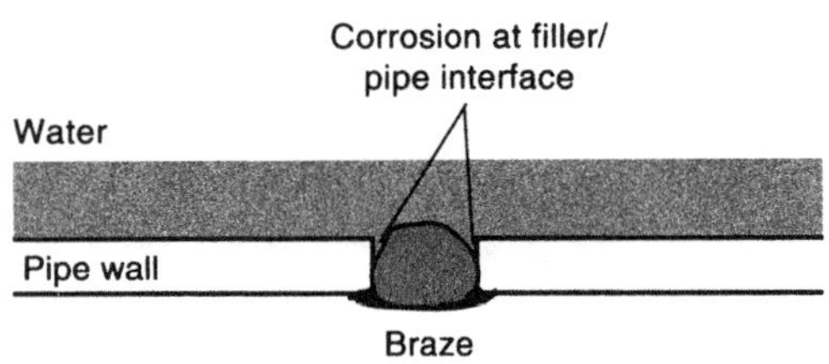

Fig. 4.44 Schematic illustration showing the corrosion of butt joints between stainless steel pipes conveying tap water, joined with an Ag-Cu-Zn braze

By way of another example, certain stainless steels, e.g., type 304, suffer from a reduction in resistance to corrosion after brazing with low-melting-point filler metals. The problem is that on heating to between 400 and 800 °C, some of the chromium combines with carbon in the steel to form chromium carbides. The loss of chromium decreases the ability of the stainless steel to form a corrosion-resistant oxide skin and is commonly referred to as "sensitization." This situation can be avoided by substituting a grade of stainless steel that is not "sensitized" in this manner, e.g., type 304L; "L" denoting low carbon content, in place of type 304. Alternatively, by using a high-melting-point braze to effect joining, the corrosion resistance can be restored by a heat treatment to redissolve the precipitated carbides. Two hours at 870 °C is often recommended as a suitable homogenization heat treatment.

REFERENCES

- Ashby, M.F., 1994. *Materials Selection in Mechanical Design,* Pergamon Press
- *ASM Handbook,* Vol 13A, ASM International. *Corrosion: Fundamentals, Testing, and Protection,* 2003
- Barranger, J., 1989. Critical Parameters of Measurement Using the Wetting Balance, *Solder. Surf. Mt. Technol.,* Vol 1 (No. 2), p 11–13
- Bever, M.B., ed., 1986. *Encyclopaedia of Materials Science and Engineering,* Pergamon Press
- *Brazing Handbook,* 4th ed., 1991. American Welding Society
- Bredzs, N., 1954. Investigation of Factors Determining the Tensile Strength of Brazed Joints, *Weld. J. Res. Suppl.,* Vol 43 (No. 11), p 545s–562s
- Breinan, E.M. and Kreider, K.G., 1969. Braze Bonding and Joining of Aluminum Boron Composites, *Met. Eng. Quart.,* Vol 9 (No. 11), p 192–202
- Cao, J. and Chung, D.D.L., 1992. Carbon Fibre Silver-Copper Brazing Filler Composites for Brazing Ceramics, *Weld. J. Res. Suppl.,* Vol 71 (No. 1), p 21s–24s
- Chekunov, I.P., 1996. A Composite Brazing Alloy for Stainless Steel and Creep Resistant Steels, *Weld. Int.,* Vol 10 (No. 9), p 735–738
- Dieter, G.E., 1976. *Mechanical Metallurgy,* 2nd ed., McGraw-Hill
- Eagles, A.M., Mitchell, S.C., and Wronski, A.S., 1995. Electrodeposition of Copper or

Nickel Barrier Layers on Steels to Produce Strong Joints Using a Copper-Phosphorus Brazing Alloy, *British Association for Brazing and Soldering,* (No. 12), p 9–12

- Eley, D.D., 1961. *Adhesion,* Oxford University Press
- Eustathopoulos, N., Nicholas, M.G., and Drevet, B., 1999. *Wettability at High Temperatures,* Pergamon
- Fernie, J.A. and Ironside, K.I., 1999. Ceramic Brazing, *Mater. World,* Vol 7 (No. 11), p 686–688
- Gale, W.F. et al., 2002. Microstructure and Mechanical Properties of Titanium Aluminide Wide-Gap Transient Liquid Phase Bonds Prepared Using a Slurry-Deposited Composite Interlayer, *Metall. Mater. Trans. A,* Vol 33A (No. 10), p 3205–3214
- Glascock, H.H. and Webster, H.F., 1983. Structured Copper: A Pliable High Conductance Material for Bonding Silicon Power Devices, *IEEE Components Hybrids and Manufacturing Technology,* Vol 6 (No. 4), p 460–466
- Hammond, J.P. and Slaughter, G.M., 1971. Bonding Graphite to Metals with Transition Pieces, *Weld. J. Res. Suppl.,* Vol 50 (No. 1), p 33s–40s
- Hanson, W. and Fernie, J., 1994. Ceramics in Turbine Applications, *TWI Bull.,* No. 5, p 103–106
- Harris, J.A. and Adams, R.D., 1984. Strength Prediction of Bonded Single Lap Joints by Non-Linear Finite Element Methods, *Int. J. Adhes. Adhes.,* Vol 4 (No. 2), p 65–78
- Haug, T., Schaefer, W., and Schamm, R., 1989. Joining Electrochemical High Temperature Components, *Proc. 3rd International Conf. Joining Ceramics, Glass and Metal,* Kraft, W., Ed., April 26–28 (Bad Nauheim, Germany), p 171–178
- Ho, C.T. and Chung, D.D.L., 1990. Carbon Fiber Reinforced Tin-Lead Alloy as a Low Thermal Expansion Solder Preform, *J. Mater. Res.,* Vol 5 (No. 6), p 1266–1270
- Huchisuka, T., 1986. Bonding of Sintered Alloys, *Met. Technol.,* Vol 56 (No. 5), p 21–27
- Iseki, T., Kameda, T., and Murayama, T., 1984. Interfacial Reactions Between SiC and Aluminium During Joining of MMCs, *J. Mater. Sci.,* Vol 19, p 1692–1698
- Jacobson, D.M., Humpston, G., and Sangha, S.P.S., 1996. A New Low Melting Point Aluminium Alloy Braze, *Weld. J. Res. Suppl.,* Vol 75 (No. 8), p 243s–249s
- Jarman, R.A., Linekar, G.A.B., and Booker, C.J.L., 1973. Interfacial Corrosion of Brazed Stainless Steel Joints in Domestic Tap Water, Part II: Metallographic Aspects, *Br. Corros. J.,* Vol 10 (No. 3), p 150–154
- Jarman, R.A., Myles, J.W., and Booker, C.J.L., 1973. Interfacial Corrosion of Brazed Stainless Steel Joints in Domestic Tap Water, *Brit. Corros. J.,* Vol 8 (No. 1), p 33–37
- Kassner, M.E. et al., 1992. Recent Advances in Understanding the Mechanical Behaviour of Constrained Thin Metals in Brazes and Solid-State Bonds, *The Metal Science of Joining,* Cieslak, M.J. et al., Ed., The Minerals, Metals and Materials Society, p 223–232
- Kinloch, A.J., 1982. The Science of Adhesion Part 2: Mechanics and Mechanisms of Failure, *J. Mater. Sci.,* Vol 17, p 617–651
- Kon-ya, S. et al., 1990. New Metallising Process of Alumina Ceramics for Hermetic Sealing, *Proc. Conf. 3rd Electronic Materials and Processing Congress,* Aug 20–23 (San Francisco, CA), p 19–24
- Kuhn, A.T., Rawlings, R.D., and May, R., 1984. A Potentiometric and Microstructural Study of the Corrosion of Silver-Brazed Stainless Steel Joints, *Brazing Soldering,* Vol 6 (No. 1), p 14–20
- Lancaster, J.F., 1965. *The Metallurgy of Welding, Brazing and Soldering,* George Allen and Unwin Ltd.
- Lea, C., 1991. Quantitative Solderability Measurement of Electronic Components, Part 5: Wetting Balance Instrumental Parameters and Procedures, *Solder. Surf. Mt. Technol.,* Vol 7 (No. 1), p 10–13.
- Levi, G. and Kaplan, W.D., 2002. Oxygen Induced Interfacial Phenomena during Wetting of Alumina by Liquid Aluminum, *Acta Mater.,* Vol 50, p 75–88
- Li, C.H., 1993. Dynamic Mismatch between Bonded Dissimilar Materials, *JOM,* Vol 45 (No. 6), p 43–46
- Li, J. and Krsulich, V., 1996. "Metal Alloy Applied in Ceramic Package Lids Reduces Stress," Semiconductor International, Feb, p 105–110
- Lim, L.C., Lee, W.Y., and Lai, M.O., 1995. Nickel-Base Wide Gap Brazing with Pre-Placement Technique, *Mater. Sci. Technol.,* Vol 11 (No. 9), p 1041–1045
- Liu, P.S. and Liang, K.M., 2000. Preparation and Corresponding Structure of Nickel Foam, *Mater. Sci. Technol.,* Vol 16 (No. 5), p 575–578

- Messler, R.W. Jr., Jou, M., and Orling, T.T., 1995. A Model for Designing Functionally Graded Material Joints, *Weld. J. Res. Suppl.,* Vol 74 (No. 5), p 160s–167s
- Miyazawa, Y. et al., 1989. Effect of Precompression on the Strength of Ceramic/Steel Joints, *Proc. Conf. Materials Research Society International Meeting on Advanced Materials* (Tokyo, Japan), Vol 8, Metal-Ceramic Joints, p 131–137
- Mizuhara, H. and Mally, K., 1985. Ceramic-to-Metal Joining with Active Braze Filler Metal, *Weld. J. Res. Suppl.,* Vol 64 (No. 10), p 27s–32s
- Moorhead, A.J., Elliott, W.H., and Kim, H.E., 1993. Brazing of Ceramics and Ceramic-to-Metal Joints, *Welding, Brazing, and Soldering,* Vol 6, ASM Handbook, ASM International
- Nielsen, K., 1984. Corrosion of Soldered and Brazed Joints in Tap Water, *Brit. Corros. J.,* Vol 19 (No. 2), p 57–63
- Nolte, H.J., 1954a. Method of Metallising Ceramic Member, U.S. Patent 2,667,427
- Nolte, H.J. 1954b. Metallised Ceramic, U.S. Patent 2,667,432
- Norton, M.G., 1993. Indirect Bonded Metallisation of Aluminum Nitride, *Proc. Symp. Materials Research Society,* Vol 314, p 223–234
- Pershall, C.S., 1949. New Brazing Methods for Joining Non-Metallic Materials to Metals, *Mater. Methods,* Vol 30 (No. 6), p 61–62
- Pincus, A.G., 1953. Metallographic Examination of Ceramic Metal Seals, *J. Am. Ceram. Soc.,* Vol 36 (No. 50), p 152–58
- Radsijewski, W.N., 1992. High Temperature Brazing of Large-Sized Constructions at Wide Joint Clearance, *Proc. Conf. Brazing, High Temperature Brazing and Diffusion Welding,* Aachen, Nov 24–26, (DVS-Berichte Band 148, Dusseldorf), p 35–37
- Schwartz, M.M., 2003. *Brazing,* 2nd ed., ASM International
- Sloboda, M.H., 1961. Design and Strength of Brazed Joints, *Weld. Met. Fabr.,* Vol 29 (No. 7), p 291–296
- Solomon, H.D., Delair, R.E., and Thyssen, J., 2003. The High Temperature Wetting Balance and the Influence of Grit Blasting on Brazing of IN718, *Weld. J. Res. Suppl.,* Vol 72 (No. 10), p 278s–287s
- Thamm, F., 1976. Stress Distribution in Lap Joints with Partially Thinned Adherends, *J. Adhes.,* Vol 7, p 301–309
- Timoshenko, S., 1925. Analysis of Bi-Metal Thermostats, *J. Optical Society of America and Review of Scientific Instruments,* Vol 11 (No. 9), p 233–255
- Totty, D.R., 1979. Brazing Cast Iron—No Longer a Problem?, Paper 7, *Proc. Conf. 3rd International Brazing and Soldering Conference,* London, UK
- Tung, S.K. and Lim, L.C., 1994. Void Formation in the Wide Gap Brazing Using Pre-Packs of the Nickel Base Braze Mixes, *Mater. Sci. Technol.,* Vol 10 (No. 5), p 364–369
- Tung, S.K. and Lim, L.C., 1995. Wide Gap Brazing with Prepacks of Nickel Base Braze Mixes, *Mater. Sci. Technol.,* Vol 11 (No. 9), p 949–954
- Twentyman, M.E., 1975. High-Temperature Metallizing, Part I: The Mechanism of Glass Migration in the Production of Metal-Ceramic Seals, *J. Mater. Sci.,* Vol 10, p 765–776
- Twentyman, M.E. and Popper, P., 1975a. High-Temperature Metallizing, Part II: The Effects of Experimental Variables on the Structure of Seals to Debased Aluminas, *J. Mater. Sci.,* Vol 10, p 777–790
- Twentyman, M.E. and Popper, P., 1975b. High-Temperature Metallizing, Part III: The Use of Metallizing Paints Containing Glass or Other Inorganic Bonding Agents, *J. Mater. Sci.,* Vol 10, p 791–798
- Werner, W.J. and Slaughter, G.M., 1968. Brazing Graphite to Hastellloy N for Nuclear Reactors, *Weld. Engineer,* Vol 53, p 65
- Wu, X.W. et al., 2001. Wide Gap Brazing of Stainless Steel to Nickel-Based Superalloy, *J. Mater. Process. Technol.,* Vol 113, p 215–221
- Xian, A. and Si, Z., 1992. Interlayer Design for Joining Pressureless Sintered Sialon Ceramic and 40Cr Steel Brazing with Ag57Cu38Ti5 Filler Metal, *J. Mater. Sci.,* Vol 27 (No. 3), p 1560–1566
- Yang, X.F. and Xi, X.M., 1995. Critical Wetting Angle for Spontaneous Liquid Infiltration into Orderly Packed Fibres and Spheres, *J. Mater. Sci.,* Vol 30, p 5099–5102
- Yu, Y.H. and Lai, M.O., 1995. Effects of Gap Filler and Brazing Temperature on Fracture and Fatigue of Wide-Gap Brazed Joints, *J. Mater. Sci.,* Vol 30, p 2101–2107
- Zhu, M. and Chung, D.D.L., 1997. Improving the Strength of Brazed Joints to Alumina by Adding Carbon Fibres, *J. Mater. Sci.,* Vol 32, p 5321–5333

CHAPTER 5

Filler Metals for Carat Gold and Hallmark Silver Jewelery

BRAZES FOR JEWELERY, silverware, and objets d'art merit separate consideration from industrial brazing alloys because, in addition to providing joints that are mechanically durable (in terms of their strength, ductility, and wear resistance), they must satisfy two additional and fundamentally different criteria specific to this usage, namely, they must match the caratage and the color of the components. The surface texture of the solidified filler metal and its resistance to corrosion also must not differ greatly from that of the joined parts so that any joints remain essentially invisible to the naked eye.

Some jewelery manufacture is highly automated, like much other industrial assembly. A good example is silver and gold chain production, in which the links are formed, brazed, and finished on fully automatic machines at many tens of links per second. In former times, chain was handcrafted, and premium-quality items are still made this way. Chain represents the oldest forms of jewelery, and foxtail-style chain dating from 2,500 B.C. has been excavated from the city of Ur. It remains one of the most popular styles, although well over 200 other varieties of chain are currently available. In mass-produced chain making, the links are fashioned from stock wire or tape. This may be a homogeneous precious alloy or cored with base metal; if the latter, during the chain-forming process, the core is dissolved out, prior to brazing, thereby reducing the weight of the item [Raw 2002]. The links are formed into chain by machines and removed as cut lengths. The brazing operation is by a powder method, which involves tumbling the chain in a nongold, silver-base braze powder. It is then retumbled in talc to remove braze powder adhering to exposed surfaces, leaving it only in the gaps of the links. The brazing operation is then performed by passing the prepared chain through a belt furnace in inert atmosphere. The stock gold wire or tape is over-carated to take the gold-free braze into account, so, on average, chain meets the caratage requirement. An alternative approach involves using stock of braze-cored wire. This needs to be formed only into links, fluxed, and heated. A cross section through the joint in a chain link made using braze cored wire is shown in Fig. 5.1 [Grimwade 2002]. The most recent technology used in chain making involves in situ welding of chain by microplasma torch or laser, which dispenses with brazes.

Brazes for carat gold jewelery must meet or exceed the fineness/caratage of the component piece parts of the assembly in order for it to meet the national fineness/caratage standards and marking or hallmarking regulations for jewelery. However, an exception is made in the United Kingdom for 22 carat jewelery, where 18 carat gold brazes are allowed. In contrast, the marking/hallmarking constraints on brazed silver items are relatively relaxed. While hallmarked sterling

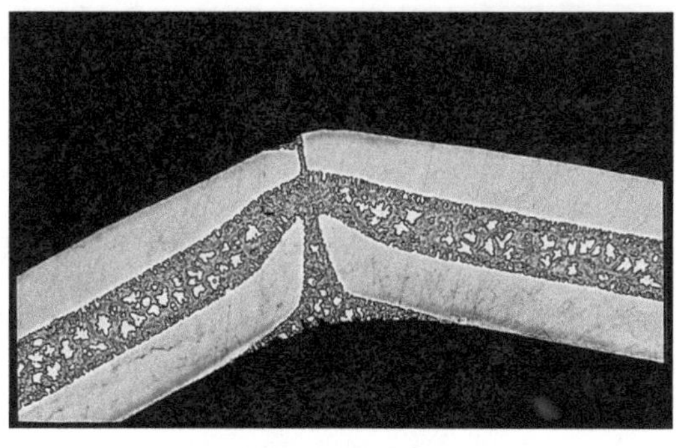

Fig. 5.1 Section through a link in a brazed carat gold chain, in which the chain stock comprises braze-cored wire. Courtesy of Cookson Precious Metals Ltd. Reprinted from Grimwade 2002

silver contains a minimum of 92.5% silver, the acceptance limit on the silver content of brazes used with sterling-grade piece parts is much lower, at 67.5%. This is in acknowledgment of the fact that, until recently, it was not possible to match the silver content of sterling silver with a brazing alloy of the same hue and of sufficiently low melting point. Relaxation of the silver hallmarking requirement means that certain members of the regular, industrial, braze alloy family, based on silver-copper-zinc, containing 67.5% silver, meet the functional requirements and are used for this purpose. Further details on these alloys can be found in Chapter 2, section 2.1.5. Recent innovation in braze alloy development has yielded silver-base brazes able to satisfy the functional requirements, including moderately low melting point, while, at the same time, fulfill the hallmarking standard. One of these is the active brazing alloy of composition Ag-5Cu-1.25Ti-1Al, which has a melting range of 860 to 912 °C (1580–1673 °F). However, hallmarking standards have not yet changed to reflect the availability of these new alloys. Hallmarking is an independent audit system to ensure caratage conformance that was introduced in the United Kingdom about 700 years ago as a means of protecting the consumer against unscrupulous manufacturers. In many countries, similar hallmarking regulations apply, but in others, including Italy, Germany, and the United States, self-marking by manufacturers is the norm. European legislation was proposed in 2001 to allow hallmarking and self-certification by manufacturers as alternative approaches. However, this proposal was formally rejected in 2004.

Because the requirements of brazes for silverware and silver jewelery are satisfied by industrial brazes, the remainder of this chapter concentrates on brazes for gold jewelery, which are specific to this application. Before embarking on a discussion of brazes for carat gold jewelery, it is necessary to have some understanding of the metallurgy of gold jewelery alloys. This is considered in the following section.

Communication between engineers and jewelery craftsmen on the subject of metallurgy and metalworking processes can give rise to confusion. This is because each profession has its own specific terms and phrases. Some surprises awaiting the engineer include:

- Solder: A braze
- Soda: Colloquial term for a solder, i.e., a braze
- Joint: Part of the structure of pin fastenings used on brooches, etc.
- Fine: Pure, e.g., fine gold is 100% gold
- Fire staining: Oxidation of base metals during heating
- Comet tails: Hard inclusions that are revealed during surface finishing

Most readers will be familiar with the term *carat* used in jewelery contexts. It is used as a unit for measuring the purity of gold and, quite separately, as the mass of precious stones. On the scale of gold purity, 24 carat (karat in the United States and Germany) represents pure gold. Thus, an 18 carat gold alloy contains 18/24 (i.e., 75%) weight fraction of the precious metal. Quality items of jewelery are made with yellow gold of 18 or higher caratage. The word *carat* actually derives from the Arabic *quirrat,* meaning seed, and the weight of a semiprecious carob seed was taken as 1 carat. Nowadays, by definition, a 1 carat gemstone has a mass of 0.2 gm (0.00705 oz).

5.1 Metallurgy of Gold Jewelery Alloys

Most metals exhibit optical reflectivity that is insensitive to the wavelength of the incident light; hence, they appear gray-white in color. Gold and copper are exceptions. Their ruddy hues arise from an asymmetrical spectral reflectance, which favors long wavelength radiation. This can be seen from reflectance curves for rhodium and gold as a function of wavelength, shown in Fig. 5.2. The process by which this

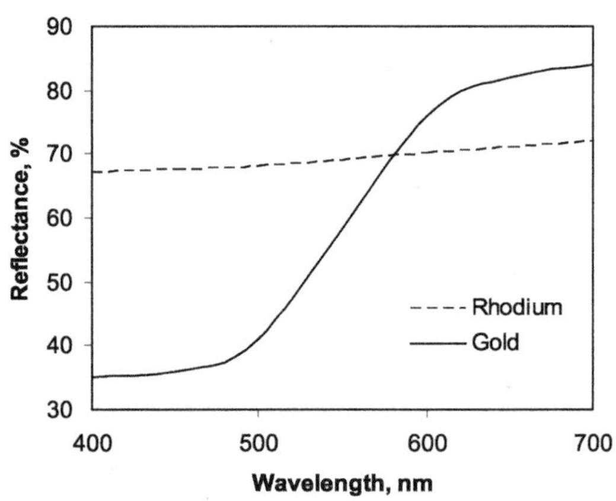

Fig. 5.2 Spectral reflectance curves for pure gold and electrodeposited rhodium. Gold reflects predominantly at longer wavelengths and therefore appears red, compared with most other metals.

occurs can be explained in terms of quantum physics, and it involves electron transitions between different atomic orbitals. Regardless of the science, the aesthetic appeal of gold relates to its unique deep yellow luster and resistance to tarnish and corrosion.

Pure gold is generally considered too soft and ductile to withstand the rigors of use, although the development of precipitation-hardened microalloyed gold alloys has led to the adoption of jewelery of 99.5% purity in China, Hong Kong, and Taiwan (called *chuk kam,* meaning "pure gold," in Chinese; explained subsequently). More commonly, substantial alloying additions are made to gold to improve the mechanical properties, reduce the melting temperature of the alloy, alter color, lower density, and make gold jewelery more affordable. The principal additions for colored carat golds are copper and silver.

The strengthening that can be achieved in the gold-rich portion of the gold-silver-copper ternary alloy system arises from three metallurgical phenomena:

- *Work hardening:* Although pure gold will recover and anneal at room temperature, alloys with copper and silver do so only at greatly elevated temperature. Therefore, any cold work imparted to the ternary alloy in the process of shaping the jewelery item will be retained unless it is heated. Typically, furnace annealing temperatures range from 500 to 700 °C (930–1290 °F), depending on the composition and cross-sectional thickness of the component. The annealing temperature may be stated as being considerably higher if torch annealing is used where the heating duration is obviously considerably shorter.
- *Precipitation hardening:* Copper and silver enter into eutectic reaction with each other and are therefore partially immiscible in the solid state, as shown in Fig. 5.3. The projection of this phase field into the ternary Ag-Au-Cu system can be exploited by solution treatment and quench-cooling of the jewelery alloy, followed by an aging heat treatment at about 300 °C (570 °F). The resulting fine distribution of second-phase particles developed in the alloy microstructure by this means roughly doubles the hardness of the gold alloy, compared with simply air cooling after annealing.
- *Ordering of the copper-rich primary phase:* Ordering has many similarities with precipitation hardening, and in gold alloys this ef-

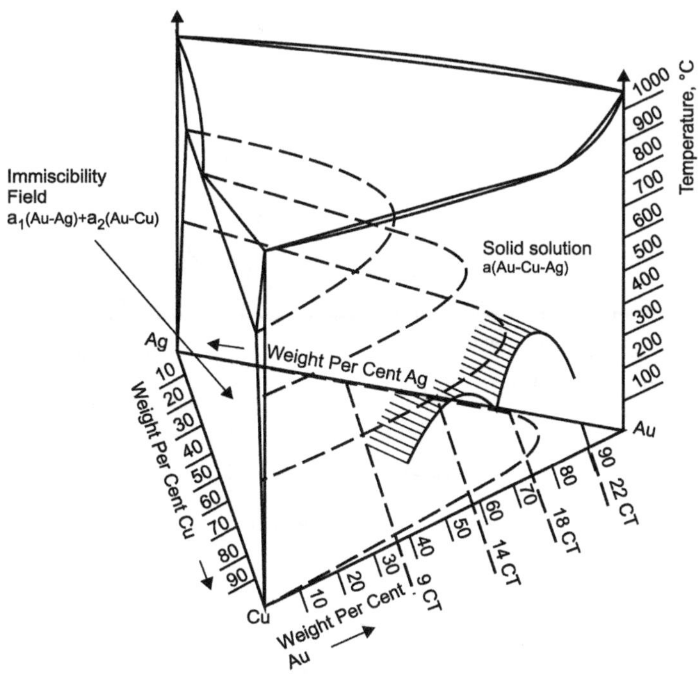

Fig. 5.3 Isometric representation of the ternary Ag-Au-Cu phase diagram, showing the field of solid solution immiscibility. Adapted from Grimwade [2002]

> fect derives primarily from the gold-copper binary system (see Fig 2.12). Although gold and copper are fully miscible in the solid state, at low temperatures, the two species of atoms prefer to adopt an ordered atomic arrangement by forming clusters of Au:Cu and Au:3Cu in the material. The associated change in lattice dimensions results in internal strains and hardening of any copper-rich solid solution phase that is present.

A minimum hardness of 175 Hv is usually considered necessary for a simple wedding band and progressively higher for gem setting and watchcases, respectively.

Because copper is a reddish metal and silver is white, addition of these metals to gold gives rise to color changes. A color map for basic silver-gold-copper alloys is presented in Fig. 5.4. Zinc is often substituted for some of the silver. The zinc addition reduces the two-phase region deriving from the silver-copper eutectic, especially in the lower carat range. This particularly benefits the malleability of 14 carat gold alloys, and the jewelery work-hardens at a more controlled rate, making it more convenient for working by hand [Rapson and Groenewald 1978]. Similarly, small additions of nickel and aluminum help simplify the process control necessary to achieve reproducible age hardening. Elements such as cobalt, iridium, and ruthenium are added as grain-refining agents to casting alloys. Other minor additions, of silicon, zinc, and boron, are introduced as deoxidizers.

In different countries, the standard caratages that are permitted or are popular for gold jewelery vary and also differ in their definition of "pure" gold. In some parts of the world, even if the metal is only 99.0% pure, it is categorized as 24 carat gold. With the advent of modern refining techniques, which can deliver 99.99% and higher-purity gold at acceptable cost, this tolerance margin in the specification is exploited in a family of high-carat gold alloys that contain up to 1% of other constituents. An example is "990 gold," which is a precipitation-hardening alloy containing 1% of titanium and exhibits exceptional yellow hue with superior wear properties and is now used mostly in watches [Gold Alloy Data 1991]. Other forms of gold, of 99.5% or higher purity, contain subpercentage additions of antimony, cobalt, calcium, or rare earths, which strengthen the precious metal by conventional precipitation hardening or through the for-

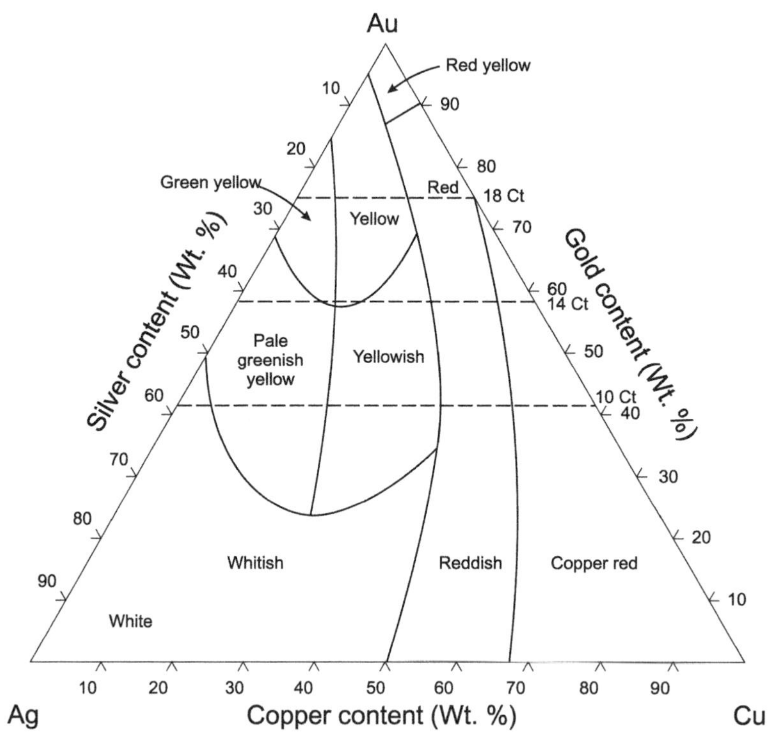

Fig. 5.4 Relationship between composition and color for silver-gold-copper alloys. Adapted from Rapson [1990]

mation of other dispersed phases, are used in chuk kam jewelery in the Far East [Corti 2001; 2002; du Toit et al. 2002].

White gold alloys were developed commercially in the 1920s to mimic the appearance of platinum—white jewelery being a fashionable setting for diamonds at that time. The addition of any white metal will tend to bleach it. Nickel and palladium have strong bleaching effects and are used most widely for this purpose [Normandeau 1992]. Nickel-containing white gold alloys may also contain copper to improve their malleability, while copper additions are added to palladium-containing white gold alloys to strengthen them and reduce their density. However, the addition must be restricted because adding copper degrades their white hue.

Gold jewelery is produced in other colors, such as purple, blue, and black. These colors are produced by one of two techniques, namely, by the formation of gold intermetallic compounds or by the application of a surface coating or patination [Corti 2004]. Purple gold is represented by the $AuAl_2$ intermetallic compound and blue gold by $AuIn_2$, which are brittle phases. On the other hand, black gold is generally produced as a surface electroplating of rhodium or ruthenium on carat golds. These colored gold compounds or layers are vulnerable to rubbing or abrasion, a characteristic that sets them apart from normal carat gold alloys.

Pickling of fabricated articles of jewelery is normal practice to remove surface scale formed during casting and heat treatment. Preparations of sodium hydrogen sulphate ($NaHSO_4$) and dilute sulfuric acid (H_2SO_4) are commonly employed for this purpose. By using more reactive agents, the base metal components can be dissolved from the surface of the alloy, and, after polishing, a deeper yellow appearance may be obtained. This "coloring" process has been practiced since ancient times, the traditional chemical for this task being either a mixture of the copiapite (a mineral of hydrated ferric sulphate) with salt and wine vinegar, or oxalic acid derived from rhubarb [Jacobson 2000]. "Bombing" in a mixture of cyanide and hydrogen peroxide (H_2O_2) is commonly used for dissolving out base metals from gold items [Faccenda 1999]; other proprietary solutions are commercially available, at least one of which is based on sulfuric acid.

Carat golds, brazes, and base metals that contain nickel and will be used in situations where prolonged skin contact is to be expected have been subjected to legislative restrictions in the EU countries since July 2001, although these do not apply to many other countries, including the United States. This is because many people become sensitized and develop an allergic reaction, known as "nickel rash," to the metal, particularly where the jewelery is inserted into open wounds, as in newly pierced earlobes. In response to this restriction, "nickel-free" gold and silver jewelery alloys and brazes have become readily available from most manufacturers.

5.2 Traditional Gold Jewelery Brazes

The popular term used in the jewelery industry, carat gold "solders," is a misnomer because these filler metals have melting points well above 450 °C (840 °F). Their working temperatures are mostly in the range 725 to 925 °C (1340–1700 °F) so that, technically, they are brazes.

The liquidus surface of the ternary gold-silver-copper phase diagram is shown in Fig. 5.5. In consequence of the fact that copper and silver enter into a eutectic reaction, a valley in the liquidus surface extends from the binary silver-copper alloy toward the gold-rich end of the ternary phase diagram. This liquidus depression has provided the basis for gold-brazing alloys and, traditionally, compositions of brazes for high-carat gold were chosen to lie in the vicinity of the eutectic valley such that their liquidus is situated below the solidus temperature of the jewelery alloy of matching caratage (see Table 5.3). Brazes in current use contain other additions to further lower the liquidus temperature and to suitably adjust their color to match the jewelery alloys, as discussed subsequently.

Tables 5.1 and 5.2 list a representative range of colored and white carat gold brazing alloys, respectively. The silver, copper, cadmium, zinc, nickel, indium and gallium additions serve to adjust both the melting temperatures and, more importantly for jewelery applications, the caratage and color of the alloys. Yellow, red and white brazes are available commercially to match the color of corresponding jewelery alloys. The carat gold alloys are adapted as brazes by adding low melting point metals, such as zinc, cadmium, indium, tin, and reducing the proportions of other alloying metals. As a consequence, the color whitens, so in the case of colored yellow-red golds, the proportion of copper tends to be increased to restore the color. The color of an alloy is primarily determined by that of the phase constituting the largest volume fraction.

Consideration of Tables 5.1 and 5.2 reveals that colored gold brazes are modified gold-silver-copper jewelery alloys, while white gold brazes, like their counterpart jewelery alloys, contain silver, nickel, zinc, copper, indium, and tin additions, or are based on gold-silver-palladium with possibly copper, zinc, and nickel additions [Normandeau 1992]. For each given carat range, there is an adjustment of the ratio of the white metal constituents on the one hand and copper on the other, to achieve the correct color balance together with the melting temperature and suitable combination of physical properties. It is possible to isolate a few characteristic features of these brazes. For example, increasing the silver-to-copper ratio in gold brazes of a fixed caratage generally increases the liquidus temperature of the alloy, except for low ratios in lower gold caratage brazes, as illustrated in Table 5.3. Conversely, brazes with a low silver-to-copper ratio are better suited to applications where gaps have to be bridged. As might be expected, the more silver-rich brazes are whiter in color, while the copper-rich filler metals tend toward red. The low silver-to-copper ratio brazes are also usually harder and more receptive to aging treatments. Copper is more effective than silver in hardening gold by solid solution strengthening, due to larger atom size differences. Silver atoms are slightly larger than those of gold, whereas copper atoms are 14% smaller. There is a tendency, albeit slight, for brazes with proportionately more silver and gold to be more resilient to tarnishing resulting from prolonged contact with sweat and general domestic chemicals (e.g., household cleaning agents, garden chemicals, and automotive fluids).

Zinc is one of the most important base metal constituents of gold brazes, being particularly effective in lowering the liquidus temperature and melting range. However, its volatility; its whitening (or bleaching) effect; and its detrimental influence on alloy ductility and malleability, as well as on the wetting characteristics of the braze, impose limits on the levels of zinc that can be tolerated. In particular, if zinc is present above about 5%, evolution of vapor becomes significant, with the consequential appearance of porosity in brazed joints.

Until recently, many of the gold brazing alloys contained cadmium as an alloying element, which offered the advantages of acting as a melting-point depressant and imparting favorable

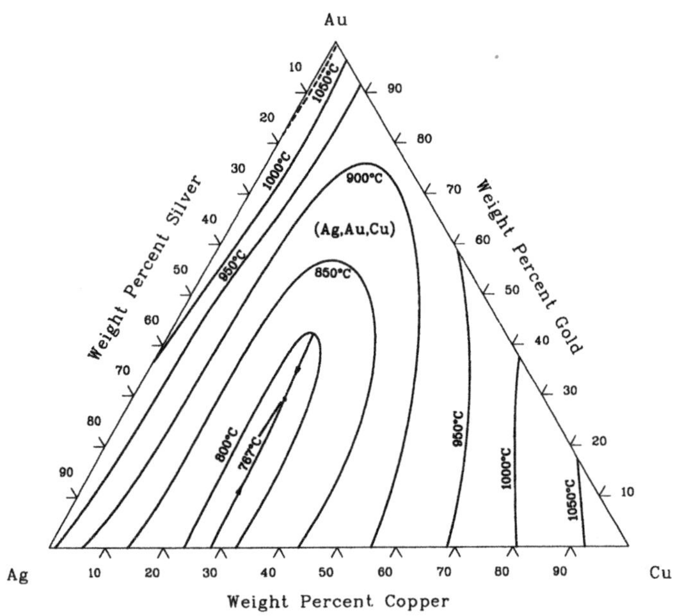

Fig. 5.5 Liquidus surface of the silver-gold-copper system (same as Fig. 2.13). From *ASM Handbook,* Vol 3 [1992]

flow and wetting characteristics of the molten braze, without strongly whitening the hue. However, the toxicity of the fume of this element is now acknowledged in legal restrictions on its use in brazes and solders in many countries, which has led to its replacement by tin, indium, and gallium [Grimwade 2002; Normandeau 1996]. Of these three base metals, gallium has the least bleaching effect [Ott 1996]. The substitution of these elements for cadmium generally results in an increase in the liquidus temperature and a widening of the melting range. Although gold brazes containing cadmium have been withdrawn from general use in most Western countries, they are still used extensively in many countries, including several major centers of jewelery production in Asia, where such restrictive legislation has not yet been introduced.

The cadmium-free jewelery brazes generally exhibit lower yield strength and enhanced ductility, which makes them more amenable to drawing to fine wire when in the form of a filler metal as well as to further mechanical working operations of a part-fabricated jewelery item. The designation range "easy," "medium," and "hard" denotes the magnitude of the temperature difference between the braze liquidus and the carat jewelery alloy solidus. The larger is this value, which is defined as "utility variance," the easier is the braze to work with, due to the greater temperature tolerance and, hence, these designations. This system of classification has significance for jewelery manufacture where multiple brazing operations need to be carried out sequentially. The operator will begin with the "hard" braze and progress down through "medium," "easy," and even "extra easy" to avoid remelting previously made joints. When carrying out repair operations, the "easy" or "extra easy" grade of brazing alloy is selected for the same reason. The fluxes used with both silver and gold jewelery brazes are standard formulations, used also in industrial brazing, as discussed in Chapter 3, section 3.2. The most common fluxes used with gold brazing alloys are based on sodium tetraborate ($Na_2B_4O_7 \cdot 10H_2O$), commonly known as borax, which is fluid above 760 °C (1400 °F), applied in the form of a paste.

The majority of jewelery brazing is still carried out by gold- and silversmiths using torches and often by hand [Grimwade 2002]. The braze is frequently applied in the form of wire, thin strip, or coupons or slugs ("paillons") cut from

Table 5.1 Composition of representative colored carat gold brazes and their melting ranges

								Melting range	
Carat designation	Gold	Silver	Copper	Zinc	Other	Grade	Color	°C	°F
22	95.0	. . .	. . .	. . .	5.0 Ga	. . .	Yellow	415–810	779–1490
	91.8	2.4	2.0	1.0	2.8 In	Easy	Yellow	850–895	1562–1643
	91.8	3.0	2.6	1.0	1.6 In	Medium	Yellow	895–900	1643–1652
	91.8	4.2	3.1	1.0	. . .	Hard	Yellow	940–960	1724–1760
	91.6	0.4	3.0	5.0	. . .	. . .	Yellow	865–880	1589–1616
	91.6	. . .	. . .	8.4	. . .	. . .	Yellow	754–796	1389–1465
21	88.0	6.0	. . .	. . .	6.0 Ga	. . .	Yellow	550–790	1022–1454
	87.5	. . .	4.5	4.0	4.0 Sn	Easy	Yellow	662–813	1224–1495
	87.5	. . .	5.5	4.8	2.2 In	Medium	Yellow	751–840	1384–1544
	87.5	4.0	3.5	5.0	. . .	Hard	Yellow	834–897	1533–1647
18	75.0	12.0	8.0	. . .	5.0 Cd	. . .	Yellow	826–887	1519–1629
	75.0	9.0	6.0	. . .	10.0 Cd	. . .	Yellow	776–843	1429–1549
	75.0	5.0	9.3	6.7	4.0 In	Easy	Yellow	726–750	1339–1382
	75.0	6.0	10.0	7.0	2.0 In	Medium	Yellow	756–781	1393–1438
	75.0	6.0	11.0	8.0	. . .	Hard	Yellow	797–804	1467–1479
	75.0	5.3	12.2	6.5	1.0 In	Hard	Yellow	792–829	1458–1524
	75.0	6.1	11.0	7.9	. . .	. . .	Red	805–810	1481–1490
14	58.5	25.0	12.5	. . .	4.0 Cd	. . .	Yellow	788–840	1450–1544
	58.5	8.8	22.7	. . .	10.0 Cd	. . .	Yellow	751–780	1384–1436
	58.3	14.4	13.0	11.7	2.5 In	Easy	Yellow	685–728	1265–1342
	58.3	17.5	15.7	6.0	2.5 Sn	Medium	Yellow	757–774	1395–1425
	58.3	20.0	18.2	3.5	. . .	Hard	Yellow	795–807	1463–1485
10	41.7	27.1	20.9	5.3	2.5 In + 2.5 Sn	Easy	Yellow	680–730	1256–1346
	41.7	29.4	22.2	4.2	2.5 Sn	Medium	Yellow	743–763	1369–1405
	41.7	33.2	23.9	1.2	2.5 Sn	Hard	Yellow	777–795	1431–1463
9	37.5	31.9	18.1	8.12	3.12 In + 1.25 Ga	Extra Easy	Yellow	637–702	1179–1296
	37.5	29.4	19.4	10.6	2.5 In + 0.6 Ga	Easy	Yellow	658–721	1216–1330
	37.5	36.3	18.2	8.0	. . .	Medium	Yellow	735–755	1355–1391
	37.5	29.8	27.5	5.2	. . .	Hard	Yellow	755–795	1391–1463
	37.5	26.1	27.4	9.0	. . .	. . .	Red	685–790	1265–1454

strip, but increasing use is being made of braze paste [Hilderbrand 1993].

Gold braze pastes are blended mixtures of the carat brazing alloy in the form of a fine powder combined and an organic binder that may or may not contain a flux, depending on whether torch heating or furnace heating with a protective atmosphere is to be used. As is normal practice with braze and solder pastes, the material is dispensed from plastic syringes using a hollow needle of appropriate size. The dosage is dispensed by depressing the plunger. In simple handcrafted jewelery manufacture, this may be done by hand, and in more highly controlled operations, an electric actuator is used to direct compressed air pneumatically onto the plunger for a predetermined time. Braze paste is available in all caratages up to and including 22 carat and in the standard colored grades.

Although braze pastes are more expensive than wire and strip, they may provide an overall cost saving through:

- Shortening the brazing operation and increasing throughput
- Deskilling of the brazing operation so that joint quality is less sensitive to craft skills of the operators
- Providing a more exact and reproducible dosage of braze to joints
- Reduction in braze scrap, which is expensive
- Offering better consistency of brazed joints, leading to higher yields
- Enabling a more precise placement of the braze, as compared with paillons

The high working temperatures required for gold brazes have two principal drawbacks. First, the elevated process temperature required for

Table 5.2 Composition of representative white carat gold brazes and their melting ranges

								Melting range	
Carat designation	Gold	Silver	Copper	Zinc	Nickel	Other	Grade	°C	°F
20	83.0	...	...	6.7	10.0	...	...	855–885	1571–1625
19	80.0	...	...	8.0	12.0	...	...	782–871	1440–1600
18	75.0	...	6.0	13.5	5.5	...	Easy	802–826	1476–1519
	75.0	...	9.0	7.0	9.0	...	Hard	843–870	1549–1598
	75.0	...	6.5	6.5	12.0	...	Easy	803–834	1477–1533
	75.0	...	1.0	7.5	16.5	...	Hard	888–902	1630–1656
14	58.33	22.0	4.42	12.0	1.25	2.0 In	Easy	695–716	1283–1321
	58.33	26.0	3.67	9.0	3.0	...	Hard	755–805	1391–1481
	58.33	15.75	5.0	15.9	5.0	...	Easy	707–729	1305–1344
	58.33	15.75	11.0	9.2	5.0	...	Hard	800–833	1472–1531
10	41.67	35.0	13.5	5.83	...	1.0 In + 3.0 Sn	Easy	715–745	1319–1373
	41.67	42.0	9.83	3.0	...	3.5 Sn	Hard	770–808	1418–1486
	41.67	28.1	14.1	6.13	10.0	2.5 Sn	Easy	763–784	1405–1443
	41.67	30.13	15.1	1.1	12.0	...	Hard	800–832	1472–1530
9	37.5	33.4	23.1	...	...	3.0 In + 3.0 Sn	...	725–735	1337–1355

Adapted from Grimwade [2002]

Table 5.3 Selected gold-silver-copper carat alloys showing the effect of varying the copper-to-silver ratio on their liquidus temperatures

					Liquidus temperature	
Carat designation	Gold	Silver	Copper	Cu:Ag ratio	°C	°F
22	91.6	6.3	2.1	1:3	1024	1875
	91.6	4.2	4.2	1:1	971	1780
	91.6	2.1	6.3	3:1	954	1749
18	75.0	21.4	3.6	1:6	976	1789
	75.0	17.0	8.0	1:2	934	1713
	75.0	12.5	12.5	1:1	882	1620
	75.0	8.0	17.0	2:1	882	1620
	75.0	3.6	21.4	6:1	881	1618
9	37.5	53.5	9.0	1:6	905	1661
	37.5	41.5	21.0	1:2	800	1472
	37.5	31.25	31.25	1:1	825	1517
	37.5	21.0	41.5	2:1	875	1607
	37.5	9.0	53.5	6:1	915	1679

Adapted from Rapson and Groenewald [1978]

carat gold brazes adds to the complexity or, more specifically, the skill required to use them. Precise and accurate temperature control is necessary because the working temperatures are very close to the melting points of jewelery alloys. For example, the commercially available, cadmium-free, 22 carat brazes have liquidus temperatures that are at most 100 °C (180 °F) below the solidus of the corresponding jewelery alloy. Second, brazing is inherently detrimental to the mechanical robustness of jewelery. This is because pure gold and high-carat gold alloys anneal and soften when heated above about 450 °C (840 °F). Figure 5.6 shows how the hardness of 18 carat yellow gold (type 750Y-3, i.e., 75Au-12.5Ag-12.5Cu) in the cold-rolled condition, drops in the course of (furnace) heat treatment 750 °C (1380 °F), as compared with the much smaller change that occurs at 450 °C (840 °F). Most of the strength generated by cold work is removed on heating for more than a few seconds to typical brazing temperatures. This also applies to electroformed (gold-silver alloy) jewelery where the high hardness of the electroform is lost on brazing the end fittings.

Currently, this second limitation is circumvented, with varying degrees of success, by making the heating cycle of the joining operation extremely rapid and designing the fabrication sequence such that additional cold working and precipitation/aging treatments can be performed after the joining process has been completed. Nevertheless, the jewelery industry could clearly benefit from the availability of lower-melting-point brazes for 18 and 22 carat gold items. Some recent work toward achieving this objective is described next.

5.3 Target Properties of Filler Metals for Carat Gold Jewelery

The principal requirements of filler metals for carat gold jewelery are that they must:

- Have a gold concentration that satisfies the fineness standards/hallmarking regulations, namely, for example, 37.5% for 9 carat gold, 75% for 18 carat gold and 91.6 wt% fineness for joining 22 carat gold in the United Kingdom; 10, 14, and 18 carat in the United States; and 8, 14, and 18 carat in Germany. In the United Kingdom, as in the Far East, the fineness standards include 990 (i.e., 99%) and 999 (99.9%) finenesses, which are classified as 24 carat. As mentioned earlier, the regulations in the United Kingdom still allow the use of 18 carat brazes for 22 carat gold jewelery.
- Be color matched to the jewelery alloy, whatever its hue, for aesthetic purposes. Most jewelery gold maintains the popular requirement for a lustrous yellow hue.
- Have a low melting point. Ideally, this temperature would be below 450 °C (840 °F) so that the filler metal may be classified as a true solder. Lower joining temperatures afford the benefits of enabling the bonding operation to be carried out without significantly softening the jewelery and requiring simpler and cheaper heating methods and finishing operations.

Other, important requirements include good wetting and spreading behavior of the filler, adequate mechanical properties (strength, ductility, and fatigue resistance), and resistance of the joints to corrosion and wear.

As a first step in examining the possibility for creating new filler metals for jewelery applications, it is instructive to consider the existing gold brazes and evaluate the constitutional basis of their formulation.

5.4 Carat Gold Braze Metallurgy

While the alloying additions made to the gold brazes shown in Tables 5.1 and 5.2 depress the

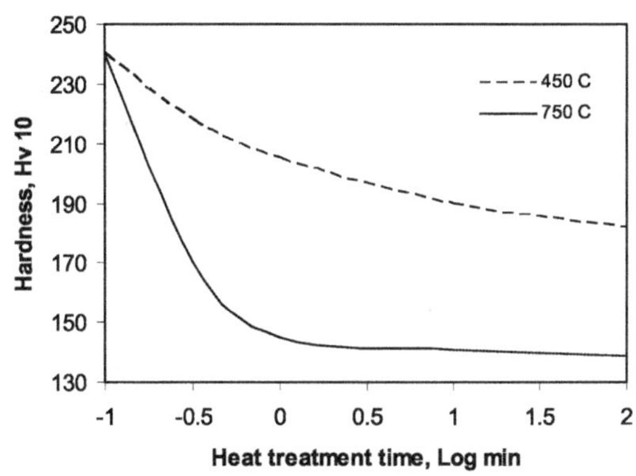

Fig. 5.6 Hardness of 18 carat yellow gold, type 750Y-3 (75Au-12.5Ag-12.5Cu) in the cold-rolled condition as a function of heat treatment time at 450 °C (840 °F) and 750 °C (1380 °F). Most of the strength generated by cold work is removed by heating for more than a few seconds to typical brazing temperatures.

liquidus temperature, the solidus temperature is depressed even more, resulting in a diverging melting range. By extrapolation, it is possible to calculate that the alloying element would have to have a melting point below absolute zero temperature in order to depress the liquidus temperature of a binary alloy containing 75% gold to below 450 °C (840 °F), in order to make it fall within the ambit of a true solder. Obviously, this is not an option. For fundamental thermodynamic reasons, multiple additions to gold-(silver)-copper alloys will be incapable of depressing the melting point below 450 °C (840 °F), as explained in Chapter 2, section 2.4. Therefore, true solders of 18 carat yellow gold must possess a different type of alloy constitution.

By definition, the temperature of a eutectic transformation must be below the melting points of the constituent phases. This type of phase change, therefore, offers scope for a large depression of the melting point of gold. Indeed, the available true "gold solders" belong to this category. Phase diagrams of binary alloys used as gold solders, which have a eutectiferous constitution, namely, gold-silicon, gold-germanium, gold-tin, gold-indium, and gold-antimony, are given in Fig. 5.7 to 5.11.

A list of the binary eutectic gold-base alloys, some of which are used in engineering applications, and melt below 450 °C (840 °F), is given in Table 5.4, which also details each individual eutectic composition, its caratage, color, and melting point. These are all gold-rich alloys, which fortuitously closely meet or exceed the minimum hallmarking specification for 18 carat gold because they contain 75% or more of gold. Gold-silicon eutectic solder would also be acceptable for use with 22 carat jewelery, having a gold content that is slightly higher than the 22 carat specification, actually achieving 23 caratage. Furthermore, the alloys listed in Table 5.4 also qualify for joining to jewelery due to the relatively high slope of the liquidus phase boundary between the various eutectic points and pure gold. In principle, this characteristic should help ensure minimal erosion of jewelery

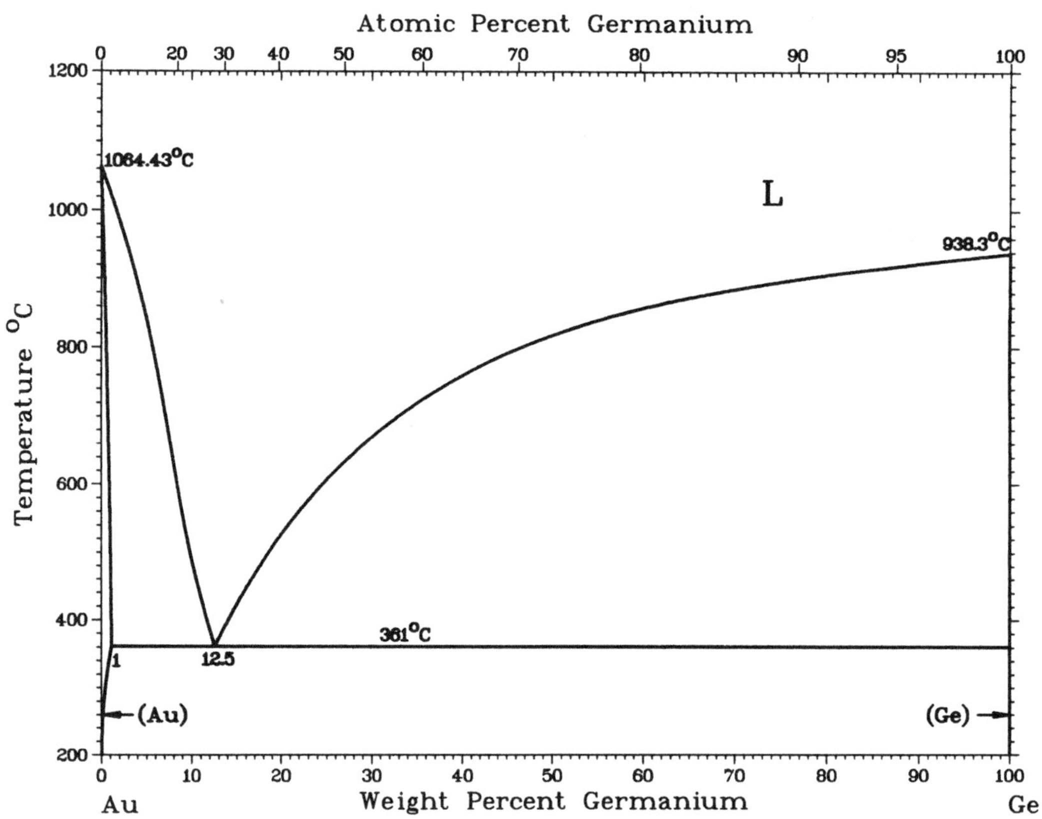

Fig. 5.7 Gold-germanium phase diagram. Adapted from Massalski [1990]

parts by the molten solder and result in good retention of the solder in the joint gap. Three of these solders, gold-germanium, gold-silicon, and gold-tin, are used widely in the electronics industry as high-temperature solders for attaching semiconductor devices, including bare chips, into packages and for building hermetic enclosures for sensitive compound semiconductors and optical devices. The high-precious-metal content of these solders means that they can be used without fluxes, provided some form of oxygen-diminished atmosphere is used to minimize the formation of scale from the base metal in the solder. Further information on solders for electronics and photonics assembly is given in the companion volume *Principles of Soldering.*

Ternary alloying of these constituents (i.e., of gold with any two of the elements, germanium, indium, antimony, silicon, and tin) mostly results in the formation of intermetallic phases that are not conducive to obtaining good-quality soldered joints. For several of the combinations, this disadvantage is compounded by the lack of a comprehensive understanding of the relevant metallurgy. The sole exception is the combination of Au-3.2Si, which is 23 carat, with Au-12.5Ge, which is 21 carat. A continuous eutectic valley extends from the Au-3.2Si composition to the Au-12.5Ge eutectic point, running in an almost straight line across the gold-germanium-silicon ternary system. This eutectic valley crosses the 22 carat fineness at the composition 91.6Au-6.8Ge-1.6Si, as shown in Fig. 5.12.

The microstructure of the gold binary eutectic alloys listed in Table 5.4 is duplex, comprising a gold-rich phase, which for the gold-germanium-silicon alloy system is essentially pure gold, interspersed with the other constituent phase of the eutectic reaction. In gold-germanium and gold-silicon binary alloys, the two constituent phases of the eutectics are virtually the pure metals owing to the very low solubility of germanium and silicon in gold, and vice-versa, in the solid state. In the gold-antimony system, the second phase is the intermetallic phase $AuSb_2$, which is rich in antimony. All of

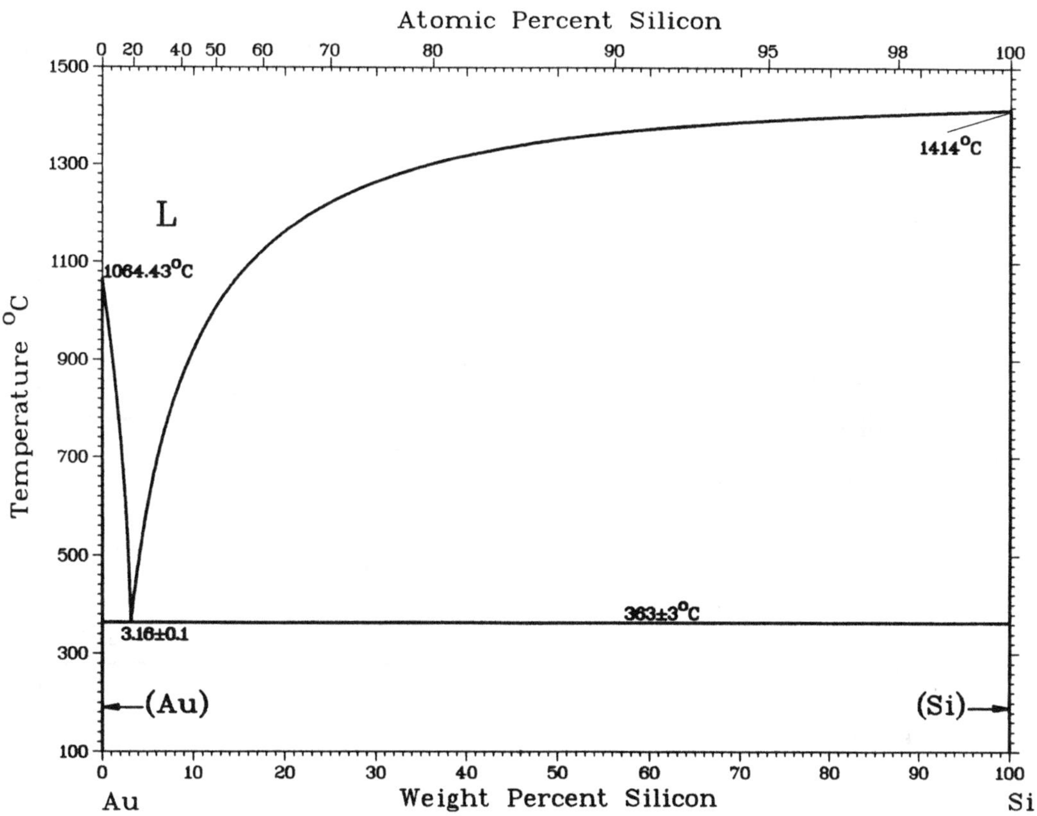

Fig. 5.8 Gold-silicon phase diagram. Adapted from Massalski [1990]

these base metal phases are silvery or metallic gray in color and will whiten the alloy overall to an extent that is largely determined by their volume proportion. Thus, the Au-3.2Si alloy contains 81.5 vol% of gold and appears pale gold, while the Au-12.5Ge alloy, with only 72% gold by volume, is even lighter in hue. The Au-25.4Sb eutectic alloy, which happens to be 18 carat, is completely gray-white in color because its duplex microstructure comprises equal proportions of gold and metallic white phases. The Au-20Sn and Au-24In eutectic alloys have no trace of golden luster whatsoever because the constituent phases are two grayish intermetallic compounds, respectively, Au_5Sn and AuSn; Au_7In_3 and AuIn.

What is clear from these considerations is that the condition needed to achieve a yellow hue in a eutectic gold alloy containing 75 wt% (18 carat) or less gold, namely, that the alloy contains a sufficient volume percentage of the yellow gold phase, is not met. This is largely a consequence of the fact that the elements that enter into eutectic equilibrium with gold have much lower density and therefore constitute a large volume fraction. For this reason, 18 carat gold alloys based on low-melting-point eutectics predominantly have a base metal color. It is not possible to offset the whitening effect of the base metal by adding the only other metal with a reddish color, namely, copper. Copper reacts with all the base metals that enter into eutectic reaction with gold to form intermetallic phases that are also dull gray or whitish in color. Hence, it is practically impossible to formulate an 18 carat gold solder, which has a melting point below 450 °C (840 °F) and simultaneously, a yellow gold hue [Humpston and Jacobson 1994]. However, there would be greater flexibility with regard to the selection of low-melting-point gold alloys as solders for white golds.

The different alloying elements used in carat gold brazes and solders have implications for recycling that need to be considered. Many base metals can cause embrittlement in gold alloys, even at low concentrations, and this problem

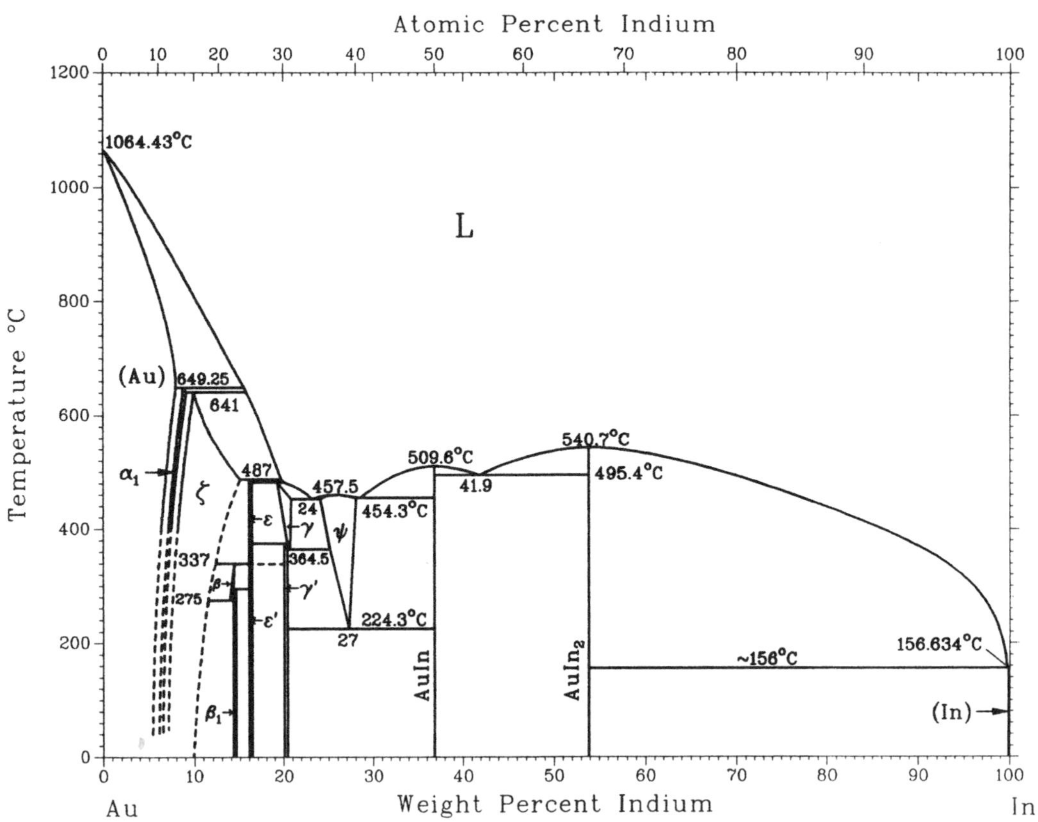

Fig. 5.9 Gold-indium phase diagram. Adapted from Massalski [1990]

would have to be recognized in the treatment of scrap. Suitable solutions, analogous to the handling of aluminum scrap, would have to be implemented.

5.5 New 22 Carat Gold Solders

The situation with regard to 22 carat solders is more encouraging, and gold solders of 22 carat composition based on the ternary combination of gold, silicon, and germanium have been produced, taking advantage of the deep eutectic valley that extends through the ternary phase diagram, which intercepts the 22 carat fineness, as mentioned previously and illustrated in Fig. 5.12 [Jacobson and Sangha 1996]. In co mmon with the industry standard gold solders used in the electronics industry, foil can be manufactured by hot rolling and wire by hot extrusion. Strip casting is another possibility (see Chapter 1, section 1.3.2.2). Strip cast foils of carat gold-germanium-silicon alloys are brittle, owing to the formation of the hard, metastable intermetallic $Au_3(Si,Ge)$, which forms when the cooling rate exceeds 5 °C/s [Johnson and Johnson 1983]. Ductility can be restored by heat treating the foil at 285 °C, or 545 °F ($\sim$0.9$T_{solidus}$), for 20 minutes or more, as can be seen from Fig. 5.13.

No satisfactory flux has been identified for use with gold-silicon-germanium alloys or, indeed, for the industry-standard gold-silicon and gold-germanium eutectics. Silicon and germanium react with oxygen when heated to form stable refractory oxides on the free surfaces, which impede wetting by these alloys. Rosin fluxes have been found to be ineffective in either dissolving or disrupting these oxides, and stronger acids or halogens corrode the alloys, owing to the large electrochemical potential difference between gold on the one hand and germanium and silicon on the other. Fluxes based on mixtures of salts and hydroxides of the alkali metals show some promise and have the merit that the residues are soluble in water. The fluxing mechanism, in this case, is believed to involve a combination of oxygen exclusion and chemi-

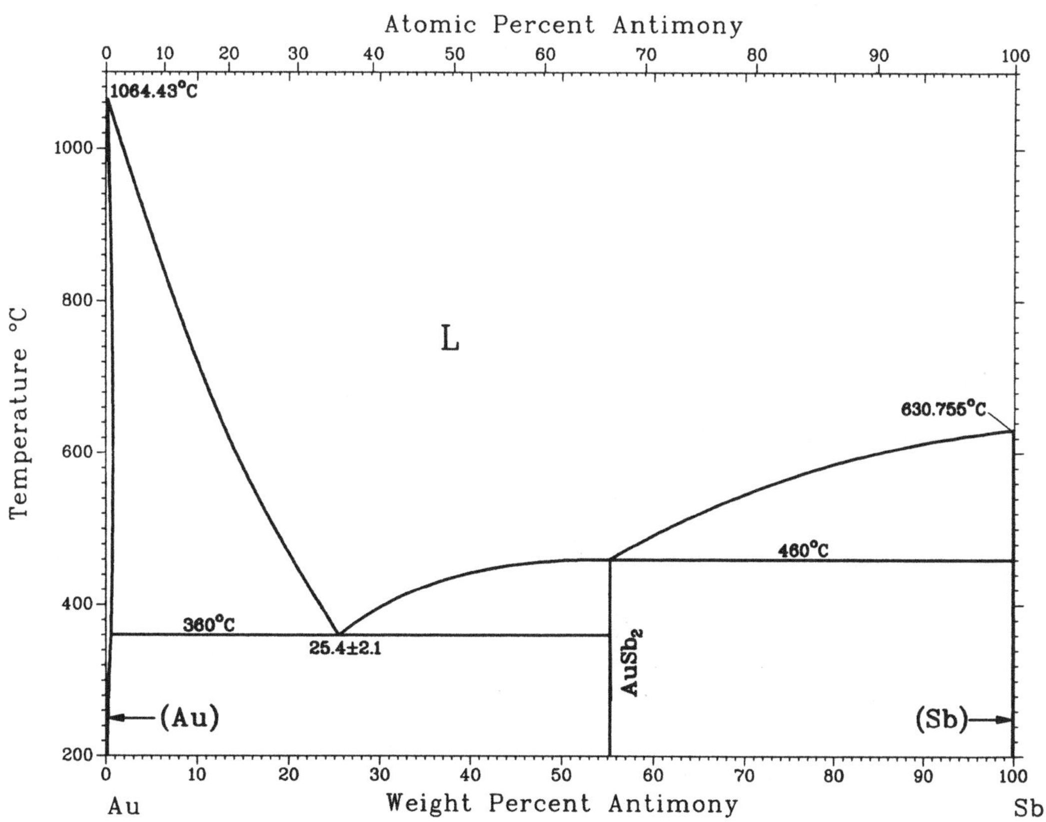

Fig. 5.10 Gold-antimony phase diagram. Adapted from Massalski [1990]

cal attack of the refractory oxides. Both the jewelery and the electronics packaging industries would benefit from more research in this area.

The binary gold-silicon and gold-germanium alloys can be used successfully as solders if the joining operations are carried out in a nitrogen atmosphere with low oxygen and water vapor content (<5 ppm in total). These and the ternary gold-germanium-silicon alloys can be made more tolerant to the joining environment by protecting the solder foil or wire with a coating of gold that is impervious to oxygen. An alternative approach that is widely practiced in the electronics industry with the constituent binary solders is to dip preforms of the filler metal in hydrofluoric acid (then rinse and dry) immediately prior to use. This treatment strips both the oxide and the near-surface nonmetal, and hence significant reoxidation does not occur until more of the nonmetal has had an opportunity to diffuse through and oxidize at the surface. The shelf life of solder so prepared is short, typically 30 min at room temperature, but is adequate for handcrafting of jewelery.

A 22 carat solder foil has been successfully developed for use with yellow gold jewelery alloys. It comprises a core that is slightly deficient in gold and also the light element, silicon, with respect to the eutectic valley linking the Au-3Si and Au-12.5Ge binary eutectics. It is coated with a gold plating to inhibit formation of dross and thereby facilitate wetting and spreading of the solder. The thinness of the plating means that preforms have a finite shelf life, although it is several months. The core alloy is of composition 90Au-8Ge-2Si, which has a melting range of

Table 5.4 Gold-base eutectic solders, some of which are used for engineering applications, that are possible candidates as filler metals for jewelery applications

Eutectic composition	Actual caratage	Color	Melting point °C	Melting point °F
Au-3.2Si	23.2	Light yellow	363	685
Au-12.5Ge	21.0	Pale yellow	361	682
Au-20.0Sn	19.2	White	278	532
Au-24.0In	18.2	Gray	458	856
Au-25.4Sb	17.9	White	360	680

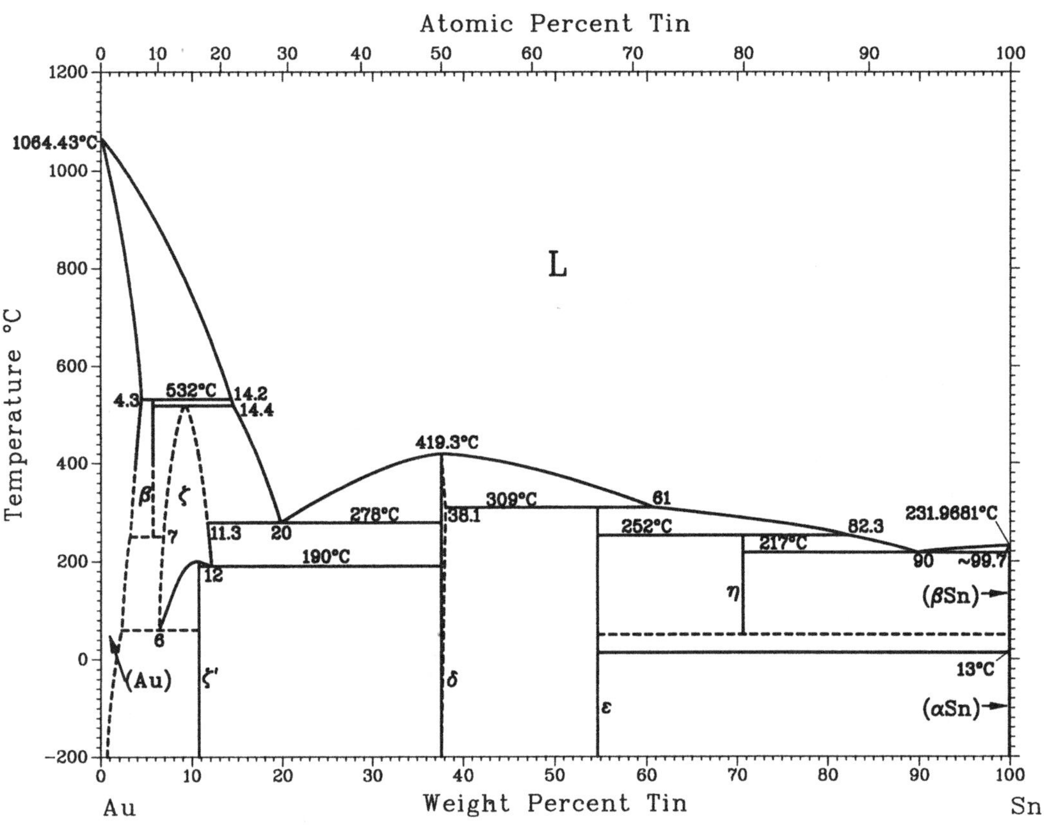

Fig. 5.11 Gold-tin phase diagram. Adapted from Massalski [1990]

362 to 382 °C (684–720 °F). The application of the gold coating of sufficient thickness raises the overall caratage to 22 and reduces the melting range to 362 to 370 °C (684–698 °F). In jewelery terminology, the two-metal structure of the preforms is referred to as a "double" or "onlay." The requisite protective atmosphere conditions for use of this solder (<5 ppm combined oxygen and water vapor) can be obtained by using nitrogen gas drawn off a liquid nitrogen tank. The inlet needs to be made leak tight, but the outlet beyond the mouth of the furnace can be left open provided the nitrogen flow rate exceeds 0.5 m/s. This gas velocity is faster than back diffusion can occur, and hence the flowing nitrogen maintains low oxygen and water vapor levels in the furnace. Because there is good heat transfer from the heating elements via the nitrogen at near-atmospheric pressure, the soldering cycle on jewelery items can be accomplished rapidly. Although the furnacing requirements are modest, they are more sophisticated and expensive in terms of capital expenditure than a simple set of torches, which means that the main area of application would be for jewelery manufacture offering sufficient profit margins.

The joint quality obtained using this plated solder foil at a process temperature of 425 °C (~800 °F) is excellent, as demonstrated by the T-joint shown in Fig. 5.14. The spread and filleting of the solder are comparable to that obtained using conventional high-temperature 22 carat braze with flux, but the low-temperature solder offers the advantages of not requiring post-process cleaning to remove flux residues and of not noticeably softening the gold jewelery items. As made, the joints appear whitish with respect to the 22 carat yellow gold jewelery, but the color match is readily restored by a modest temperature heat treatment, at 285 °C (545 °F), maintained for at least 120 min, in a shroud of nitrogen. The yellowing effect of the heat treatment correlates well with the joint microstructure. On heating, the morphology of the

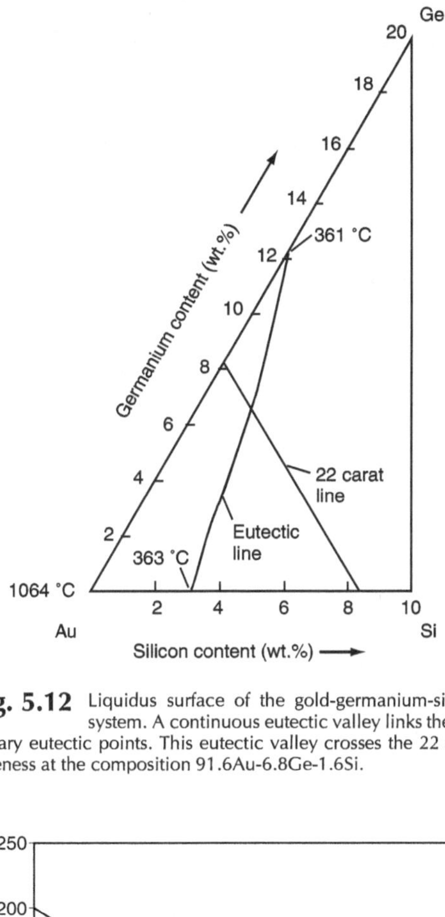

Fig. 5.12 Liquidus surface of the gold-germanium-silicon system. A continuous eutectic valley links the two binary eutectic points. This eutectic valley crosses the 22 carat fineness at the composition 91.6Au-6.8Ge-1.6Si.

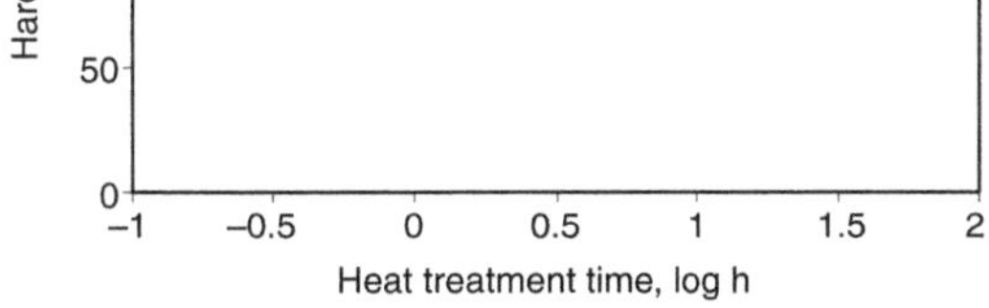

Fig. 5.13 Progressive recovery of the ductility of strip-cast foils of the 22 carat gold solder 91.6Au-6.8Ge-1.6Si, during heat treatment at 285 °C (545 °F)

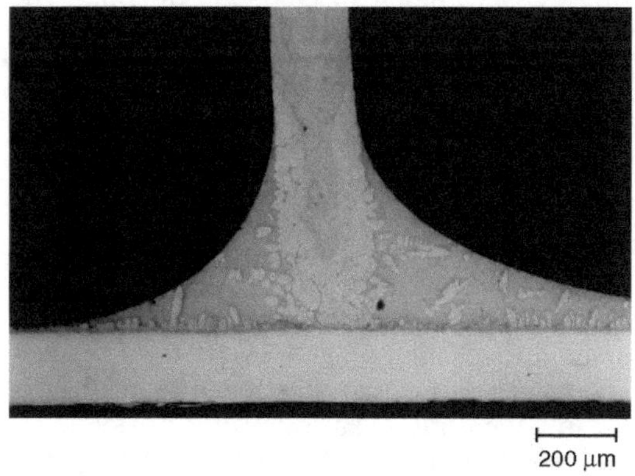

Fig. 5.14 Metallographic cross section of a T-joint made to 22 carat gold jewelery using a true gold solder (i.e., melting point <450 °C, or 840 °F)

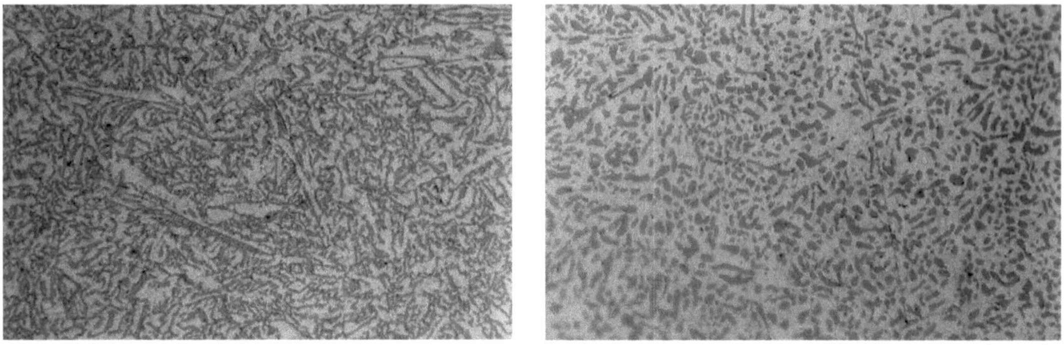

Fig. 5.15 Microstructure of the 92.5Au-6Ge-1.5Si alloy. (a) Before heat treatment at 285 °C, showing dendritic form of the germanium-silicon precipitates and (b) after heat treatment at 285 °C, showing spheroidal form of the germanium-silicon precipitates

germanium-silicon precipitates throughout the solder, changing from dendritic to spheroidal, so that the same proportion of this phase accounts for a smaller proportion of the area of the free surface of the alloy than it does prior to the heat treatment. In consequence, the yellow-gold matrix becomes dominant, and the overall color of the alloy favorably alters accordingly (Fig. 5.15a,b).

The mechanical integrity of joints made to 22 carat gold substrates has been assessed in lap-shear and peel resistance tests, with joint areas in the waisted region of 2 mm × 2 mm. Failure always occurred in the 22 carat gold substrate rather than through or adjacent to the joint. For the reasons explained in Chapter 4, section 4.3.3, this result should not be taken as evidence that the joint is stronger than the parent materials. Such a presumption neglects the role of stress concentration in this style of joint. The important metric is that joint shear strengths typically exceed 210 MPa (30 ksi), with good peel resistance and fracture toughness and are therefore adequate for jewelery applications. Likewise, corrosion tests, designed to assess the susceptibility of joints to degradation from skin acids and household chemicals, did not reveal any deficiencies in the joints.

There exists one other low-temperature method for joining 18 and 22 carat gold jewelery, namely, diffusion soldering. This process is the low-temperature analogue of diffusion brazing, which is described in detail in Chapter 6. The successful application of this technique to gold jewelery alloys is discussed in the companion volume *Principles of Soldering*.

REFERENCES

- *Alloy Phase Diagrams,* 1992. Vol 3, *ASM Handbook,* ASM International, p 3.5
- Corti, C.W., 2001. Strong 24 Carat Golds: the Metallurgy of Microalloying, *Gold Technology* (No. 33), p 27–36
- Corti, C.W., 2002. Micro-Alloying of 24 Carat Golds: Update, *Gold Technology* (No. 36), p 34
- Corti, C.W., 2004. Blue, Black and Purple! The Special Colours of Gold, *Proc. Conf. 18th Annual Santa Fe Symposium,* May 23–26 (Albuquerque, NM)
- Faccenda, V., 1999. *Handbook on Finishing in Gold Jewellery Manufacture,* World Gold Council
- Grimwade, M.F., 2002. *Handbook on Soldering and Other Joining Techniques in Gold Jewellery Manufacture,* World Gold Council
- Gold Alloy Data, 1991. Au 990–Ti 10 ("990 gold"), *Gold Bull.,* Vol 24 (No. 1), p 15–19
- Hilderbrand, H.H., 1993. Gold Solder Pastes, *Gold Technology* (No. 9), p 8–12
- Humpston, G. and Jacobson, D.M., 1994. Do 18 Carat Gold Solders Exist?, *Gold Bull.,* Vol 27 (No. 4), p 110–116
- Jacobson, D.M., 2000. Corinthian Bronze and the Gold of the Alchemists, *Gold Bull.,* Vol 33 (No. 2), p 60–66
- Jacobson, D.M. and Sangha, S.P.S., 1996. A Low Melting Point Solder for 22 Carat Yellow Gold, *Gold Bull.,* Vol 29 (No. 1), p 3–9
- Johnson, A.A. and Johnson, D.N., 1983. The

Room Temperature Dissociation of Au_3Si in Hypereutectic Au-Si Alloys, *Mater. Sci. Eng.,* Vol 61 (No. 3), p 231–235

- Massalski, T.B., Ed., *Binary Alloy Phase Diagrams,* 2nd ed., Vol 1, ASM International, 1990
- Normandeau, G., 1992. White Golds: A Review of Commercial Material Characteristics and Alloy Design Alternatives, *Gold Bull.,* Vol 25 (No. 3), p 94–103
- Normandeau, G., 1996. Cadmium-Free Gold Solder Alloys, *Gold Technology* (No. 18), p 20–24
- Ott, D., 1996. Development of 21 Carat Cadmium-Free Gold Solders, *Gold Technology* (No. 19), p 2–6
- Rapson, W.S., 1990. The Metallurgy of the Coloured Carat Gold Alloys, *Gold Bull.* Vol 23 (No. 4), p 125–133
- Rapson, W.S. and Groenewald, T., 1978. *Gold Useage,* Academic Press
- Raw, P., 2002. Hollow Carat Gold Jewellery from Strip and Tube, *Gold Technology* (No. 35), p 3–10
- du Toit, M. et al., 2002. The Development of a Novel Gold Alloy with 995 Fineness and Increased Hardness, *Gold Bull.* Vol 35 (No. 2), p 46–52

CHAPTER 6

Diffusion Brazing

In brazing, wetting of the component surfaces is not always easy to achieve and when it does occur, the resulting alloying between the filler and components can cause excessive erosion of the parent materials, embrittlement of joints due to the formation of phases with inferior mechanical properties, and other undesirable effects. Thereafter, the upper working temperature of the assembly is also compromised by the presence of the lower-melting-point filler metal. These problems not withstanding, brazes have the singular merit of being able to fill joints of irregular dimensions and produce well-rounded fillets at the edges of the joint.

Diffusion bonding sidesteps the need for wetting and spreading by a filler metal (see Chapter 1, section 1.1.7.3). Once formed, diffusion-bonded joints are stable to high temperatures so that the service temperature of the assembly can actually exceed the peak temperature of the joining process without risk of the joint remelting. While the formation of undesirable intermetallic phases can also occur in diffusion bonding, because there are usually fewer constituents involved, it is easier to select a safe combination of materials. However, diffusion bonding tends to be limited as a production process because it is not tolerant of joints of variable width, and moreover, its reliability is highly sensitive to surface cleanliness. High loads (typically 10 to 100 MPa (or 1.4 to 14 ksi) have to be applied during the bonding cycle to ensure good metal-to-metal contact across the joint interface. Also, the duration of the heating cycle is typically hours, compared with seconds for brazing, because solid-state diffusion is much slower than wetting of a solid by a liquid. These factors, and the absence of any significant fillets to minimize stress concentrations at the edges of joints (see Chapter 4, section 4.2.4), considerably limit the applications for diffusion bonding.

There exists a hybrid joining process that combines the good joint filling, fillet formation, and tolerance to surface preparation of conventional brazing, together with the greater flexibility with regard to service temperature and metallurgical simplicity that is obtainable from diffusion bonding. This process uses a molten filler metal to initially fill the joint gap, but during the heating stage, the filler diffuses into the material of the components to form solid phases and, in consequence, the remelt temperature of the joint is raised. This process is called diffusion brazing, sometimes also referred to as transient liquid-phase (TLP) joining, or transient liquid phase bonding (TLPB) [Nicholas 1998, 91–102]. It is an established joining and repair process that has been used for decades in the aerospace industry. Further information on its low temperature analogue, diffusion soldering, can be found in the companion volume *Principles of Soldering*. It is noteworthy that although diffusion soldering processes are, by definition, performed at process temperatures below 450 °C, the remelt temperature of joints can exceed the solidus temperature of common brazes.

6.1 Process Principles

The steps involved in making a diffusion-brazed joint are illustrated schematically in Fig. 6.1. The joint configuration generally comprises the two component parts and a filler preform inserted between them, and heating is commenced (stage 1). The latter is usually much thinner than 50 μm (2 mils), which is thinner than a typical braze preform. The components are clamped together with a compressive stress of a few megapascals and the assembly is then usually heated to above the liquidus temperature of the filler metal, which melts, wets the joint surfaces, fills the joint gap, and forms small edge fillets (stage

2). If the filler and base material are elemental metals and enter into a simple eutectic reaction, then melting can be brought about by heating the assembly to a temperature above the eutectic point. If this process temperature is below the melting point of the pure filler, then initially interdiffusion occurs in the solid state, which forms part of the first stage of the process. This results in depression of the melting point of the filler so that melting and wetting occur (stage 2).

Due to the generation of liquid in the joint, the necessary applied pressures are generally much less than those required for normal diffusion bonding, and typically in the range of 0.5 to 1 MPa (70 to 145 psi), but greater than those normally used during brazing (0–0.1 MPa, or 0–15 psi). This is important not only because it enables the joining operation to be less demanding in equipment terms but also because the pressure needed is only a small fraction of the yield stress of the parent material so that macroscopic deformation is avoided [Askelsen 1992].

If the assembly were then cooled, the joint would resemble a thin, but conventional, brazed joint in its properties and metallurgical characteristics. However, by maintaining the assembly at elevated temperature, the composition of the filler metal will change with time as alloying elements in the braze, added as melting-point depressants, diffuse away from the joint and into the parent metals. In consequence, the solidus temperature of the filler rises and solidification occurs isothermally (stage 3). Ultimately, the joint becomes essentially homogeneous with the parent metal (stage 4).

Diffusion brazing provides a means to fill joints that are not perfectly smooth or flat (a feature of liquid-phase joining), while offering greater flexibility with regard to joint configurations and mechanical properties than can be obtained from diffusion bonding. Owing to the features of the process, including the high degree of control applied to the process parameters, the following additional advantages may be obtained:

- Good reproducibility of the process
- High dimensional control of the joint width, i.e., narrow tolerances
- Facilitation of exceptionally good joint filling in large area joints, which ensures leak tightness. This is especially important where the joint forms part of a hermetic enclosure.
- Attainment of very narrow joints, typically less than 50 μm (2 mils), which benefits properties such as mechanical properties and thermal conductivity
- Ability to tightly control edge spillage of the molten filler and even prevent it altogether

An alloy system suitable for diffusion brazing should have the following characteristics:

- It should be based on a simple alloy binary or ternary system to keep the joint design and joining process as simple as possible.
- It should have a phase constitution that includes a relatively low-melting-point eutectic reaction to initiate the melting process.
- It should have as few intermetallic compounds as possible. Those that do exist should be stable over a wide range of composition and ideally melt at temperatures below or comparable with the joining temperature. This reduces the propensity for diffusion barriers to become established that would impede the isothermal solidification process and the proliferation of brittle phases in the joint.
- The terminal primary metal phase should possess a wide range of solid solubility of the other constituents to minimize the risk of

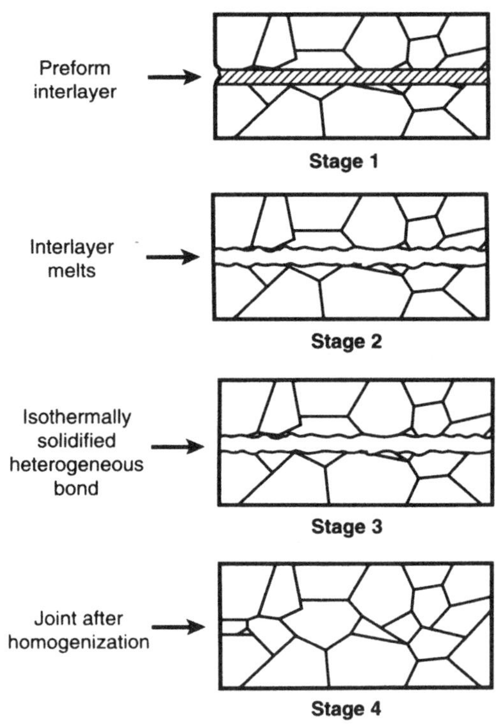

Fig. 6.1 Schematic illustration of the steps involved in making a diffusion-brazed joint

intermetallic phases precipitating during cooling of the assembly from the processing temperature and to provide a greater tolerance of the process to the amount of filler metal introduced into the joint.

A selection of representative examples of alloy systems that satisfy these conditions and lend themselves to viable diffusion brazing processes is shown in Table 6.1.

6.2 Examples of Diffusion Brazing Systems

One of the first diffusion brazing processes to be exploited was developed for joining nickel-base superalloys using the Ni-4B eutectic braze (melting point = 1090 °C, or 2084 °F) [Duvall, Owczarski, and Paulonis 1974]. Nickel-12P eutectic braze (melting point = 880 °C, or 1616 °F) has also been used in the same application [Ikawa, Nakao, and Isai 1979].

Both boron and phosphorus play a dual role, being very effective melting-point depressants and fast-diffusing elements. This second characteristic is especially advantageous for diffusion brazing because boron and phosphorus diffuse rapidly out of the molten filler at the joining temperature and become widely dispersed in the parent metals. In consequence, isothermal solidification takes place rapidly at the joining temperature as the composition of the nickel filler tends toward pure nickel. Because boron embrittles many filler metals, this makes the production of ductile preforms difficult. One solution to this problem has been the development of surface enriched brazing foils. These are produced free of boron and are therefore ductile, and the boron is added to the surface region by a vapor phase process. Ductile nickel-base preforms for diffusion brazing are now normally made by rapid solidification technology. The formation of eutectic nickel and iron alloys with boron and phosphorus having deep eutectic wells makes them amenable to foil manufacture

Table 6.1 Selected material combinations used for diffusion brazing

		Process temperature		Remelt temperature		
Substrate	Filler metal	°C	°F	°C	°F	Ref
Alumina	Cu/Ni/Cu interlayers	1150	1920	1400	2550	Shalz et al. [1994]
Alumina/kovar	Ni/Ti/Ni interlayers	980	1800	984–1310	1805–2390	Zhang, Qiao, and Jin [2002]
Cobalt alloys	Ni-4B	1175	2145	1475	2685	Duvall, Owczarski, and Paulonis [1974]
Nickel alloys	Ni-4B	1175	2145	1450	2640	Duvall, Owczarski, and Paulonis [1974]
	Ni-4.5Si-3.2B	1065	1950	1450	2640	Gale and Orel [1996]
	Ni-12P	1100	2010	1450	2640	Ikawa, Nakao, and Isai [1979]
Inconels (Ni-Cr-Fe alloys)	Ni-7Cr-3Fe-3.2B-4.5Si (AMS 4777)	1150	2100	1400	2550	Wu, Chandel, and Li [2001]
	Al	700	1290	1133	2070	Khanna et al. [2000]
Ferritic superalloys	Ni-4.5Si-3.2B	1100	2010	1400	2550	Khan and Wallach [1995]
Steel (including stainless)	Fe-12Cr-4B	1050	1920	1400	2550	Nakahashi et al. [1985]
	Ni-20Cr/Ni-B powder compacts	1000	1830	Not determined		Zhuang and Eagar [1997]
Copper	Sn	690	1275	1080	1975	Sangha, Jacobson, and Peacock [1998]
Silver alloys	Ag-30Cu	825	1515	950	1740	Tuah-Poku, Dollar, and Massalski [1988]
Tantalum	Ti-30V	1675–1760	3050–3200	>2095	>3800	Schwartz [1987]
Titanium aluminide	Titanium aluminide/Cu powder mixture	1150	2100	>1350	>2460	Gale et al. [2002]
Titanium alloys	Cu-50Ni	975	1785	1700	3090	Norris [1986]
	Ag-15Cu-15Zn	700	1290	700(a)	1290(a)	Elahi and Fenn [1981]
Aluminum alloys	Cu	550	1020	550(a)	1020(a)	Niemann and Wille [1978]
	Zn-1Cu	525	975	525(a)	975(a)	Ricks et al. [1989]
	Ag-29Cu	510	950	660	1220	Bushby and Scott [1995]

(a) Estimated, but not measured

by melt spinning (see Chapter 1, section 1.3.2.2). This process excels at producing thin (20–50 μm, or 0.8–2.0 mils) and highly uniform foil preforms that are ideal for diffusion brazing processes.

Another melting-point depressant that is widely used in melt-spun foils is silicon. However, silicon diffuses much more slowly in nickel than do boron and phosphorus. As a result, relatively persistent silicides tend to form in joints during diffusion brazing operations, which can cause embrittlement [Khan and Wallach 1995]. Nevertheless, it has been shown that nickel silicide can be completely dispersed and solid solution achieved in a diffusion-brazed joint between nickel alloy (Inconel) components made with a nickel-iron-silicon-boron alloy held at 1150 °C (2100 °F) for 120 minutes [Wu, Chandel, and Li 2001].

Diffusion brazing of aluminum alloys that contain silicon can be accomplished using a copper or brass preform [Timisit and Janeway 1994]. The brazing temperature needs to be above 530 °C (990 °F) for the copper preform and 510 °C (950 °F) for the brass preform. The role of zinc is as a melting-point depressant. Joint strengths above 100 MPa (15 ksi) can be obtained. Aluminum-lithium alloys, which have high specific stiffness, are difficult to join by brazing because the lithium stabilizes the native oxide in the form of lithium-rich spinels. However, they may be joined by diffusion brazing using copper interlayers and applied pressures in the region of 5 MPa (0.7 ksi) at a process temperature of 530 °C (990 °F). The partial oxygen pressure in the process chamber needs to be below 2 MPa (3×10^{-7} psi) and the faying surfaces scrupulously cleaned for joining to be successful [Urena et al. 1996]. Diffusion brazing of aluminum/boron fiber composites 6061 Al-50%B has been achieved using Al-12Si brazing foils for use on the space shuttle [Schwartz 2003]. The process temperature must be kept low to prevent reaction between the aluminum and boron and reduction of mechanical properties. The boron fibers or tapes will often be coated with silicon carbide to aid protection in this regard.

Titanium alloys can be diffusion brazed with copper. Copper plays the role of a temporary melting-point depressant, diffuses readily in titanium, and can be applied as a thin layer to faying surfaces by electroplating. Some care may be needed in the applicability of this approach because copper stabilizes the beta phase in titanium and decreases the beta-to-alpha transition temperature. Tantalum alloys are often joined by diffusion brazing as a means of achieving high-temperature service without using an exceptionally high-melting-point braze. Ti-30V braze, when used with an extended heating cycle at 1675 to 1760 °C (3050–3200 °F), to form a diffusion-brazed joint to tantalum alloys, results in a joint remelt temperature that exceeds 2095 °C (3800 °F) [Schwartz 2003, 98]. The high reactivity to tantalum, when hot, with most gases means that vacuum is required for brazing and a thin surface coating of copper or nickel is often required as an additional measure to slow reaction with residual species in the vacuum.

A number of interesting variants of diffusion brazing have been reported. One of these makes use of a filler of near-eutectic composition Ag-29Cu for joining aluminum. The unusual feature here is that the filler metal has a higher melting point (779 °C, or 1434 °F) than the parent metal, aluminum [Bushby and Scott 1995]. In this case, the process relies on the formation of a low-melting-point ternary eutectic alloy between silver, copper, and aluminum at 505 °C (941 °F). When the assembly is heated above this temperature, solid state interdiffusion across the interface between the three metal constituents results in a ternary alloy, which melts. The copper diffuses out of the joint faster than the silver. The latter forms the intermetallic compound Ag_2Al at the center of the joint, which remains solid at the joining temperature. Raising the holding temperature to the melting point of the Ag_2Al intermetallic compound (567 °C, or 1053 °F), but still below the melting temperature of the parent metal, removes it from the joint. Aluminum casting alloys and aluminum-boron composites can be diffusion brazed by plating commercial Al-7.5Si filler metal with copper. Provided the copper is sufficiently thin and well diffused after brazing, the mechanical and corrosion resistance of the aluminum parts are not compromised. The joining process temperature can be as low as 540 °C (1000 °F). Using the same idea, the filler metal foil or coating can be dispensed with altogether if certain combinations of different metals or alloys are being joined. This approach has been used in the development of a diffusion brazing process for joining Zircaloy 2 (Zr-1.5Sn-0.25(Fe,Ni,Cr)) to stainless steel at approximately 950 °C (1740 °F). The process relies on a eutectic reaction between iron and zirconium at 934 °C (1713 °F) [Owczarski 1962].

Diffusion brazing is not limited to metal components but can also be applied to ceramic parent materials. A nickel foil, 100 μm (4 mils) thick and coated both sides with a 3 μm (120 μin.) layer of copper has been used as an interlayer in diffusion brazing components of alumina [Shalz et al. 1994]. The joining operation was carried out in vacuum at 1150 °C (2100 °F), which is above the melting point of copper (1085 °C, or 1985 °F). At this temperature, the copper melts and alloys with the nickel, leading to isothermal solidification. At the same time, the nickel diffuses toward the alumina interface where it forms $NiO-Al_2O$ spinel, which provides a strong bond. An applied pressure of 5 MPa (700 psi) ensures good initial interfacial contact, so although the volume of liquid is small, it is still sufficient to fill the joint gap. A lower-temperature process for joining alumina to Kovar (Fe-29Ni-17Co) makes use of a titanium foil 0.3 mm (12 mils) thick sandwiched between 15 μm (0.6 mil) thick nickel foils [Zhang, Qiao and Jin 2002]; alternatively, the three foils could be replaced by a so-rolled trifoil or a titanium foil plated both sides with nickel. The trifoil is held between the aluminum and Kovar components under a comparatively low uniaxial pressure of 30 to 40 kPa (4.3–5.7 psi). When the assembly is heated in vacuum, titanium diffuses across the interface with the nickel foils to form a eutectic that melts at 942 °C (1728 °F). The nickel layers are progressively replaced by layers of Ti_2Ni intermetallic, which melts at 984 °C (1803 °F), and the thickness of the titanium core is correspondingly reduced. Heating the assembly above 980 °C (1796 °F) for at least 20 min is sufficient to form a joint with a shear strength of about 80 MPa (11 ksi), provided the joint is stress relieved by a subsequent heat-treatment at 400 °C (750 °F) [Qiao, Zhang and Jin 2003]. Bonding to the alumina is accomplished by the formation of an NiTiO interfacial phase. This process is a hybrid of active brazing, involving titanium, and diffusion brazing.

In place of foils, powder mixtures can be used to affect diffusion-brazed joints. This opens the possibility of applying the filler metal by other means such as screen printing and syringe dispensing (when in a suitable binder). This has been demonstrated for an active diffusion-brazed joint made between Fecralloy (Fe-30Cr-5Al-0.06Y_2O_3) and calcia-stabilized zirconia (CSZ) [Li and Xiao, 2001]. Starting with a powder mixture of equivalent composition Fe-20Cr-5Al and heating to 1000 °C (1830 °F), the aluminum melts, dissolves some chromium, and is then able to wet the ceramic. Chromium is consumed by the formation of chromates at that interface, while both aluminum and chromium dissolve in iron, forming a solid solution. Consequently, the braze solidifies at the joining temperature. Further information on active brazes is given in Chapter 7.

Another embodiment of diffusion brazing involves conducting the joining operation with a temperature gradient imposed. Aluminum alloys have been joined using copper foils in this manner. The temperature gradient forces the liquid interface to move slowly through the joint and into the parent materials until it runs out of copper and solidifies. For reasons not fully understood, as the interface moves, it tends to adopt a rippled, sine wave profile, which greatly reduces any tendency of the joint to fail when stressed in shear [Shelley 1998]. Unfortunately, the complexity of the equipment required to develop the necessary controlled temperature gradient limits the applicability of this novel approach.

The minimum suitable temperature for a diffusion brazing process is not simply determined by the melting point of the filler or that of the alloys generated by reaction. If it is desired that the end product of reaction is a solid solution across the joint and that there are no residual interfacial intermetallic phases present, then the solubility limit of the minor constituents in the primary metal phase is also crucially important. This issue was highlighted in the development of a fluxless copper-tin diffusion brazing process, which makes use of tin-coated copper foils (100 μm, or 4 mils, thick), sandwiched between copper-coated components [Sangha, Jacobson, and Peacock 1998]. With the aim of avoiding the formation of the brittle Cu_3Sn (ε) intermetallic phase by reaction of tin with copper, a process temperature exceeding the melting point of this phase of 676 °C (1249 °F) was chosen. The effect of brazing temperature on joint strength is shown in Fig. 6.2. However, in addition, it was found that only when the maximum thickness of the tin layer was restricted to 2.5 μm (100 μin.) was it possible to consistently suppress the formation of the Cu_3Sn intermetallic phase. This is due to the substantial decline in the solubility of tin in copper as the temperature falls toward ambient room temperature. Whereas copper can accommodate up to 16% tin at 520 °C (970 °F), the solubility reduces to a mere 1% at 200 °C (390 °F). Figure 6.3 shows a section through an optimized joint made at 690 °C (1275 °F), using the copper-tin diffusion brazing process, with 2.0 μm (80 μin.) of tin, showing an absence of visible particles of the intermetallic Cu_3Sn intermetallic phase.

The diffusion brazing process based on copper-tin has been successfully employed in producing heat exchangers for Sterling engines by a layer manufacturing technique, as described in section 6.5.

6.3 Modeling of Diffusion Brazing

Attempts have been made to model diffusion brazing processes in order to understand the significance of the various process parameters and their interrelationship [Tuah-Poku, Dollar and Massalski 1988; Nakagawa, Lee, and North 1991; MacDonald and Eagar 1992; 1998; Zhou, Gale and North 1995]. The analysis is simplest for binary alloy systems comprising solid solutions or simple eutectics, which do not include intermetallic compounds. Some of the limits of these models are highlighted in Nicholas [1998, 94–101]. Even so, further refinements to the models continue to be made, including the treatment of ternary systems [Sinclair, Purdy, and Morral 2000].

In their review of the subject, Zhou, Gale, and North [1995] split the process of diffusion brazing (involving the relatively simple configuration of a homogeneous filler between two components of the same base material) into four stages, which are identical to those enumerated in section 6.1:

1. Heating of the assembly from room temperature to the filler melting temperature, during which significant interdiffusion occurs between the parent material and the filler metal. This part of the process is particularly important when the filler metal layer is very thin.
2. Dissolution of the base material into the molten filler, resulting in the width of the liquid zone increasing. At the same time, the temperature increases from the melting point to the joining temperature.
3. Isothermal solidification at the joining temperature, as a result of diffusion of one or more constituents of the filler into the base metal (and vice versa).
4. Homogenization of the joint and base metal.

In the simplest case considered, the parent material and filler are both elemental metals and enter into a single eutectic reaction, and no intermediate intermetallic compounds form. The diffusion brazing process can then be represented on the generic eutectic phase diagram shown in Fig. 6.4.

Analytical models of diffusion brazing make use of the following assumptions:

- There are no initial barriers to direct contact between the parent material and filler metal

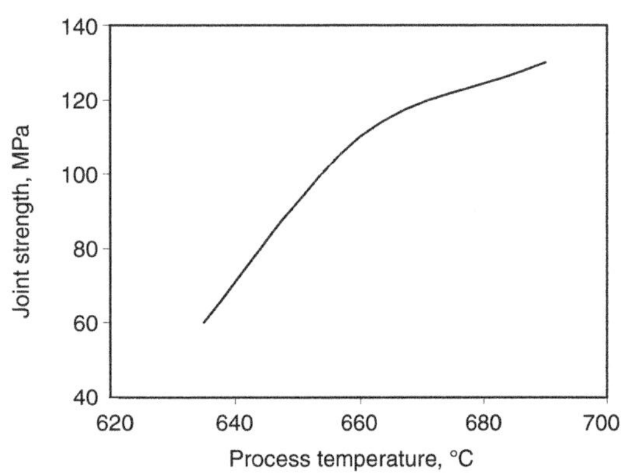

Fig. 6.2 Plot of shear strength as a function of brazing temperature for diffusion-brazed assemblies, each comprising a foil of copper plated on both sides with a 2 μm (80 μin.) thick layer of tin sandwiched between copper-plated CuCrZr test pieces. The assembly was held for 5 min at the brazing temperature under a compressive load of 2.5 ± 0.5 MPa (360 psi). The variation in shear strength is consistent with the progressive dissolution and dispersion of tin in copper and the concomitant reduction in the formation of the Cu_3Sn intermetallic phase.

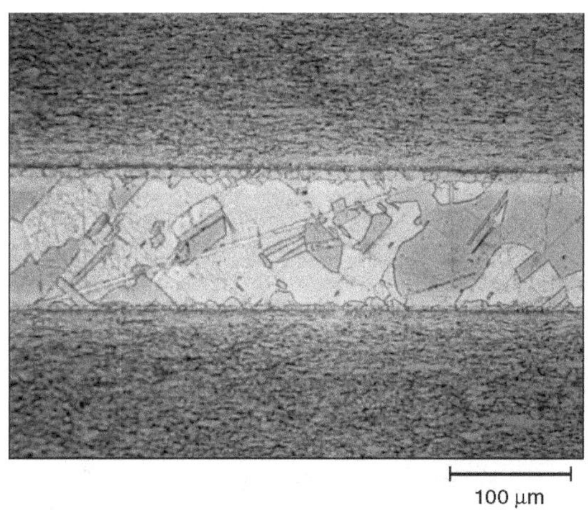

Fig. 6.3 Micrograph of a well-reacted joint in a dispersion-strengthened copper assembly formed at 690 °C (1274 °F). There is no visible evidence of residual intermetallic Cu_3Sn phase at the interface between the reaction zone and the copper components in this sample.

so that metal-to-metal interface formation can occur instantaneously when the components are brought into contact. This means that the effect of surface oxides in inhibiting or retarding diffusion across interfaces is neglected.

- It is implicitly assumed that the components and filler are pressed together under sufficient applied stress, that any asperities yield so that contact is rapidly established across the entire surface areas of abutment.
- The parent and filler metals are initially homogeneous.
- The classical Fick's diffusion equations apply.
- The parent metal has a semi-infinite surface at the joint interface, in the simplest case.
- Mass conservation applies; in particular, there is no loss of the low-melting-point constituent from the edges of the joint.
- The influence of grain boundaries is not taken into account.

If intermetallic phases form between the filler and that of the final primary metal solid solution, the situation becomes more complicated. In particular, the intermetallic phases hinder the dissolution process because the low-melting constituent then has to diffuse through the intermetallic phase. Diffusion in intermetallic compounds is generally much slower than in pure metals.

Applying this approach, but not taking into account solute dissolution during the heating stage (stage 1 above) nor density changes on solidification, Tuah-Poku, Dollar, and Massalski derived a relatively simple relationship between the reaction time (t) on the one hand and the thickness of the low-melting-point interlayer (W_0); the diffusivity of the filler metal B in the base metal A (D_α); the initial concentration of the filler metal (C_B, normally unity); and the concentration of the filler metal at the solidifying interface ($C_{\alpha L}$, which is identical to the solubility limit of the filler in the base metal at the process temperature). This situation can be described by:

$$t = \frac{\pi W_0^2}{16 D_\alpha}\left(\frac{C_B}{C_{\alpha L}}\right)^2 \qquad \text{(Eq 6.1)}$$

A refinement to Eq 6.1 has been made by MacDonald and Eagar [1998], which takes into account movement of the solid/liquid interface and density changes on solidification. This adds the following terms to the above equation: ρ_L, ρ_S (density of the liquid and solid, respectively, on either side of the solidifying interface) and $\beta = M/2\sqrt{D_\alpha t}$, where M denotes the position of the interface. The modified equation is:

$$t = \frac{W_0^2}{16 D_\alpha}\left(\frac{\rho_L C_B}{\beta \rho_S C_{\alpha L}}\right)^2$$

According to both equations, the layer of liquid in a stable diffusion brazing process should reduce in thickness with time, in proportion to $t^{1/2}$. This square root dependence has been observed when diffusion brazing nickel and nickel alloy components with an Ni-B interlayer [Nicholas 1998, 96–98].

The simpler equation (Eq 6.1) has been applied in the analysis of the copper-tin diffusion brazing at 700 °C (1290 °F), i.e., in the temperature range between 676 °C (1249 °F), the decomposition temperature of the Cu_3Sn phase, and 756 °C (1393 °F), the decomposition temperature of the γ phase, for which the diffusion data are available. In this case, there is a complication because, in order for the fully reacted end product to be primary copper, tin has to diffuse through the intervening γ and β copper-tin phases (see Fig. 6.5). This not only limits the

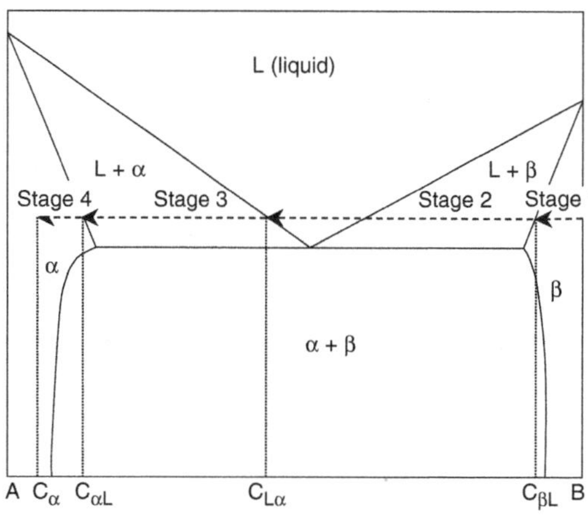

Fig. 6.4 Sequential stages in diffusion brazing for a parent metal A and filler metal B that enter into a single eutectic reaction and do not form intermediate intermetallic compounds. In stage 1, at the commencement of heating, some interdiffusion occurs until melting commences at composition $C_{\beta L}$. Dissolution of some parent continues through stage 2, and the zone of liquid widens, until composition $C_{L\alpha}$ is reached, when isothermal solidification begins. In stage 3, isothermal solidification proceeds and is completed when the last increment of liquid transforms to solid with the composition $C_{\alpha L}$. After solidification is complete, the remaining β phase can be dissipated by a homogenization heat treatment (stage 4).

reaction rate but adds complexity to the theoretical analysis. Furthermore, the rapidly declining solubility of tin in copper as the temperature is reduced toward room temperature promotes re-precipitation of Cu_3Sn and introduces a further issue for consideration.

In order to use the previously described analytical models, an average diffusivity, $D_{\bar{\alpha}}$, is defined, which takes account of the diffusion of the low-melting-point filler through any intervening intermetallic compounds to primary phase. Furthermore, the term $C_{\alpha L}$ is replaced in Eq 6.1 by $C_{\alpha\beta}$, the limit of solubility of the filler in the base metal at the joining temperature. As shown by MacDonald and Eager [1992], this equation can be represented in nomograph form, reproduced in Fig. 6.6. The value of $D_{\bar{\alpha}}$ is not readily available from the literature but can be calculated from the variables in Eq 6.1 after some limited experimental work to determine the time required for complete solidification to occur for the given thickness of the low-melting-point phase.

The estimated value of $D_{\bar{\alpha}}$ for tin in copper, corresponding to a tin layer thickness of 2 μm (80 μin.) and a bonding time of ~10 min at 680 to 690 °C (1256–1274 °F), is 2×10^{-13} cm²/s. Using this value of $D_{\bar{\alpha}}$, the bonding time required to fully react tin layers 5 μm (200 μin.) and 10 μm (400 μin.) thick are 90 and 360 min, respectively, which are far in excess of the bonding time used in the copper-tin diffusion brazing process development. These long diffusion times help to explain why it was not possible to completely react and disperse the tin into copper solid-solution when tin coatings of these thicknesses were tried. There is another contributory factor responsible for a residual copper-tin phase, Cu_3Sn, to be present at the end of the bonding cycle. The diminishing solubility of tin in copper as the diffusion-brazed assembly is cooled down to room temperature will cause this phase to re-precipitate out of copper solid solution.

It is worth pointing out that the aggregate diffusivity value, $D_{\bar{\alpha}}$, calculated from Eq 6.1 lies

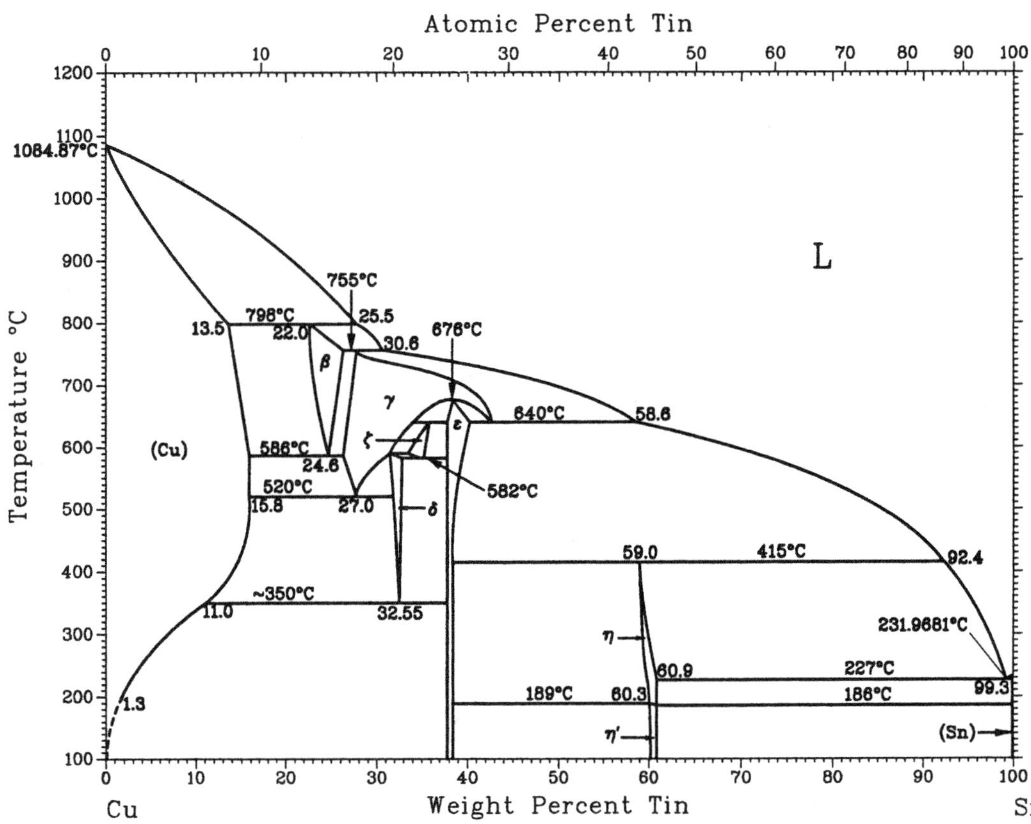

Fig. 6.5 Copper-tin phase diagram. Source: Saunders and Miodownik [1990]. Published in Massalski [1990]

almost exactly between the measured value of ~5 × 10^{-11} cm^2/s for the diffusion of tin in Cu_3Sn at 707 °C (1305 °F) and the corresponding value of ~9 × 10^{-16} cm^2/s for tin diffusion in pure copper [Brandes and Brook 1992]. These values of diffusivity are therefore consistent with one another, especially taking into account the respective melting points of pure copper (1085 °C, or 1985 °F) and of the γ phase (676–756 °C, or 1249–1393 °F). As a general rule for metals, diffusivity generally decreases as the melting point increases.

This example shows the value of the simple analytical model for providing general guidelines. However, as Zhou, Gale, and North [1995] have pointed out, some caution needs to be observed in the use of the analytical models of transient liquid-phase bonding that have been proposed, and they cannot be relied on for quantitative purposes. In the first place, the resolution of such a joining process into a number of discrete processes is at variance with what actually occurs in practice. Classical diffusion theory yields a parabolic relationship for base metal dissolution as a function of time at constant temperature. Such behavior is not generally observed.

Grain boundaries are well known to play a major influence on the rate of diffusion and, therefore, on the kinetics of the diffusion brazing process. Grain boundaries are high-diffusivity paths, with diffusivity values typically four orders of magnitudes higher than those for volume diffusion for the same element close to the melting point, although the aggregate rate of material transport along grain boundaries is limited by virtue of their narrow width, which is typically ~1 nm (0.04 μin.). Grain boundaries will also enhance diffusion rates by increasing the area of a solid surface because they form grooves where they emerge at the free surface. The density of grain boundaries is directly related to the grain size of a material, so the grain boundary effect on diffusion brazing can be seen in the enhancement of the diffusion brazing process as grain size is reduced, all other parameters remaining unchanged. This effect has been confirmed in the diffusion brazing of silver and nickel components [Nicholas 1998, 97–98]. It is widely held that future work on the modeling of TLP bonding should focus on a numerical model that encompasses all the mass transport mechanisms that occur during the process, namely, volume diffusion in each of the phases present, grain boundary diffusion, and interfacial diffusion between the constituent phases. This will be a formidable undertaking, all the more so where there are more than two elemental constituents and where transitory phases form and dissociate during the bonding process. Such an approach has been investigated by Ohsasa, Shinmura, and Narita [1999] for analyzing the TLP bonding of nickel using the Ni-15.2Cr-4.0B ternary alloy braze.

6.4 Application of Diffusion Brazing to Wide-Gap Joining

Diffusion brazing has been effectively adapted to bridging wide gaps. As detailed in Chapter 4, section 4.3.4.2, powder mixtures comprising the braze and a "gap filler" are used mostly for wide gap brazing and require the application of pressure to minimize gross voiding or porosity. If these constituents and the process conditions are judiciously chosen, it is possible

$D_{\bar{\alpha}} = 2 \times 10^{-13}$ m^2/s

Tin thickness, μm	Process time, h
2	0.2(a)
5	1.5
10	6

(a) measured

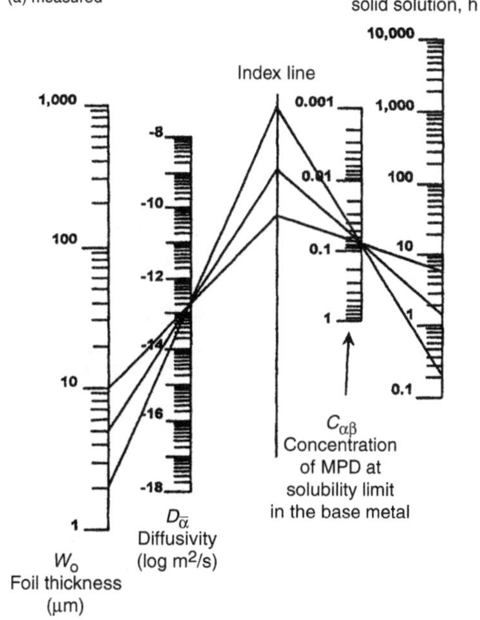

Fig. 6.6 Nomograph based on Eq 6.1 in the text defining the relationship between brazing time (t), tin thickness (W_0), and the diffusivity ($D_{\bar{\alpha}}$) of the solute in the base metal, according to the model of Tuah-Poku, Dollar, and Massalski 1988 [adapted from MacDonald and Eagar 1992]. Results for copper-tin diffusion brazing are shown on this diagram. Here tin is the melting point depressant, MPD.

to achieve a reaction between the braze and "gap filler," which leads to isothermal solidification and complete dispersion of the braze into the gap filler and parent materials.

One approach that has been followed for joining titanium aluminide is to apply a composite mixture of the aluminide and copper powder (with particle sizes less than 44 μm, or 1.7 mils, and 53 μm, or 2.1 mils, respectively), in the form of a slurry to the joint [Gale et al. 2002]. The components of titanium aluminide with the slurry interlayer were then pressed together under a compressive load of 4 MPa (580 psi) and heated to 1150 °C (2100 °F) for 10 min. This process temperature is above the melting point of copper (1085 ° C, or 1985 °F), but over 200 °C (360 °F) below that of the titanium aluminide. In consequence, the copper melts and diffuses into the aluminide, solidifying at the process temperature. As applied, the layer of slurry was 500 μm (20 mils) thick but contracted to a final joint width of about 200 μm (8 mils). As bonded, the joints registered a four-point bend strength of 580 MPa (85 ksi). Gale [1999] has also reviewed the application of diffusion brazing (TLPB) to the joining of other structural intermetallic compounds.

A different scheme, but still involving powders, has been applied to large root-opening 304 stainless steel joints [Zhuang and Eagar 1997]. In this case, individual particles of nichrome (Ni-20Cr) powder, less than 44 μm (1.7 mils) in size, were coated with a layer of Ni-10P by electroless deposition, with the deposit typically making up 16 wt% of the processed powders. The powder was compacted into 10 mm discs in a die under 350 MPa (51 ksi) compression. Subsequent heating of the 314 stainless steel assemblies at 1000 °C (1830 °F) with compacts held in place under a compressive load of 0.29 MPa (42 psi) and protected in a vacuum of better than 4×10^{-1} mPa (3×10^{-5} torr) led to isothermal solidification and achieved full densification. The coating ensured that liquid existed between individual particles of nichrome, and this considerably aided the formation of fully dense joints, with tensile strengths that approached that of the stainless steel base metal (~580 MPa, or 85 ksi), for joints 1 to 4 mm (40–160 mils) wide. While the phosphorus evidently provided fluxing of the stainless steel surfaces, the resulting phosphides that formed represented the weakest link in the joints. Trials using coated 304L stainless steel powder in place of the nichrome gave poorer results, namely, significant porosity and weaker joints. It is believed that, being soluble in iron, both the nickel and phosphorus in the coating dissolved into the steel particles and isothermal solidification occurred before most of the shrinkage could occur.

Zorc and Kosec [2000] have used a system of parallel wires to reinforce a joint gap with a braze used to wet and fill the interstices. By operating a diffusion brazing cycle, under compressive loads, continuous joints between the parent material and the reinforcement wires were achieved. Because the joints between the parent materials and the wires were not interrupted by layers of relatively weak braze, the resulting joints between common engineering materials (copper, steels) proved to be exceptionally tough and reasonably strong. The wire reinforcements, in all cases, 2 mm (80 mils) in diameter, were either copper or copper-plated steels used in conjunction with copper parent material or of various steels to match steel components. A selection of silver- and nickel-base brazing alloys and copper-phosphorous eutectic (Cu-7.3P, or BCuP-2) were used and the joining operations, which were carried out above their respective melting (solidus) temperatures. Extremely high pressures were applied for the duration of the diffusion brazing operation ranging between 2.4 GPa (350 ksi) for the copper combination, to 4.1 GPa (590 ksi) for the steel parent materials and reinforcements. The high pressures noticeably flattened the wires and punched them into the surfaces of the parent materials. These are somewhat extreme conditions, which are likely to greatly limit the practical application of this approach.

6.5 Application of Diffusion Brazing to Layer Manufacturing

Diffusion brazing offers an effective means for manufacturing parts with geometries that are difficult, if not impossible, to achieve using conventional manufacturing technology. This has been convincingly demonstrated for heat exchangers designed for use with Sterling engines.

A Sterling engine makes use of an external source of heat, which is supplied to a gas. The hot gas drives a piston, which converts the heat into useful work. Residual heat must then be rapidly removed from the gas in readiness for the next cycle. The heat removal is achieved by passing the gas through specially designed cooler elements. The design required 1800 oval tubes 0.7 mm × 0.9 mm (28 × 35 mils) in an annular zone of a finned copper cylinder 68 mm (2.7 in.) long for carrying the gas to be cooled. The central hollow space of the copper cylinder was to contain the water coolant. This heat exchanger module is shown in plan and cross section in Fig. 6.7. The manufacture of these items presented severe problems in respect of the oval holes. Even circular holes of this fine size would be difficult to drill conventionally because of their length. A possibility that was considered was machining a series of notched annular rings, each 1 mm (40 mils) thick and of increasing diameter that would have to be shrink fitted together. However, this approach was ruled out as difficult, expensive, and unlikely to be technically successful. The solution was to fabricate the coolers using a layer manufacturing process involving bonding copper sheets or laminations together by diffusion brazing [Bocking, Jacobson, and Bennett 2000].

The laminations, of oxygen-free high conductivity (OHFC) copper, were 0.455 mm thick, and 185 such sheets were required for each assembly (which included blanking plates at either end for spigots). It was decided to produce the holes by photochemical machining, which is eminently suitable for cutting fine detail in thin sheets. In all, six coolers were required for each engine. Accordingly, they were all built together in a single operation, and each sheet was etched with the layer patterns for all six coolers, as shown in Fig. 6.8. Following photochemical machining, the sheets were plated on each side, with 2.0 ± 0.5 μm (8 ± 2 μin.) of pure tin. The sheets were then stacked in a jig, provided with three alignment pins to ensure that they registered correctly, as shown in Fig. 6.9. The stack was then pressed between a pair of steel plates at a pressure of approximately 3 MPa (440 psi). The pressure was applied by means of steel bolts positioned along the central cavity in each of the six heat exchanger modules. An appropriate torque was applied to give a compressive loading of ~90 kN (10 tons) on each bolt. This gave an overall pressure of close to 2 MPa (290 psi). A further 1 MPa (145 psi) of compression was provided by the thermal expansion mismatch between the tool steel press and the copper as the assembly was heated to a bonding temperature of 820 °C (1510 °F). The bonding operation was

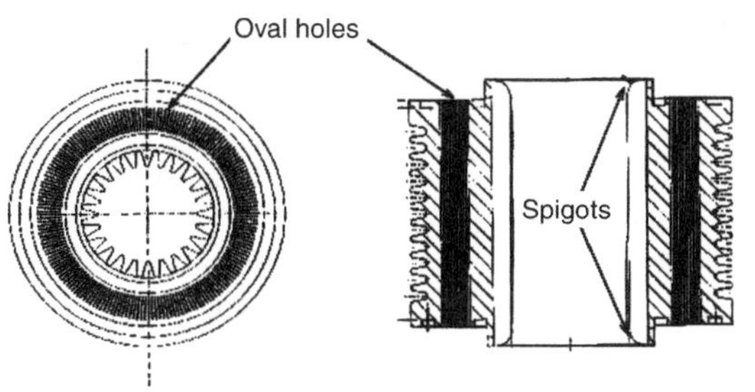

Fig. 6.7 (a) Plan and (b) cross-sectional views of a heat exchanger module. The number (1800), aspect ratio (~85:1), and size of the oval holes, each measuring 0.7 mm × 0.9 mm (28 × 35 mils) in diameter by 68 mm (2.7 in.) long, would make manufacture of these parts from solid an expensive proposition.

Fig. 6.8 Single sheet of copper patterned by photochemical etching to contain a single plan section through six heat exchanger modules. To achieve the total component height, 185 identical sheets were required. The diffusion braze, in this case, 2 μm (8 μin.) of tin, was applied to both sides of each sheet by electroplating.

Fig. 6.9 Patterned, plated, and stacked sheets, aligned in a jig ready for application of the compressive load and diffusion brazing in a vacuum oven. The diffusion brazing conditions were a compressive stress of 3 MPa (440 psi) and a process temperature of 820 °C (1510 °F), sustained for 10 h.

Fig. 6.10 Fully machined heat exchangers fabricated by copper-tin diffusion brazing

carried out in a vacuum oven with a residual pressure of 10 Pa (1.5 × 10^{-4} psi). The heating rate used was 5 °C/min (9 °F/min), and the assembly was held at the peak joining temperature for 10 hours to ensure sustained thermal equilibrium throughout. Four of the six fully machined heat exchanger modules demonstrated good hermeticity and absence of blockages of any of the fine holes through the seepage of filler (Fig. 6.10). The two units that were not leak-tight had two small leaks, which were ascribed to inadvertent contamination of the tin plating from residual photoresist. These leaks were subsequently sealed with a solder applied to the inner walls of the two heat exchangers and were perfectly usable thereafter.

This successful trial clearly demonstrated the capabilities of layer manufacturing combined with diffusion brazing for manufacturing metal products with complex internal features. It is envisaged that this approach could be developed into an important rapid manufacturing method [Bocking et al. 1997].

REFERENCES

- Askelsen, O.M., 1992. Diffusion Bonding of Ceramics, *J. Mater. Sci.,* Vol 27, p 569–579
- Bocking, C. et. al., 1997. Rapid Prototype Tools Using Layer Manufacturing Technology, *GEC J. Technol.,* Vol 14, p 110–14
- Bocking, C., Jacobson, D., and Bennett, G., 2000. Layer Manufacturing of Heat Exchange Elements Using Photochemical Machining, Electroplating and Diffusion Brazing, *Trans IMF,* Vol 78 (No. 6), p 243–246
- Brandes, E.A. and Brook, G.B., Ed., 1992. *Metals Reference Book,* Butterworth-Heinemann
- Bushby, R.S. and Scott, V.D., 1995. Liquid Phase Bonding of Aluminium and Aluminium/Nicalon Composite Using Interlayers of Cu-Ag Alloy, *Mater. Sci. Technol.,* Vol 11 (No. 7), p 643–649
- Duvall, D.S., Owczarski, W.A., and Paulonis, D.F., 1974. TLP Bonding: A New Method for Joining Heat Resistant Alloys, *Weld. J. Res. Suppl.*, Vol 53 (No. 4), p 203s–214s
- Elahi, M. and Fenn, R., 1981. The Joining of a Titanium Alloy Using a Copper/Silver Intermediate Layer, *Proc. Conf. Joining of Metals: Practice and Performance,* April 10 (Coventry, UK), p 137–144
- Gale, W.F., 1999. Applying TLP Bonding to the Joining of Structural Intermetallic Compounds, *JOM,* Vol 51 (No. 2), p 49–52
- Gale, W.F. and Orel, S.V., 1996. A Microstructural Investigation of NiAl/Ni-Si-B/NiAl Transient Liquid Phase Bonds, *J. Mater. Sci.,* Vol 31, p 345–349
- Gale, W.F. et al., 2002. Microstructure and Mechanical Properties of Titanium Aluminide Wide-Gap, Transient Liquid-Phase Bonds Prepared Using a Slurry-Deposited Composite Interlayer, *Met. Mat. Trans. A,* Vol 33A, p 3205–3214
- Ikawa, H., Nakao, Y., and Isai, T., 1979. Theoretical Considerations on the Metallurgical Processes in TLP Bonding of Nickel-Based Superalloys, *Trans. Jpn. Weld. Soc.*, Vol 10 (No. 1), p 24–29
- Khan, T.I. and Wallach, E.R., 1995. Tran-

sient Liquid-Phase Bonding of Ferritic Oxide Dispersion Strengthened Superalloy MA957 Using a Conventional Nickel Braze and a Novel Iron-Base Foil, *J. Mater. Sci.*, Vol 30, p 5151–5160

- Khanna, P.H. et al. 2000. Novel Ni/Al/Ni Diffusion Soldered Joints for High Temperature Applications, *Proc. Conf. EuroMat 2000,* Sept 27–30 1999 (Munich, Germany) p 1–6
- Li, J. and Xiao, P., 2001. Joining Ceramic to Metal Using a Powder Metallurgy Method for High Temperature Applications, *J. Mater. Sci.,* Vol 36 (No. 6), p 1383–1387
- MacDonald, W.D. and Eagar, T.W., 1992. Transient Liquid Phase Bonding Processes, in Cieslak, M.J. et al., Ed., *The Metal Science of Joining,* TMS, p 93–100
- MacDonald, W.D. and Eagar, T.W., 1998. Isothermal Solidification Kinetics of Diffusion Brazing, *Met. and Mater. Trans. A,* Vol 29A, p 315–325
- Nakagawa, H., Lee, C.H., and North, T.H., 1991. Modelling of Base Metal Dissolution Behavior during Transient Liquid-Phase Brazing, *Metall. Trans. A,* Vol 22A (No. 2), p 543–555
- Nakahashi, M. et al., 1985. Transient Liquid Phase Bonding for Heat Resistant Steels, *J. Jpn. Inst. Met.,* Vol 49 (No. 4), p 285–290
- Nicholas, M.G., 1998. *Joining Processes, Introduction to Brazing and Diffusion Bonding,* Kluwer Academic
- Niemann, J.T. and Wille, G.W., 1978. Fluxless Brazing of Aluminum Castings, *Weld. J. Res. Suppl.,* Vol 57 (No. 10), p 2855s–2915s
- Norris, B., 1986. Liquid Interface Diffusion (LID) Bonding of Titanium Structures, *Proc. Conf. Designing With Titanium,* July 7–9 (Bristol, UK), p 83–86
- Ohsasa, K., Shinmura, T., and Narita, T., 1999. Numerical Modeling of the Transient Liquid Phase Bonding Process of Ni Using Ni-B-Cr Ternary Filler Metal, *J. Phase Equilibria,* Vol 20 (No. 3), p 199–206
- Ornellas, D.L. and Catalano, E., 1974. Diffusion Bonding of Gold to Gold, *Rev. Sci. Instrum.,* Vol 45 (No. 7), p 955
- Owczarski, W.A., 1962. Eutectic Brazing of Zircaloy 2 to Type 304 Stainless Steel, *Weld. J. Res. Suppl.,* Vol 41 (No. 2), p 78s–83s
- Qiao, G.J., Zhang, C.G., and Jin, Z.H., 2003. Thermal Cyclic Test of Alumina/Kovar Joint Brazed by Ni-Ti Active Filler, *Ceram. Industry,* Vol 29, p 7–11
- Ricks, R.A. et al., 1989. Transient Liquid Phase Bonding of Aluminum-Lithium Base Alloy AA8090 Using Roll-Clad Zn Based Interlayers, *Proc. Conf. Aluminum Lithium Alloys,* March 27–31 (Williamsburg, VA), p 441–449
- Sangha, S.P.S., Jacobson, D.M., and Peacock, A.T., 1998. Development of the Copper-Tin Diffusion-Brazing Process, *Weld. J. Res. Suppl.,* Vol 77 (No. 10), p 432s–438s
- Saunders, N. and Miodownic, A.P., 1990. Cu-Sn (Copper-Tin) *Binary Alloy Phase Diagrams* 2nd Edition, Massalski, T.D. ed. ASM International, 1990, p 1482
- Schwartz, M.M., 1987. *Brazing,* ASM International
- Schwartz, M.M., 2003. *Brazing,* 2nd ed., ASM International
- Shalz, M.L. et al., 1994. Ceramic Joining II: Partial Transient Liquid-Phase Bonding of Alumina via Cu/Ni/Cu Multilayer Interlayers, *J. Mater. Sci.,* Vol 29, p 3200–3208
- Shelley, T, 1998. Aluminium Alloys Can Get it Together, *Eureka;* Vol 18 (No. 6), p 51
- Sinclair, C.W., Purdy, G.R., and Morral, J.E., 2000. Transient Liquid-Phase Bonding in Two-Phase Ternary Systems, *Met. and Mater. Trans. A,* Vol 31A, p 1187–1192
- Timisit, R.S. and Janeway, B.J., 1994. A Novel Brazing Technique for Aluminum and Other Metals, *Weld. J. Res. Suppl.,* Vol 73 (No. 6), p 119s–128s
- Tuah-Poku, I., Dollar, M., and Massalski, T.B., 1988. A Study of the Transient Liquid Phase Bonding Process Applied to a Ag/Cu/Ag Sandwich Joint, *Metall. Trans. A,* Vol 19A (No. 3), p 675–686
- Urena, A. et al., 1996. Diffusion Bonding of an Aluminium-Lithium Alloy (AA8090) Using Aluminium Copper Alloy Interlayers, Part 1: Microstructure, *J. Mater. Sci.,* Vol 31, p 807–818
- Wu, X.W., Chandel, R.S., and Li, H., 2001. Evaluation of Transient Liquid Phase Bonding between Nickel-Based Superalloys, *J. Mater. Sci.* Vol 36, p 1539–1546
- Zhang, C.G., Qiao, G.J., and Jin, Z.H., 2002. Active Brazing of Pure Alumina to Kovar Alloy Based on the Partial Transient Liquid Phase (PTLP) Technique with Ni-Ti Interlayer, *J. European Ceram. Soc.* Vol 22, p 2181–2186

- Zhou, Y., Gale, W.F., and North, T.H., 1995. Modelling of Transient Liquid Phase Bonding, *Int. Mater. Rev.,* Vol 50 (No. 5), p 181–196
- Zhuang, W.D. and Eagar, T.W., 1997. Transient Liquid-Phase Bonding Using Coated Metal Powders, *Weld. J. Res. Suppl.,* Vol 76 (No. 4), p 157s–162s
- Zorc, B. and Kosec, L., 2000. A New Approach to Improving the Properties of Brazed Joints, *Weld. J. Res. Suppl.,* Vol 79 (No. 1), p 24s–31s

CHAPTER 7

Direct Brazing of Nonmetals

Frequently there is a requirement to braze together components, one or more of which is nonmetallic, in order to achieve assemblies with specific combinations of properties.

As pointed out in earlier chapters, for a molten alloy to wet and spread over the joint surfaces, a degree of chemical interaction between the filler metal and the parent materials is necessary. Where the parent materials are metals with clean surfaces, the interaction generally means alloying with some associated degree of erosion of the joint surfaces and the formation of intermetallic compounds. The wetting of nonmetallic components by filler metals is more difficult but can be accomplished by three principal routes. One of these is to apply metallizations to the joint surfaces so as to render them essentially metallic in character. The chosen metallization must obviously not be significantly soluble in the filler alloy or dewetting can occur if the molten braze erodes through it and comes into contact with the nonmetal. The design and application of metallizations that are wettable by filler metals is discussed in Chapter 4, section 4.1.2.1. Another approach involves incorporating small quantities of elements that are highly active into the filler metal. Provided that at least one of the products of reaction with the base material is metallic in character and remains as a layer on the surface of the nonmetal (even if this is extremely thin), then the filler alloy can wet it and form sound joints. In other words, the reaction results in the formation of a metallization layer as part of the brazing cycle. Some examples of active brazes are described briefly in Chapter 4, section 4.1.2.2, and further details are given in section 7.2 of this chapter. The third method, which is also discussed in the same section of Chapter 4, is to select a braze that is able to wet the nonmetal by virtue of chemical bonding between them. Copper wetted on to alumina via the copper-cuprous oxide eutectic is an example of the latter.

Direct brazing of nonmetals, i.e., without first applying a surface metallization, differs from the brazing of metals in three important respects:

- The principles underlying wetting are fundamentally different. Molten brazing alloys wet metals owing to a degree of mutual intersolubility between the constituents. Wetting occurs essentially instantaneously so that a relatively short heating cycle (up to a few minutes), provided by a torch or a furnace, that accomplishes reflow of the brazing alloy, is usually sufficient to achieve good wetting and spreading. The same is not true for brazing to nonmetals. Figure 7.1 shows the effect of time at temperature on the contact angle for several braze/nonmetal combinations. Successful brazing of nonmetals, where this is possible, normally requires time scales of more than an order of magnitude longer than for brazing metals. Typically, dwells of tens of minutes at process temperature are necessary to establish a reasonably low contact angle.

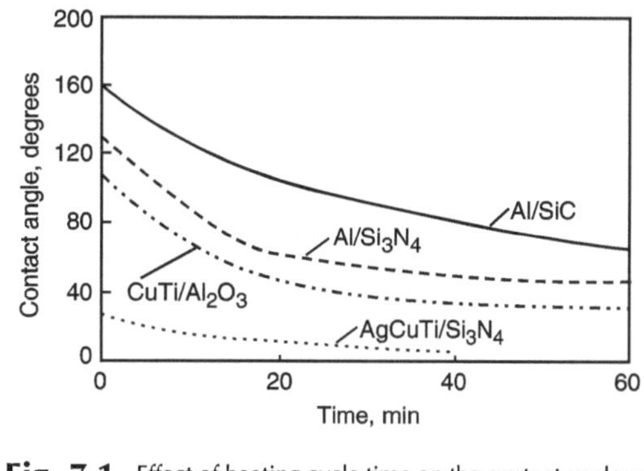

Fig. 7.1 Effect of heating cycle time on the contact angles of four braze/nonmetal combinations: Al/SiC, Al/Si_3N_4, $Cu\text{-}Ti/Al_2O_3$, and $Ag\text{-}Cu\text{-}Ti/Si_3N_4$. The time required to establish a low-contact angle is more than an order of magnitude slower than for braze/metal combinations.

- Identifying which brazes are compatible with particular nonmetals cannot be easily deduced but requires detailed knowledge of chemical thermodynamics. For example, the ability to successfully braze silicon nitride depends critically on whether the ceramic surface is virgin nitride or is oxidized. Nonoxide ceramics such as silicon nitride, silicon carbide and aluminum nitride can be endowed with a thin oxide surface film by exposure to air while the ceramics are sufficiently hot. Oxide ceramics are not wetted as readily as carbides and nitrides by reactive metals such as aluminum and silicon, and hence, without knowledge of the surface state of the nonmetal, joining can be difficult to achieve reliably [Eustathopoulos, Nicholas, and Drevet 1999].
- Nonmetals have radically different mechanical properties from metals; notably, they possess low-fracture toughness and are weak in tension. Consequently, metal/nonmetal joint configurations are more restricted, and residual stress arising from thermal expansion mismatch is an important factor that must always be taken into consideration. On the other hand, most nonmetals are stable in chemical environments and have much higher elastic moduli and wear resistance than metals. Ashby plots provide a convenient method of comparing the properties of materials. These plots were originally devised to establish a materials performance index based on specific design objectives. For example, a meaningful material performance index (M) of a stiff beam for a lightweight load-bearing application would be $M = E/\rho$, which is defined as the specific modulus. Then, all materials represented on an Ashby plot of elastic modulus (E) versus density (ρ), which fall on a line of slope equal to unity, would have an equal value of M, and therefore, equally satisfy the design criterion of equivalent performance. The use of Ashby plots to highlight some differences between nonmetals and metals are illustrated in Chapter 4, Fig. 4.4 and 4.5, and Fig. 7.2 to 7.4 (Fig. 4.4, coefficient of thermal expansivity (CTE) vs. thermal conductivity (λ); Fig. 4.5, CTE vs. elastic modulus (E); Fig. 7.2, normalized strength vs. CTE; Fig. 7.3, λ vs. E; Fig. 7.4, fracture toughness (K_{1C}) vs. E).

The joining of ceramics is treated fairly extensively in specialist books on the subject (e.g., Nicholas 1990 and Schwartz 1990), and only a brief resume of some of the critical considerations follows.

Glazing uses filler glasses rather than filler metals and, in many practical aspects, resembles brazing. Because glasses are based on oxides, they will wet and spread directly on many oxide ceramics and oxidized metals. Interestingly, active glasses, incorporating titanium oxide, are being developed to join nonoxide ceramics and nonoxidized metals [Nicholas 1993].

7.1 Wetting, Spreading, and Chemical Interaction

For a braze to wet and spread on a nonmetal requires some form of chemical interaction to occur so that the composition of the surface of the nonmetal is altered to make it sufficiently metallic in character and conducive to metallic bonding. Therefore, in brazing to nonmetals, diffusion and chemical reaction are inextricably linked to wetting and spreading by the filler.

The interaction between a braze and a nonmetal can take one of two forms. It may be classified as either chemical bonding or chemical reaction [Howe 1993a]. Which of these two types of interaction occurs for a given combination of materials can be established from a study of the contact angle of the braze with the surface of the nonmetal as a function of temperature. For a braze melted on the surface of a nonmetal where the adhesion is achieved by chemical bonding, the contact angle will decrease slowly as a function of temperature. The contact angle usually remains quite high (well above 45°). A sudden change in contact angle and/or a progressive reduction in the contact angle is indicative of chemical reaction at the interface. Thus, for example, the contact angle of aluminum on silicon carbide is approximately 160° from 660 to about 900 °C (1220–1650 °F), whereupon the contact angle drops rapidly to about 45° [Levi, Abbaschian, and Mehrabian 1978]. This fairly abrupt change is due to decomposition of the surface of the ceramic and reaction with the aluminum to form aluminum carbide at the nonmetal/braze interface, which is more readily wetted by the braze.

7.1.1 Chemical Bonding

Chemical bonding may be defined as bonding in which there is only charge transfer or sharing across the metal/nonmetal interface as the two components are brought into contact and the separation between the facing surfaces approaches the order of interatomic distances (see Chapter 1, section 1.2.8). In the ideal case, there is no accompanying reaction between constituents of the materials on either side of the original interface. Some texts refer to this situation as physical bonding. For a review of theoretical modeling of metal-ceramic chemical bonding, see Finnis 1996.

The driving force for the formation of a metal/nonmetal chemical bond is the work of adhesion, given by the the Dupré equation (Eq 1.2 in Chapter 1), which equals the work required to incrementally increase liquid-vapor and solid-vapor interfaces from a liquid-solid interface:

$$W_a = \gamma_{LV} + \gamma_{SV} - \gamma_{SL}$$

where γ_{SL} is the surface energy, or surface tension, between the solid and liquid; γ_{LV} is the surface energy between the liquid and vapor; and γ_{SV} is the surface energy between solid and vapor:

$$\gamma_{SL} = \gamma_{LV} + \gamma_{SV} - W_a$$

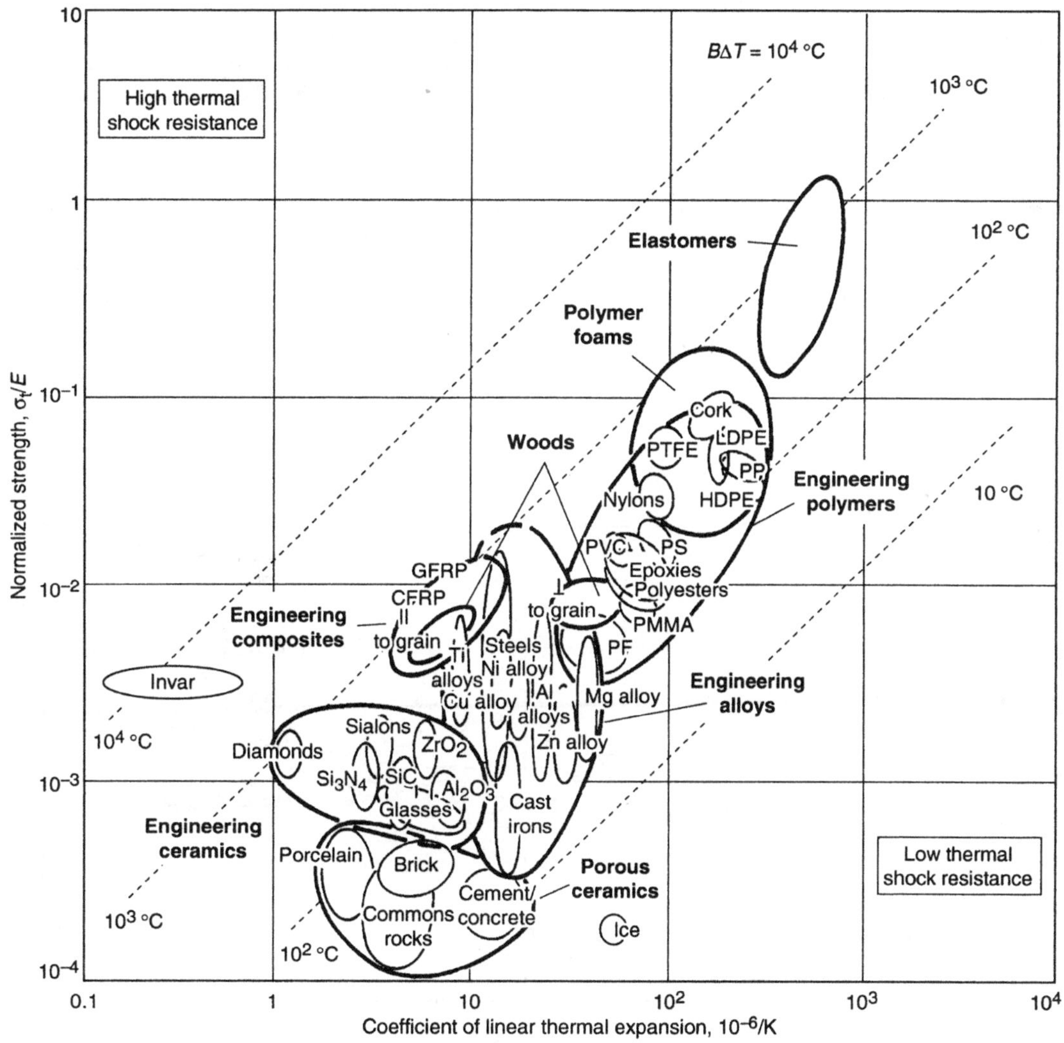

Fig. 7.2 The normalized tensile strength, σ_t/E, plotted against coefficient of linear thermal expansion, α. The contours show a measure of the thermal shock resistance, ΔT. Corrections must be applied for constraint and to allow for the effect of thermal conduction during quenching. Source: Ashby [1997]

In other words, the lower the solid-liquid interfacial energy γ_{SL} is, the higher the work of adhesion, W_a, and therefore the greater is the cohesion between the nonmetal and the metal (i.e., braze), which generally makes for stronger joints. This relationship between work of adhesion and joint strength is essentially observed in practice and is illustrated in Fig. 7.5 for different metals wetted on alumina.

There is, in theory, a direct correlation between the work of adhesion and the equilibium contact angle of the braze. The higher the work of adhesion, W_a, between the filler and the substrate material, the lower is the contact angle, as can be seen from the Young-Dupré equation presented in Chapter 1, section 1.2.2, which was obtained by combining the Young equation (Eq 1.1) with the Dupré equation (Eq 1.2):

$$W_a = \gamma_{LV}(1 + \cos\theta)$$

However, the contact angle is also affected by the surface tension of the molten filler as well as other factors, and its dependence on composition may not be straightforward. Some of the complexities involved have been highlighted by studies of the wetting of gold-silicon and aluminum-silicon binary alloys on silicon carbide, where unexpected minima in the wetting angle have been measured at intermediate compositions; see Fig. 7.6 (contact angle measured at a partial oxygen pressure of 10^{-7} mPa) [Naidich, Zhuravlev, and Krasovskaya 1998; Drevet, Kalogeropoulou, and Eustathopoulos 1993]. The contact angle may also be affected by non-equilibrium mechanisms, and in consequence vary over time, as exemplified by the wetting of molten aluminum on sapphire (α-Al_2O_3) mentioned in Chapter 1, section 1.2.2. The possibility raised by Drevet, Kalogeropoulou, and Eustathopoulos [1993] that the presence of small

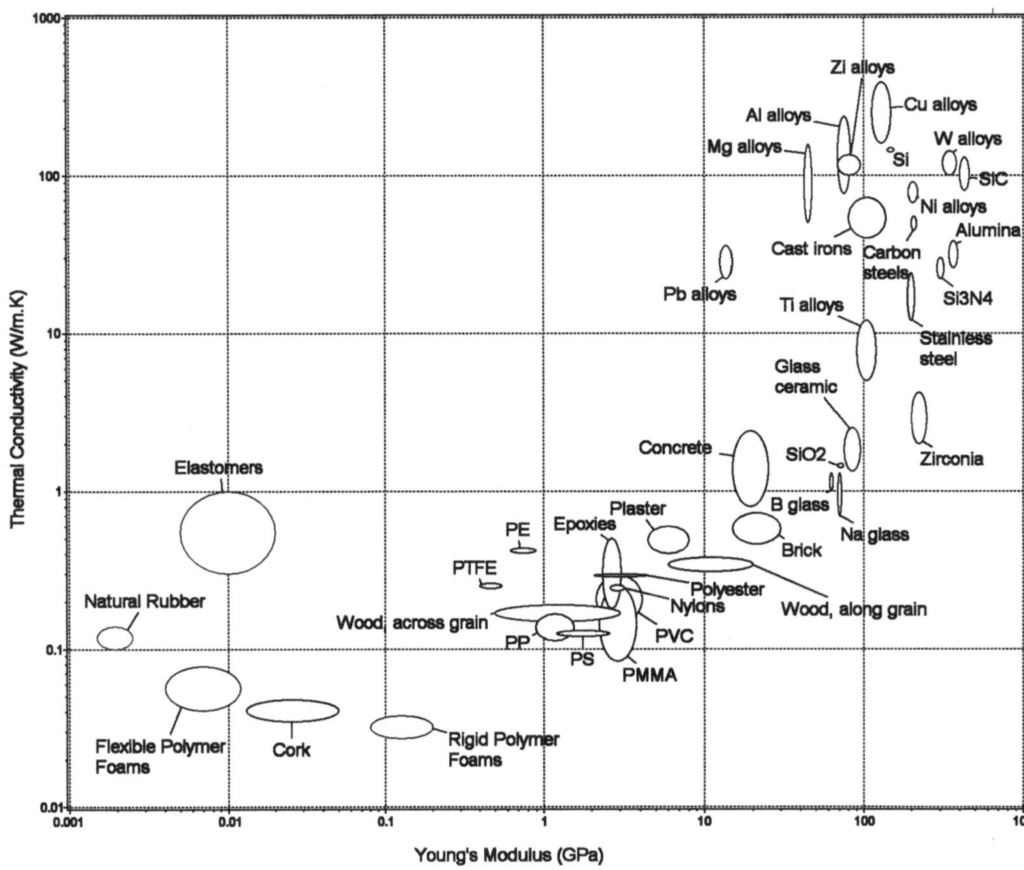

Fig. 7.3 An Ashby materials selection chart showing groups of materials plotted in terms of their thermal conductivity and elastic modulus. Courtesy of Granta Design Ltd. This figure was plotted with Cambridge Engineering Selector 4.5 [Granta 2004]

amounts of oxygen in the process atmosphere improves wetting (as compared with a high vacuum; see Fig. 7.6) has been queried. In the light of their own experimental findings, Naidich, Zhuravlev, and Krasovskaya [1998] point out that the various factors that influence chemical bonding and wetting of ceramic materials by molten metals are still far from clear and require further study.

Ionic materials are more difficult for brazes to wet than are covalent materials because the electrons are tightly bound within an ionic material so their propensity to share electrons with the electron cloud in metals is correspondingly lower. Table 7.1 illustrates this point for aluminum, copper, and silicon wetted on to alumina, silicon carbide, and graphite, with regard to the wetting angle, which is lower for the same metal on ceramics of lower ionicity [Warren and Andersson 1984]. Oxide ceramics, in particular, are predominantly ionic, and any bonding to a braze is governed by the interaction between the metal ions in the braze and the oxygen anions in the ceramic. In this case, the degree of wetting and strength of bonding is determined by the affinity of the liquid metal for oxygen.

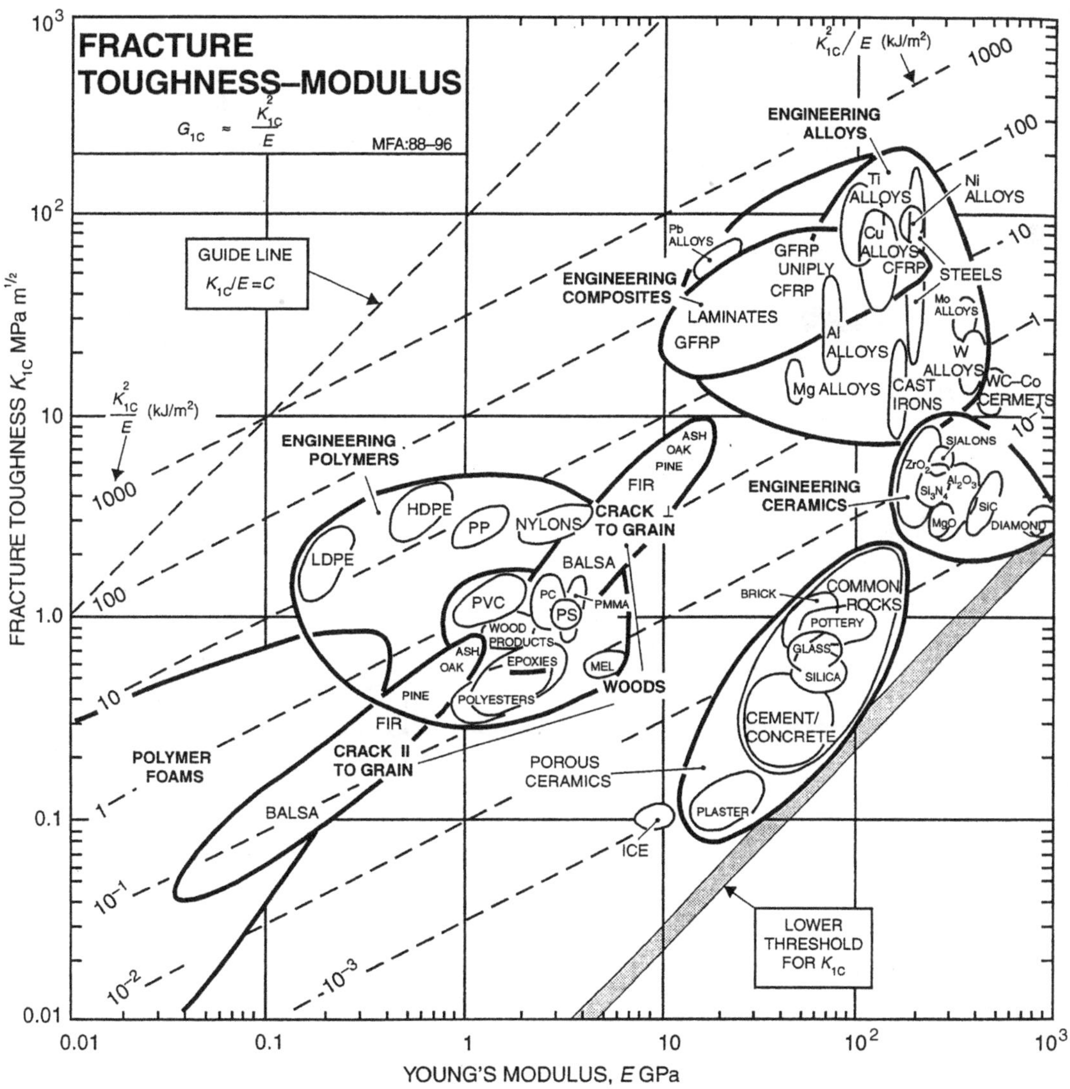

Fig. 7.4 An Ashby materials selection chart showing groups of materials plotted in terms of their fracture toughness and elastic modulus. Source: Waterman and Ashby [1997]

By contrast, covalent materials, such as the ceramics silicon carbide and aluminum nitride, are more readily wetted by metals, even when the systems are ostensibly nonreactive combinations. Thus, molten aluminum wets aluminum nitride, and molten silicon wets silicon carbide with contact angles of approximately 40 to 50° [Eustathopoulos, Nicholas, and Drevet 1999].

The effect of ionicity on wetting is evident in the effect of varying the carbon content (x) of titanium carbide (TiC_x) on the wetting angle of chemically bonded brazes as shown in Fig. 7.7. Clearly, the higher the proportion of the titanium in relation to carbon, i.e., of "free" titanium metal in the carbide, the more readily a molten metal will wet it. The same principle has been applied to achieve wetting of titanium carbide components by nonreactive metals (copper, silver, tin, and gold). It involves subjecting the surface of the titanium carbide to a partial oxidation treatment so as to denude carbon from the surface of the ceramic and leave its surface metal-rich [Froumin et al. 2000]. This treatment is equivalent to incorporating titanium in the braze, as shown in Fig. 7.8 [Frage, Froumin, and Dariel 2002].

Chemical bonds form joints of acceptable strength for many applications. Figure 7.9 illustrates the strength of alumina/metal couples that bond by this type of mechanism, measured at room temperature [Crispin and Nicholas 1976]. As might be expected, metals with a higher oxygen affinity (i.e., more "active" in this context) make for superior brazes than more noble metals in this case.

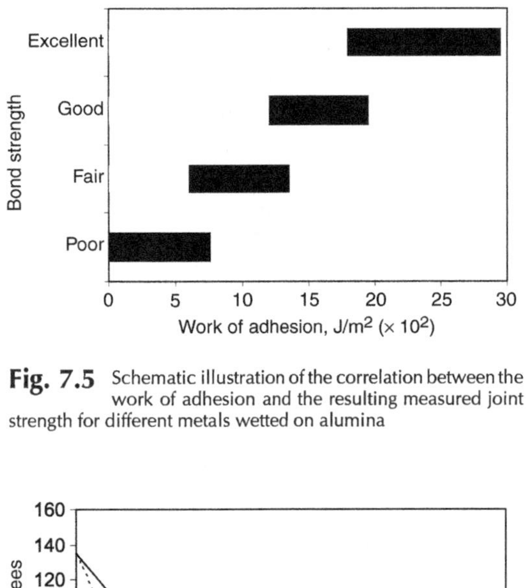

Fig. 7.5 Schematic illustration of the correlation between the work of adhesion and the resulting measured joint strength for different metals wetted on alumina

7.1.2 Chemical Reaction

Chemical reaction accounts for the bonding mechanism in the overwhelming majority of industrial brazing processes to nonmetals. In the

Table 7.1 Contact angles of brazes wetted onto substrates of differing ionicity

The more covalent in character is the substrate, the lower the resulting contact angle tends to be.

	Contact angle, degree				
	Test temperature		Substrate		
Braze	°C	°F	Al_2O_3	SiC	Graphite
Aluminium	1250	2280	48	42	39
Copper	1100	2010	155	140	140
Silicon	1450	2640	82	37	15

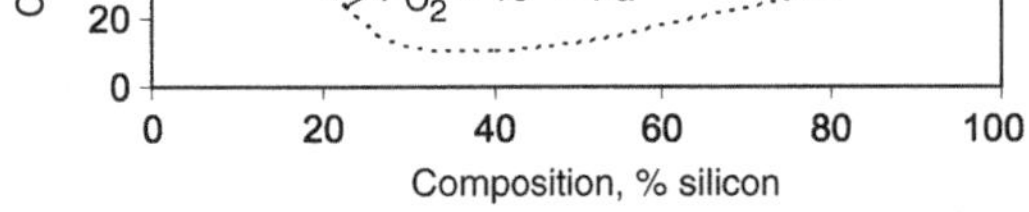

Fig. 7.6 Variation of contact angle for gold-silicon alloys on silicon carbide at 1200 °C (2190 °F) as a function of alloy composition in relatively high- and low-oxygen partial pressure atmospheres (the silicon-rich end of the curve is derived by extrapolation because silicon is not molten at the test temperature used). The observed minimum at the higher oxygen partial pressure is thought to be due to a nonequilibrium mechanism.

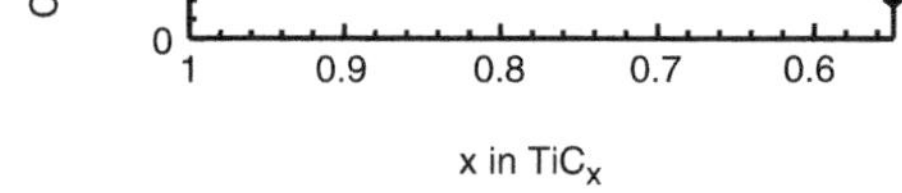

Fig. 7.7 The dependence of the equilibrium contact angle of selected molten metals on titanium carbide as a function of the carbon content of the substrate. In all cases, the contact angle decreases as the carbon deficiency in the titanium carbide, with respect to its stoichiometric value of unity, increases; i.e., the metallic character of the titanium carbide increases.

context of brazing, chemical reaction can be defined as the occurrence of some mass transport (i.e., diffusion) across the interface of a joint that accompanies a brazing operation. This process often leads to the formation of interfacial reaction layers with properties that differ from those of the braze and nonmetallic components. In this situation, the braze is said to be active toward the nonmetal, and the situation is somewhat analagous to the formation of interfacial intermetallic compounds in brazed and soldered joints in purely metal assemblies. It is to be noted that active solders, discussed in the companion volume *Principles of Soldering,* operate in a similar manner to active brazes because they utilize similar process temperatures, even though the solidus temperatures of the filler metals are very different in the two cases.

A chemical reaction is represented by an incremental Gibbs free energy ΔG_r to the classical wetting equation, which drives down the wetting angle from θ^0 before the reaction to θ as a result of it, according to the expression given in Eq 1.4:

$$\cos\theta = \cos\theta^0 + \frac{\gamma_{SL} + \gamma_{SL}^0}{\gamma_{LV}} - \frac{\Delta G_r}{\gamma_{LV}}$$

where γ_{SL} is the solid-liquid interfacial energy after reaction and γ_{SL}^0 is the interfacial energy before reaction.

The composition and pressure of the surrounding gas atmosphere can also influence the formation of interfacial phases, especially when one of the products of reaction between braze and nonmetal is gaseous. Brazing to nonmetals is normally carried out without a liquid (i.e., molten) flux because such materials tend to be ineffective in promoting wetting of nonmetals by brazing alloys. Their use is limited to the rare occasions where the brazing alloy itself needs to be protected against atmospheric degradation.

Chemical reactions between brazes and nonmetal can take any number of forms, but two are particularly prevalent. These are the "exchange reaction," where the brazing alloy dissociates the nonmetal and reacts with one or more of the constituents liberated, and the "gaseous reaction," where one or more of the products of reaction are volatile at the process temperature. The principal difference between them is that in gaseous reactions, adjustment of the process atmosphere can provide a ready means of controlling the joining process. Molten aluminum wetted onto silicon carbide above 1220 °C (1650 °F) gives rise to the exchange reaction:

$$3Al + 4SiC = Al_3C_4 + 4Si$$

and on silicon nitride, a gaseous reaction occurs, with the liberation of nitrogen:

$$Al + 2Si_3Ni_4 = Al + 6Si + 4N_2$$

In both cases, the "free" silicon dissolves in the molten aluminum braze, thereby altering its melting range, as well as the mechanical and physical properties of the resulting joint. Such chemical reactions permit the braze to properly wet the nonmetal. Spreading of the braze on the surface of the nonmetal will depend on how the reaction is sustained at the contact line between the braze front and the nonmetal. Two limiting mechanisms governing chemically reactive

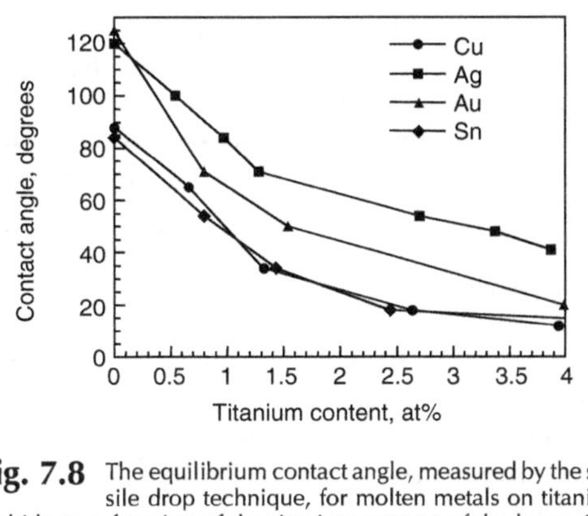

Fig. 7.8 The equilibrium contact angle, measured by the sessile drop technique, for molten metals on titanium carbide as a function of the titanium content of the braze. Test temperatures used were 1150 °C (2100 °F) for gold-, copper-, and tin-base alloys and 1050 °C, or 1920 °F for silver-base alloys. In all cases, the contact angle declines toward an asymptotic value; i.e., wetting improves as the titanium content of the braze increases.

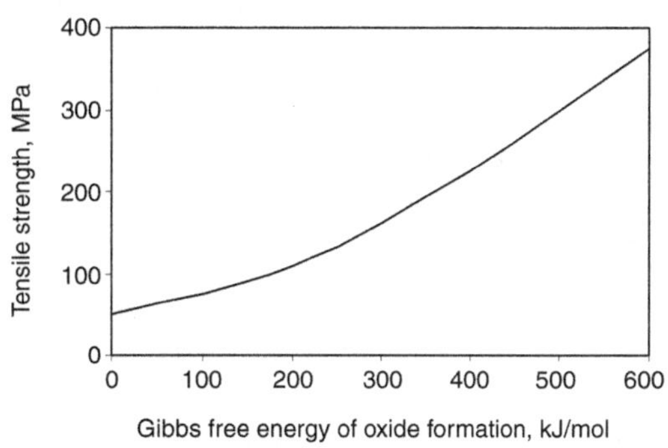

Fig. 7.9 Maximum strength of metal/alumina couples plotted as a function of the oxygen affinity of the braze

spreading have been identified, namely, diffusion controlled and reaction controlled, as described in the following sections.

7.1.2.1 Diffusion-Controlled Spreading

In diffusion-controlled spreading, expansion of the wetted area of the nonmetal by the braze is limited by the rate of arrival of the reacting species by diffusion. An example is provided by copper braze, containing titanium, wetted onto alumina, where the spreading velocity is initially rapid, but decreases with time as concentration gradients develop in the molten braze droplet.

The effect of diffusion on spreading has been modeled. To a first approximation, the fourth power of the braze droplet radius, r, varies in proportion with the product of the droplet volume, V, and reaction time, t, [Asthana and Sobczak 2000]:

$$r^4 = KVt$$

where K is a constant for the process temperature and materials combination.

7.1.2.2 Reaction-Controlled Spreading

In reaction-controlled spreading, there is a linear relationship between the radius of the braze droplet, r, and the process time, t:

$$r = K't$$

where K' is a constant for the process temperature and materials combination [Asthana and Sobczak 2000]. Aluminum wetted onto graphite at high temperature is typical of reaction-controlled spreading. Spreading is not instantaneous as in the manner of a braze wetted onto a metal, but the molten aluminum creeps slowly over the surface of the graphite at a steady rate for a considerable period (>3 h), forming Al_3C_4 at the joint interface [Landry and Eustathopoulos 1996].

Several different spreading models have been developed to explain the dynamics of spreading of reactive molten metals on ceramic surfaces, and these are conveniently reviewed in Meier, Javernick, and Edwards [1999]. In these models, the driving force for spreading is based variously on:

- The progress of the reaction between the molten reactive metal and the ceramic at the circumference of the melt pool, i.e., the triple boundary where the liquid is bounded by solid and vapor,
- The overall progress of the reaction beneath the melt pool
- More complex situations, which take account of the observation that a ridge forms at the triple boundary in the course of the reaction at the solid-liquid interface.

The driving force for spreading that pertains in practice may be a combination of several of these mechanisms. So far, the dearth of comprehensive and definitive data relating to the reaction kinetics, interfacial energy, and diffusion data has impeded elucidation of the spreading behavior.

The growth of the reaction layers formed between a braze and a nonmetal can generally be approximated by a simple parabolic growth law, indicative of diffusion-controlled growth [Torvund et al., 1996]. That is, the thickness of the reaction layer, x, is proportional to the square root of the process duration, t:

$$x = kt^{1/2}$$

where k is a constant that depends on the material system and is a function of temperature. However, in the very early stages of formation of the reaction layer, this parabolic growth has not been observed for the brazing of Ticusil (Ag-26.7Cu-4.5Ti) to alumina, and unusual phases have been identified at the growing interface [Shiue, Wu, and Wang 2000]. These phenomena point to the existence of an initial growth regime that differs from the subsequent rate-controlling mechanism, at least in this case. Here, parabolic, diffusion-controlled, growth of the reaction layer is established only after the first two minutes during brazing at 900 °C (1650 °F). Also, an unusual reaction phase was identified at the interface in this initial stage of reaction: $Ti_3(Cu,Al)_3O$ intermetallic, together with significant amounts of free aluminum.

It is perhaps not surprising that, once the interfacial layer is established, growth of the reaction layer is diffusion controlled because further increases in thickness can occur only if at least one of the reactants diffuses through the solid reaction layer to interact with the other species [Saiz, Cannon, and Tomsia 2001].

At temperatures where the braze is molten, the thickness of the reaction layer tends to have a linear dependence on temperature. Appropriate adjustment of the two variables, process temperature and time, therefore, permits some degree of control over wetting and spreading by the molten filler metal and the thickness of the reaction layer.

The strength of bonds produced by chemical reaction depends critically on the thickness and morphology of the interfacial layer formed. Optimal mechanical properties are generally obtained when the reaction products form a continuous layer that is very thin and possesses a textured interface (i.e., not perfectly smooth). In some braze/nonmetal couples, the reaction layer can be extremely thin (<0.1 μm, or 4 μin.) and is therefore difficult to detect, as, for example, in the case of copper-base brazes containing chromium, titanium, or vanadium wetted onto diamond [Scott, Nicholas, and Dewar 1975]. It has been observed that increasing the thickness of the reaction product zone at the interface by raising the brazing temperature or lengthening the time at peak temperature progressively compromises the apparent strength of the brazed joints because the reaction products tend to have low fracture toughness [Meier, Javernick, and Edwards, 1999].

7.2 Active Brazes

The economics of manufacturing dictate that process cycle times need to be short. It is apparent from the preceding discussion that reasonably rapid brazing of nonmetals can be achieved only by using filler metals that wet and spread by chemical reaction. This requirement has given rise to two classes of brazes for joining nonmetals, namely, active brazes, which incorporate elements capable of wetting nonmetals, and high-temperature brazes, in which the same elements are primary constituents of the filler alloy.

Active brazing alloys have already been introduced in Chapter 2, section 2.2.2. Analogous active solders are discussed in the companion volume *Principles of Soldering*. Active brazes are now commonly used and this section is devoted to a description of this area of brazing technology.

In active brazing, one or more constituents in the braze reacts with the ceramic to produce an interfacial layer that is sufficiently metallic in character that it is wettable by the braze. Titanium is the most commonly used activating element because it reacts with oxides, carbides, and nitrides to form wettable layers of TiO, TiC_{1-x}, and TiN_{1-x}. It is necessary for the joining atmosphere to be highly inert so as to prevent the formation of nonmetallic films or skins on the surface of the braze, which will defeat the intended role of the active constituent. A vacuum of 10 mPa (10^{-6} psi) is normally required to achieve this condition. Other atmospheres that have been used successfully include argon, helium, dry hydrogen, and mixtures of these gases. Nitrogen is ruled out because it reacts with the titanium to form the nitride. The reactivity of active constituents, including titanium, has made it impossible to use chemical fluxes with this class of brazing alloys.

Less-reactive elements such as zirconium and more-reactive elements such as hafnium are also used as active constituents, although much less so than titanium. The choice depends on the parent material and other factors, as is explained later [Loehman 1994; Lugscheider and Tillmann 1991].

Silver-copper-titanium alloys form the basis of many active brazing alloys and processes [Eustathopoulos, Nicholas, and Drevet 1999; Selverian, O'Neil, and Kang 1992; Selverian and Kang 1992; and Suganuma 1990]. One of the earliest active metal brazes based on this alloy system comprised a composite wire, in which a core of titanium was encased in a sheath of silver-copper eutectic. It was produced as a partitioned composite because the relatively high titanium content would produce a brittle product if the constituents were alloyed together. Using rapid solidification, homogeneous foils can be manufactured as ductile preforms of homogeneous alloy. Examples of brazes with the designation ABA (active brazing alloys) are given in Table 2.17. Active brazes are preferably preplaced between the mating components and reflowed in a suitable protective atmosphere. Preplacement minimizes the need for the filler metal to spread to obtain complete joint filling and thereby shortens the process cycle. Mechanically sound joints to the majority of engineering ceramics including alumina, silicon carbide, silicon nitride, aluminum nitride, graphite, and diamond can be produced by this method.

Many, but not all, active brazing alloys are based on silver-copper and the other precious metals listed in Table 2.17. A particular limitation of the copper- and silver-base alloys is the relatively low service temperatures that they can survive, particularly in oxidizing environments, where it is necessary for the service temperature to be below 500 °C (930 °F). This limitation does not suit the application requirements for many ceramics. Therefore, active brazes allowing higher service temperatures have had to be developed. A notable example is a brazing alloy composed of nickel-chromium-phosphorus, where chromium is the active constituent [Peteves et al. 1996]. This braze has been used at 900 to 1000 °C (1650–1830 °F) to produce fairly strong joints to silicon nitride, suitable for service at elevated temperature.

The relatively high expense of the silver-containing active brazes has provided another spur to the development of silver-free active brazes. Copper-nickel brazes, which are less expensive, cannot be satisfactorily activated with titanium because copper itself, as well as nickel additions to copper-titanium alloys, reduces the activity of titanium, and an increase in the concentration of titanium to compensate for this reduction results in a mechanically unworkable alloy. One reported solution to this problem has been achieved by limiting the nickel and titanium contents and adjusting the liquidus temperature of the braze through the addition of boron [Xiong, Wan, and Zhou 1998]. Although the exact alloy composition is not disclosed, the liquidus temperature appears to be close to 1000 °C (1830 °F), and respectable joint strengths to silicon nitride are reported to have been obtained.

The specialist equipment required to fabricate active brazing alloys means that they command a premium price. Special crucible materials need to be used because the molten charge will wet most materials, and a high-quality protective gas, or a vacuum facility, is required, all of which tend to be expensive.

Active filler alloys are used primarily for joining metals to engineering ceramics. They are also finding use with oxide-dispersion-strengthened materials (metals and nonmetals), which are mechanically alloyed, and metal-matrix composite materials in which the nonmetallic reinforcement phase, often silicon carbide or carbon/graphite, can constitute more than 50% by volume fraction [Lugscheider, Bürger, and Broich 1997]. The refractory reinforcement phase inhibits wetting by conventional filler alloys. Active brazes are also used for brazing refractory metals, including stainless steel and nickel-base alloys, due to their ability to wet surface oxides and either reduce or convert them to a mechanically robust form.

7.2.1 Spreading on Nonmetals

The Gibbs free energy of reaction is one possible driving force for spreading by an active braze on a nonmetal. However, just because the energy change is favorable does not mean that spreading will automatically follow, as it almost invariably does on metal components. For nonmetallic components, an additional consideration is the volume change that accompanies conversion of the surface region of the nonmetal into the reaction product.

In the case of silica wetted by aluminum, chemical reduction of silica leads to a 38% decrease in volume. This shrinkage causes cavities to develop at the contact line, which hampers spreading, because repeated rewetting is necessary to establish a continuous interfacial film on the ceramic [Zhou and De Hosson 1995]. On the other hand, for titanium-activated brazes wetted on alumina, the volume change is positive. This situation improves wetting because the reaction product will always extend just ahead of the braze contact line. These two conditions are illustrated schematically in Fig. 7.10. Table 7.2 provides data on the relative volume change for selected active metal/nonmetal couples.

As mentioned earlier, the slow rate of spreading by active brazes on nonmetals means that it is common practice to preplace the filler metal in the joint rather than rely on filling by capillary action. This means that gap widths are strongly influenced by the thickness of the preforms that are used. Also, because the molten filler metal can dewet from the nonmetal before the interfacial layer becomes established, it is again common practice to apply modest pressure to the joint during active brazing (1–5 kPa, or 0.15–0.7 psi). The pressure also assists in ensuring that the braze fills any gross surface topography on the nonmetal components but must not be so high as to extrude molten braze from the joint gap.

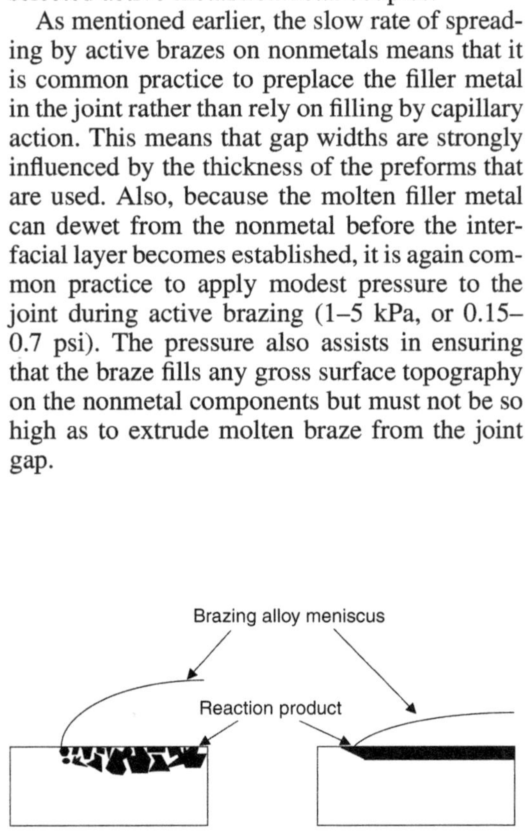

Fig. 7.10 Schematic illustration of the effect of (left) volume contraction and (right) volume expansion, as an interfacial reaction layer is formed, on the spreading behavior of a molten braze on a nonmetal substrate

7.2.2 *Influence of Concentration of the Active Constituent*

From the preceding discussion, it might appear that a high concentration of the active element is desirable. However, for many filler alloys, there are limits on the concentration. The reasons for this are various. In the case of silver-copper filler metals, alloying with titanium increases the hardness of the braze to a point where it becomes unworkable, as shown by Fig. 7.11. This limitation can be overcome by either preparing the alloy in a ductile form by rapid solidification, producing a composite preform comprising a core of titanium and a cladding of silver-copper alloy, or applying titanium as a metallization to the component surfaces prior to joining. These preparation techniques are described in detail in Chapter 4, section 4.1.5.

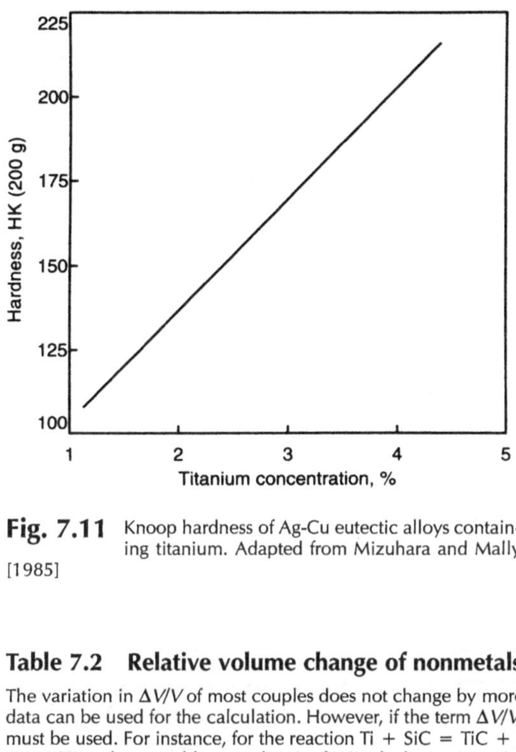

Fig. 7.11 Knoop hardness of Ag-Cu eutectic alloys containing titanium. Adapted from Mizuhara and Mally [1985]

With silver-copper-titanium brazes there are additional reasons for restricting the titanium concentration to about 2%. The first of these is the effect on the liquidus temperature of the alloy; the addition of more than 2% titanium substantially widens the melting range of the alloys, which is generally an undesirable characteristic for a filler alloy. The Ag-Cu-5Ti composition braze has a melting range of 775 to 927 °C (1427–1700 °F). Moreover, it can be seen from the silver-copper-titanium phase diagram shown in Fig. 7.12 that at concentrations of titanium above 2%, the molten braze separates into two distinct liquid compositions, with the maximum solubility of the active metal in the liquid mixture of about this percentage [Kubaschewski 1988; Villars, Prince, and Okamoto 1995]. The solidified filler alloy will then not be homogeneous when produced by conventional casting methods, which will result in the composition and properties of the joint varying in an unpredictable manner. However, rapid solidification can be used to extend the titanium content without producing deleterious segregation in the braze or in joints made with the silver-copper-titanium alloy. The 68.8Ag-26.7Cu-4.5Ti (Ticusil) filler alloy (melting in the range 780–900 °C, or 1435–1650 °F) is produced commercially by rapid solidification for brazing refractory materials.

At first sight, a 2% titanium concentration in a filler metal might appear to be too low for the braze to be effective with many ceramics. The

Table 7.2 Relative volume change of nonmetals ($\Delta V/V$) resulting from wetting by active brazes

The variation in $\Delta V/V$ of most couples does not change by more than 5% over a wide range of temperature so that room temperature data can be used for the calculation. However, if the term $\Delta V/V$ is close to zero, then data appropriate to the brazing temperature must be used. For instance, for the reaction Ti + SiC = TiC + Si, $\Delta V/V = -2.5\%$ at room temperature, but +0.01% at 1000 °C (1830 °F) and reasonable spreading is obtained when appropriate process conditions are used.

Braze	Nonmetal	Reaction product	Process temperature, °C (°F)	Volume change of nonmetal $\Delta V/V$, %	Spreading quality
Al	BN	AlN	1100 (2010)	+14	Fair
	Si_3N_4	AlN	1100 (2010)	+23	Fair
	SiO_2	Al_2O_3	800 (1470)	−38	Poor
	SiC	Al_4C_3	1200 (2190)	+30	Fair
	CaO	Al_2O_3	900 (1650)	−50	Poor
AgCuTi	SiC	TiC	1050 (1920)	+0.01	Fair
CuTi	Al_2O_3	Ti_xO_y	1025 (1880)	+10 to +50	Good
	C	TiC	1075 (1970)	+128	Excellent
NiPdTi	Al_2O_3	Ti_xO_y	1250 (2280)	+10 to +50	Good
Si	C	SiC	1450 (2640)	+134	Excellent
Fe	C	Fe_3C	1550 (2820)	+337	Excellent
	SiC	C	1350 (2460)	−57	Poor
FeSi	SiC	C	1350 (2460)	−57	Poor

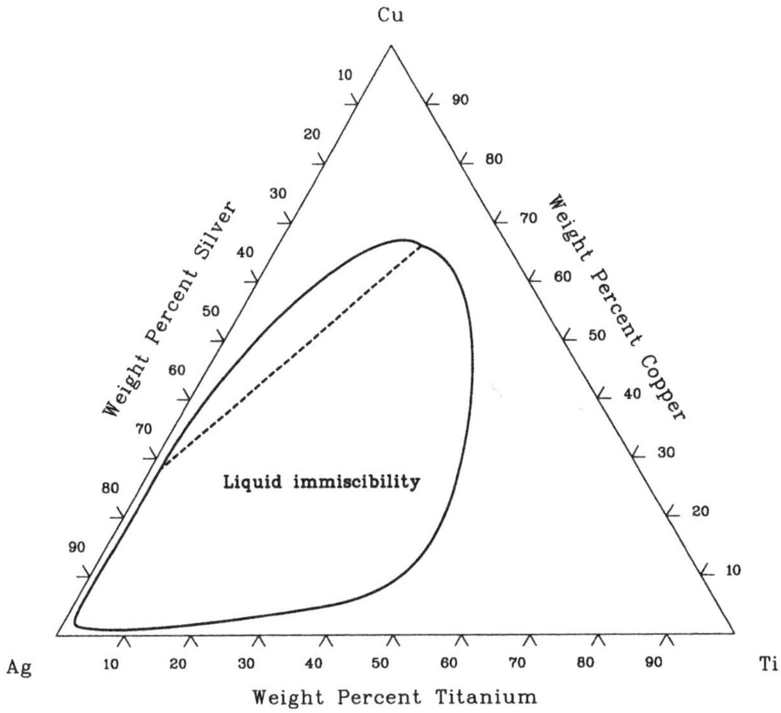

Fig. 7.12 Liquidus surface of the Ag-Cu-Ti phase diagram showing the region of liquid immiscibility. The critical tie line that links the two liquid phases of Ag-27Cu-2Ti and Ag-66Cu-22Ti is marked.

data in Fig. 7.13 would appear to indicate this to be the case with regard to copper-titanium alloys. In fact, it is not the percentage of the reactive element per se that governs the efficacy of the filler metal, but its overall chemical activity.

It is well established that the chemical activity of titanium is much less than unity in active brazes such as the silver-copper-titanium alloys in which the active element is only a minor constituent. Copper dissolves up to 67 wt% titanium, at 1150 °C (2100 °F), which denotes a strong attraction between these elements. This attraction concomitantly reduces the activity of titanium. On the other hand, certain ternary additions such as silver, indium, and tin, in which titanium has a much lower solubility, help to partially restore the activity of this active element [Nicholas and Peteves 1994]. With respect to the silver-copper-titanium alloys, it should be pointed out that the solubility limit of titanium in silver is only 3% at 1150 °C (2100 °F). In a similar manner, titanium has a solubility limit of only 6.7% in tin at the same temperature.

Besides helping to recover the activity of titanium in copper-titanium alloys, the addition of silver has the advantage of depressing the melting point of braze through the silver-copper eutectic reaction. Furthermore, as noted previously, at a concentration above 2%, titanium

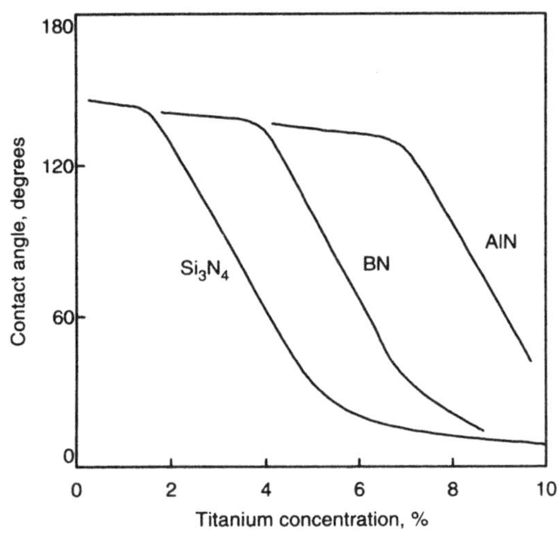

Fig. 7.13 Effect of titanium concentration on the wetting of some nitride ceramics by Cu-Ti-activated brazes, as measured by the contact angle. Adapted from Nicholas [1989a]

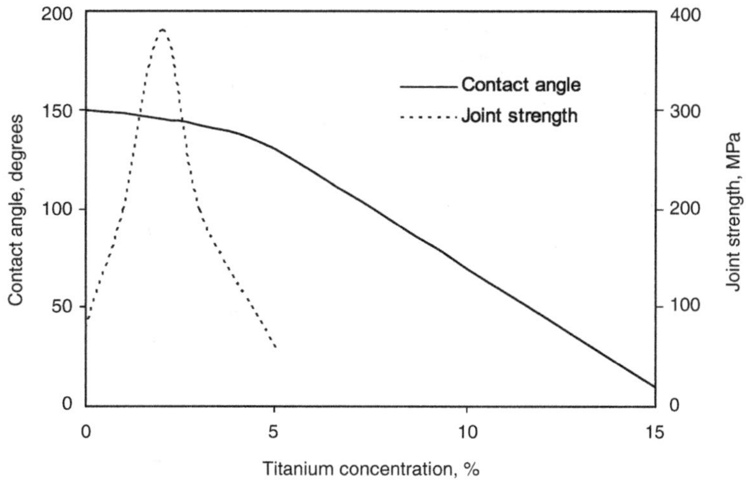

Fig. 7.14 Wettability (contact angle) and room temperature strength of alumina brazed with copper containing varying amounts of titanium, prepared by brazing in vacuum at 1150 °C (2100 °F) for 15 min

causes the molten silver-copper alloy to separate into two fractions. One of these fractions has the composition Ag-22Cu-2Ti and the other, Ag-66Cu-22Ti. The 22%Ti fraction is more than adequately active to ensure wetting of even highly refractory ceramics such as alumina, whereas the other alloy fraction is not [Paulasto and Kivilahti 1998]. It has been found that the wetting efficacy of silver-copper-titanium alloys falls away sharply with reduction in the titanium concentration much below 1.5%.

Increasing the concentration of the active ingredients promotes wetting by reducing the contact angle, but for many active brazes the concentration required for optimal bond strength is much lower than required to effect a significant reduction in the contact angle, as illustrated in Fig. 7.14 [Nicholas and Crispin 1986]. As pointed out in section 7.1.2.2 of this chapter, optimal mechanical properties are generally obtained when the reaction products form a continuous layer that is, nevertheless, very thin. Increasing the thickness of the reaction product zone at the interface will generally reduce the contact angle but degrade the strength of the brazed joints because the same reaction products that improve wetting by the braze also tend to be brittle. It has been demonstrated that adding elements such as indium and tin to silver-copper-titanium alloys, which boost the activity of the titanium, permit a reduction in the concentration of the active element [Nicholas 1988].

These points are illustrated by Fig. 7.15, which shows the progressive reduction in contact angle of copper-titanium resulting from the addition of silver and then additionally of tin, and supported by the solubility data in Table 7.3. These beneficial effects of silver, tin, and indium in restoring the activity of titanium when alloyed with copper account for the successful exploitation of silver-copper-titanium brazes containing tin or indium (see also Chapter 2, Table 2.17).

It is also possible to boost the effectiveness of the reactive constituent by enhancing its natural tendency to concentrate at the filler/ceramic in-

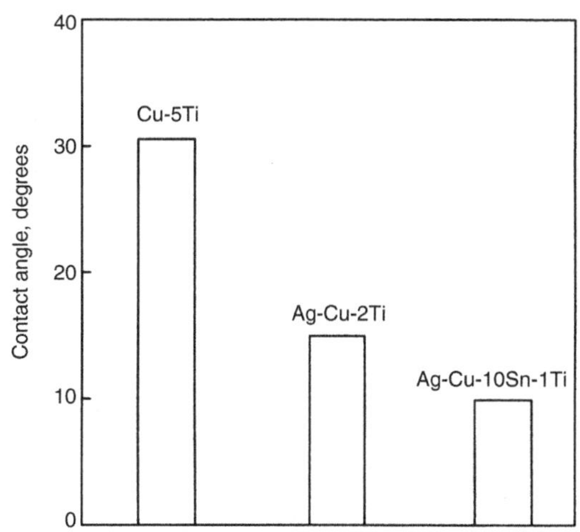

Fig. 7.15 Comparison of the influence of composition on the wetting of Si_3N_4 ceramic by titanium-activated brazes under comparable process conditions, as measured by the contact angle. Adapted from Nicholas and Peteves [1991]

terface. This can be achieved by applying an electric field across the joint gap while the braze is molten [Minegishi, Sakurai, and Morozumi 1991]. Under the influence of the applied voltage, ions of the element in the braze with the highest induced charge will tend to migrate toward the cathodic (negatively biased) side of the joint. Because the principal constituents of brazes intended for use at elevated temperatures tend to be relatively noble elements, it will be ions of the active constituent that are driven toward the cathode. This process, known as field assisted brazing, requires the ceramic to be electrically conductive, such as zirconia and silicon nitride. Here, it should be noted that many ceramics that are considered to be electrical insulators at room temperature have significant conductivity when heated to 1000 °C (1830 °F), as their volumetric resistivity reduces by six orders of magnitude over that temperature range [Morrell 1985]. Unlike the electrical resistivity of metals, which generally increases linearly with rising temperature, the resistance of ceramics decreases rapidly. At normal ambient temperatures (~20 °C, or 70 °F), the electrical resistivities of ceramics are several orders of magnitude higher than those of metals.

Table 7.3 Solubility of some reactive metals in brazing alloy constituents at 1100 °C (2010 °F)

	Solubility, wt%		
	Solute		
Solvent	Titanium	Zirconium	Hafnium
Silver	5	12	10(a)
Gold	1.5	3.5	9.5
Copper	61.5	18.5	43
Indium	3	5(a)	5(a)
Tin	18	15	12

(a) Reliable phase diagram data not available. The solubility was estimated from consideration of similar binary systems.

7.2.3 Formation and Nature of the Reaction Products

Reaction between an active filler alloy and a nonmetal results in modification of the wetted surface, with the formation of one or more interfacial compounds. As discussed in Chapter 2, when designing a joining process, due consideration must be given to all of the reaction products. Volatile elements can generate voids in a joint through the evolution of vapor, while other products of reaction can, either individually or in combination with other species, form low-melting-point phases, or produce liquid immiscibility and other undesirable features.

The variation in thickness of an interfacial layer formed between the Cu-5Ti braze on silicon nitride as a function of the brazing cycle duration and peak temperature is shown in Fig. 7.16 and Fig. 7.17, respectively. As is true of most chemical reactions, the thickness of the one or more layers of interfacial reaction products increases with the available thermal energy. The initially linear slope of the graphs presented in

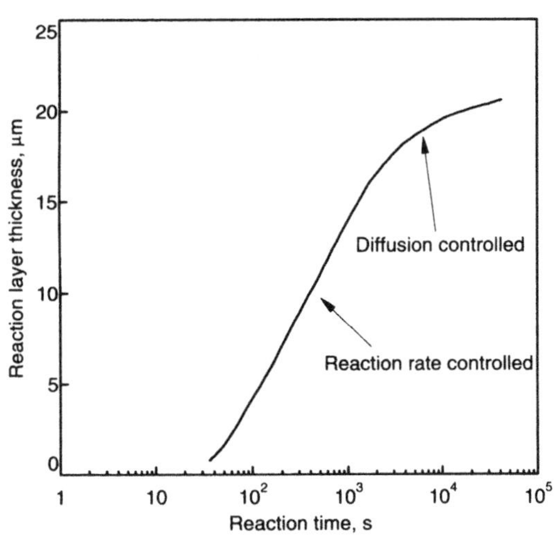

Fig. 7.16 Reaction layer thickness as a function of brazing time for Si_3N_4 wetted by Cu-5Ti at 1125 °C (2055 °F). Adapted from Nakao, Nishimoto, and Saida [1989]

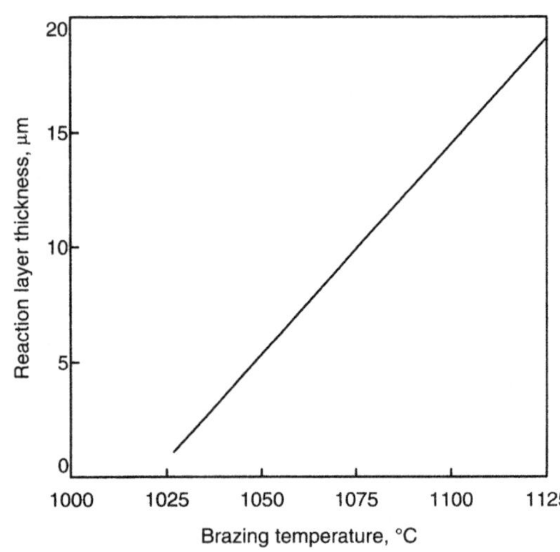

Fig. 7.17 Reaction layer thickness as a function of the brazing temperature for Si_3N_4 wetted by Cu-5Ti for 1000 s. Adapted from Nakao, Nishimoto, and Saida [1989]

Fig. 7.16 indicates that the reaction between the molten braze and the ceramic, in this case, is, at first, reaction-rate controlled. When the reaction product zone has reached a certain thickness, solid-state diffusion through it then determines the rate of subsequent growth of the layer. At this crossover, the growth slows down significantly, as can be seen from Fig. 7.16. Often, the transition between these two growth regimes is not sharp and, overall, the thickness of the reaction layer tends to increase asymptotically with time. The concentration of the active constituent also affects the thickness of the reaction layer formed under fixed process conditions, as can be seen from Fig. 7.18.

In some active braze/nonmetal combinations, a somewhat complex sequence of reaction layers can form adjacent to the interface [Meier, Javernick, and Edwards 1999]. It should be added that reactions occurring in active brazing are not confined to the immediate interface between the active braze and the nonmetal components. The reactions, which have a bearing on the integrity of the joints, also encompass intermetallic products that form as sequential layers within the braze, toward the interface with the nonmetal. In the case of titanium-activated nickel used with components of alumina, a complex sequence of no less than six intermetallic layers has been proposed, which are then followed by very thin (submicron) reactive layers of the oxide TiO_2 and $Al_2O_3 \cdot TiO_2$ spinel, the final layer forming against the alumina surface. In such cases, the thickness of the various layers, their mechanical properties, the CTE match between the layers, and their adhesion to one another and to both the nonmetal and the body of the braze will together determine the aggregate physical properties of the joint.

"Low-temperature" brazing of silicon carbide with a titanium-activated alloy will lead to the formation of titanium carbide and a joint with poor mechanical properties. Increasing the brazing temperature stabilizes a more complex carbide, Ti_3SiC_2, which has a similar CTE, but a better lattice fit with silicon carbide and hence more favorable structural properties result [Lugscheider and Tillmann 1993].

From free-energy considerations of exchange reactions, hafnium is the most active member of Group 4 of the periodic table toward nonmetals. Silver-copper alloys activated with hafnium are generally preferred to their titanium equivalents for joining to nitride ceramics because the reaction product, HfN, is a barrier to silicon and nitrogen diffusion, whereas TiN is not [Lugscheider and Tillmann 1991]. Therefore, once a continuous layer of HfN has formed, the reaction effectively ceases so that there is no fall-off in the mechanical properties through a thickening of the reaction zone on extended heating. However, there is a penalty to be paid for obtaining this benefit, namely, that more stringent furnace atmospheres are necessary due to the greater reactivity of hafnium with oxygen, and this makes brazing processes using hafnium-activated brazes relatively difficult to implement successfully. The influence of hafnium content on joint strength, measured by four-point bend tests, is shown in Fig. 7.19. There is clearly a preferred concentration of hafnium, which is in the range 3 to 5%. Lower hafnium contents do not permit the formation of continuous interfacial layers. On the other hand, if the hafnium content of the braze is too high, the interfacial layers formed are thicker and have a lower fracture toughness. Also, hafnium alloys are not suitable for all applications. For example, in joining to silicon carbide, the HfC reaction product does not make an effective barrier to further reaction so that thick interfacial layers form rapidly, resulting in a bond strength below 40 MPa (5.8 ksi), which is too low to be of commercial relevance [Lugscheider and Tillmann 1993].

The sensitivity of active brazes to the brazing atmosphere is reflected by the data given in Fig. 7.20, which shows the strength of ceramic-metal assemblies brazed using a silver-copper-titanium

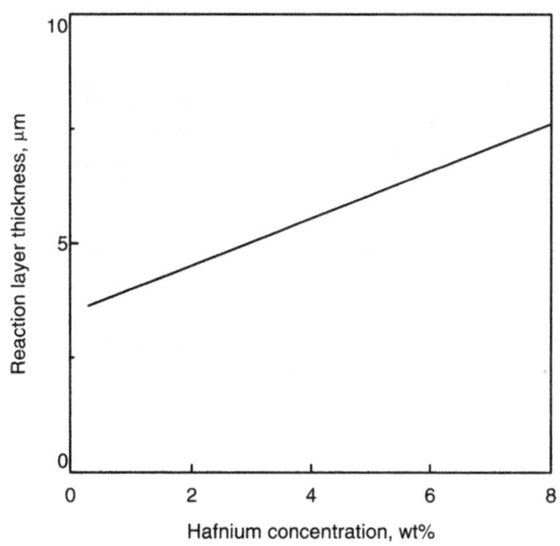

Fig. 7.18 Reaction layer thickness as a function of the concentration of the active metal for SiC brazed with Ag-Cu-Hf alloys. Adapted from Lugscheider and Tillmann [1991]

alloy in different atmospheres. The highest and most consistent joint strengths have been achieved in atmospheres of high-purity nitrogen and argon.

The thickness of the reaction layer formed between titanium-containing brazes and nitride ceramics can be greatly reduced by selected alloying additions. Kuzumaki, Ariga, and Miyamoto [1990] have shown that the addition of niobium to silver-copper-titanium alloys is effective in restricting the width of the interfacial layer. The reasons for this is not entirely clear, and there are at least two possible explanations. One possibility stems from the fact that titanium and niobium form a solid solution so that the TiN reaction layer is replaced by $(Ti_xNb_{1-x})N$, which may be a more effective barrier to silicon and nitrogen diffusion than is TiN. Alternatively, the presence of the niobium may reduce the activity of titanium in the alloy, due to the solubility of

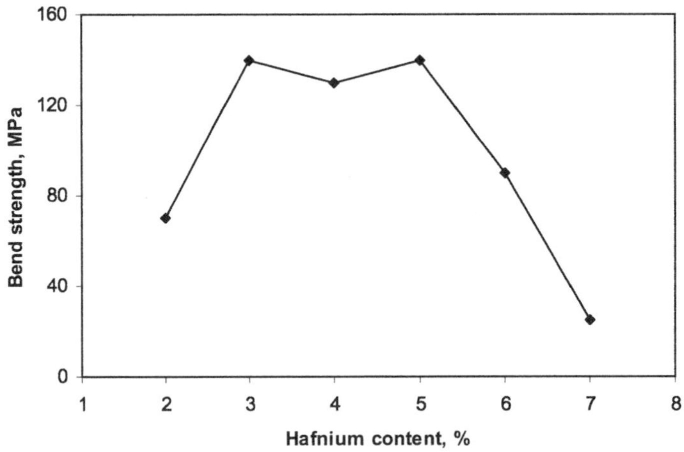

Fig. 7.19 Four-point bend strength of joints made to silicon nitride with a silver-copper-hafnium braze as a function of hafnium content. The optimum concentration for formation of a continuous, but thin, layer of reaction product appears to be in the range 3–5%.

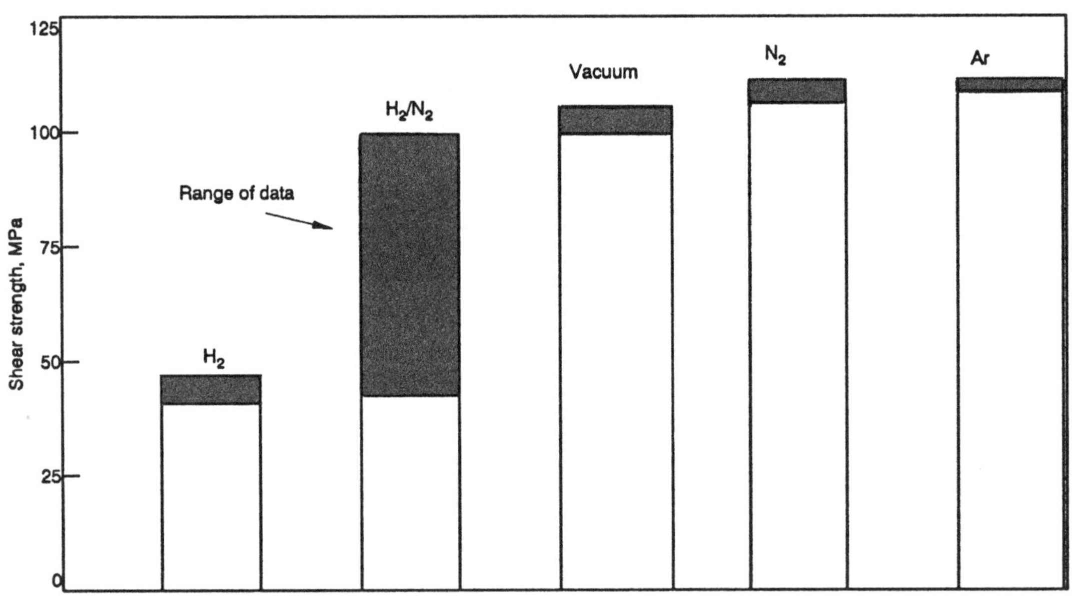

Fig. 7.20 Influence of the brazing atmosphere on the shear strength of ZrO_2/mild steel joints made with Ag-Cu-3Ti filler alloy. Adapted from Weise, Malikowski, and Krappitz [1989]

titanium in niobium [Akselsen 1992]. The influence of niobium additions on the thickness of the reaction layer can be seen in Fig. 7.21.

The factors that govern the formation of joints to ceramics with active brazes are considered next in specific cases, using as examples silicon nitride and alumina ceramics with titanium-bearing brazing alloys.

7.2.4 Active Brazes with Silicon Nitride

Nonactivated brazes do not wet silicon nitride well, even in the presence of fluxes: measured contact angles are typically greater than 130 °C (265 °F). The addition of titanium promotes wetting and spreading with low-contact angles (<15°) provided the concentration is sufficiently high. This is shown by the data given in Fig. 7.13, which relates to the wetting of three nitride ceramics by copper with different titanium contents. Because the reaction layer has to become fully established on the nonmetal before spreading can proceed, this slows the spreading of active brazes on nonmetals to rates that are typically one-tenth or less of that achieved by conventional brazes on metals [Nicholas 1993] (see Fig 7.1).

Thermodynamic analysis shows that free titanium reacts with silicon nitride to form titanium nitride and a titanium silicide:

$$1/9Si_3N_4 + Ti \rightarrow 4/9TiN + 1/9Ti_5Si_3 \qquad \text{(Eq 7.1)}$$

This is the preferred reaction at all temperatures, insofar as it results in the largest reduction in Gibbs free energy, but there is a minimum temperature threshold below which the reaction slows down considerably, as shown by the plot in Fig. 7.22, relating contact angle to temperature.

The empirical data represented in Fig. 7.22 was obtained in wetting experiments. The particular wetting test that was used in this case was the sessile drop test. This involves placing a measure of a solid alloy onto a substrate of interest, heating it to above its melting point and then measuring the contact angle visually.

The reaction scheme represented by Eq 7.1 is consistent with the formation of a layer of the compound TiN adjacent to the interface with the ceramic component and an accompanying reaction layer of Ti_5Si_3 forming alongside it, as earlier reported by Nicholas and Peteves [1991]. However, detailed microstructural studies of interfaces between silicon nitride and silver-copper-titanium braze have revealed a much more complex picture of the phases that form. In particular, Ti-Si-Cu-N compounds have been observed in the interface region, and it is not entirely certain that stoichiometric TiN forms immediately against the silicon nitride surface [Peteves et al. 1996; Paulasto and Kivilahti 1995].

These findings are not surprising in view of the fact that the reaction described by Eq 7.1

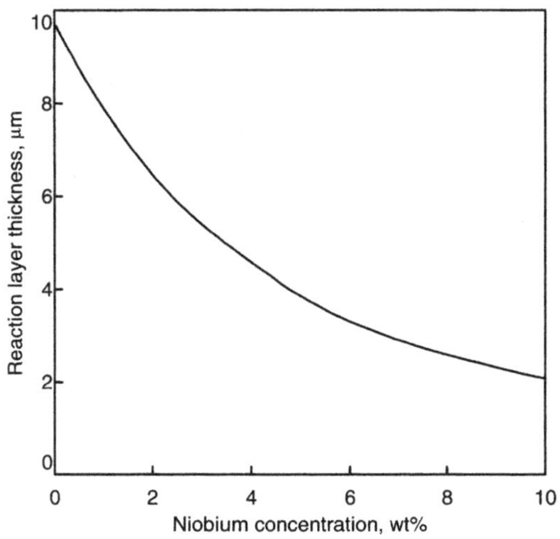

Fig. 7.21 Reduction in the thickness of the reaction layer formed by the addition of niobium to the Ag-Cu-5Ti braze wetted onto aluminum nitride under similar process conditions. Adapted from Kuzumaki, Ariga, and Miyamoto [1990]

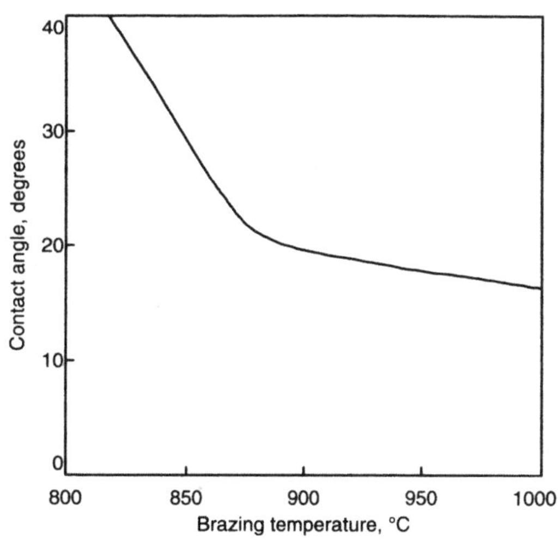

Fig 7.22 Influence of brazing temperature on the wetting of Si_3N_4 by the Ag-27Cu-2Ti alloy, as measured by the contact angle. Adapted from Nicholas and Peteves [1991]

pertains to a thermodynamic condition where the activity of titanium is unity. However, in brazing alloys in which titanium is only a minor constituent, the activity of this element is much less than unity, as pointed out in section 7.2.2 of this chapter. The presence of both silicon and nitrogen as decomposition products also clearly influence the reaction products that form between silver-copper-titanium brazes and silicon nitride [Peteves et al. 1996]. In the absence of copper, titanium, in a silver-titanium braze, segregates to the silicon-nitride interface with a high activity, and the reaction described by Eq 7.1 applies [Paulasto and Kivilahti 1995]

The formation of titanium silicides above the nitride layer is beneficial from the point of view of the brazing process. This is because titanium nitrides are essentially nonmetallic and are not readily wetted by molten filler metals. On the other hand, titanium silicides have a strong metallic character and are readily wetted by silver-copper and other brazing alloys [Nicholas 1989b].

The metallic character of the interfacial reaction products in contact with the braze is important in ensuring wetting. This can be seen from the decrease in wetting angle that occurs as the heating cycle is extended (Fig. 7.23), which results from the formation and growth of titanium silicides that become progressively titanium-rich as the reaction proceeds. A similar effect is achieved by increasing the brazing temperature, as indicated in Fig. 7.12.

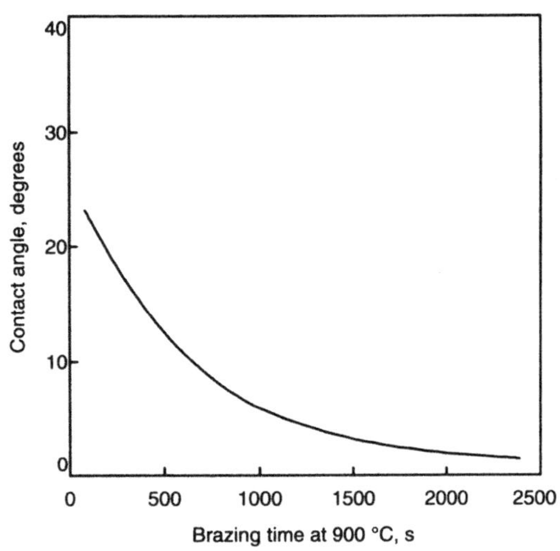

Fig 7.23 Variation in contact angle with brazing time for Ag-27Cu-2Ti on Si_3N_4. Adapted from Loehman [1988]

A more dramatic illustration of this point is provided by the contact angle of silver-copper-titanium brazes on substrates of silicon nitride, and also boron nitride and aluminum nitride, respectively, under identical conditions. Figure 7.13 shows that much higher concentrations of titanium are required to wet boron nitride; neither of the reaction products—namely, TiN or TiB_2—are wetted by plain silver-copper eutectic. It is only when titanium is added and raised to a concentration that is sufficient to force the composition of these reaction products off stoichiometry and make them titanium-rich that wetting occurs. An even higher concentration of this element is needed to induce wetting of aluminum nitride. In all cases, the titanium concentrates at the bridging compound/braze interface, leaving the solidified filler denuded of the reactive constituent.

In the case of silicon nitride, the choice of compatible active brazes is quite sharply prescribed by the relatively low decomposition temperature of this ceramic. Above 1100 °C (2010 °F), silicon nitride is unstable in an atmosphere with low nitrogen and oxygen partial pressures. Figure 7.24 shows the tendency for silicon nitride to decompose as a function of temperature and partial pressure of nitrogen in the vicinity of the joint. This limitation is graphically illustrated by a comparison of joints made to silicon nitride with a nickel-chromium-phosphorus and a nickel-chromium-silicon braze [Peteves et al. 1996]. Joints made with nickel-chromium-phosphorus braze in the temperature range 900 to 1000 °C (1650–1830 °F) are fairly strong (up to 170 MPa, or 25 ksi), whereas those made similarly in vacuum with the higher-melt-

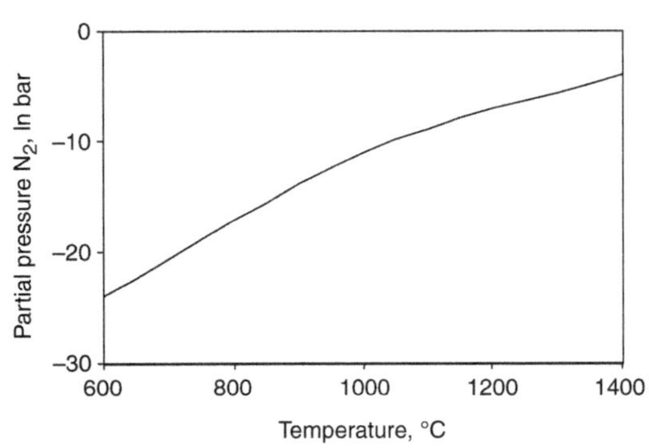

Fig. 7.24 Nitrogen partial pressure as a function of temperature for the decomposition of silicon nitride. Adapted from Lugscheider and Tillmann [1993].

ing-temperature nickel-chromium-silicon braze at 1150 to 1200 °C (2100–2190 °F) are weak.

7.2.5 Active Brazes with Alumina

The approach of using the Gibbs free energy change accompanying exchange reactions to predict which products are formed is satisfactory for many applications, but as noted in the previous section, a detailed analysis often reveals a far more complex picture.

This is true for a silver-copper braze activated with titanium used on alumina. The following reaction between titanium and alumina:

$$3Ti + Al_2O_3 = 3TiO + 2Al$$

has a positive free energy and should therefore not occur [Moorhead, Henson, and Henson 1987]. However, the silver-copper-titanium braze wets and spreads on alumina very well, decomposing the surface of the latter in the process. Chidambaram, Edwards, and Olson [1991] explain the observed reduction of alumina by considering the surface thermodynamics of the nonmetal, for which the free energy of reaction is negative, instead of limiting consideration to bulk properties.

The wetting of titanium-activated silver-copper brazes on oxide ceramics has been the subject of much study, and a thermodynamic explanation that is consistent with experimental observation is beginning to emerge [Paulasto and Kivilahti 1998]. It transpires that the reaction product depends not only on the composition of the braze but also on the amount of oxygen present, both dissolved in the braze and also a contribution from the process atmosphere. In the absence of oxygen, titianium dissolves alumina to form an intersititial solid solution Ti(Al,O). Over this forms a second reaction product $(Ti,Al)_4Cu_2O$ and, after extended wetting times, a third layer Ti(Al,Cu,O), which is also a mixed oxide, forms. In silver-rich brazes, the silver repels oxygen from the melt (silver oxide is not stable at these temperatures) and results in binary titanium oxides forming in contact with the alumina, over which a second layer of titanium aluminides then develops. When more oxygen is made available, the reaction products change to contain more oxygen than is provided by the dissociation of stoichiometric alumina so that altering the process atmosphere provides a means for controlling spreading and the reaction products [Camilo de Camargo 1995]. The role of oxygen in controlling the development of interfacial layers has also been observed in the active brazing of alumina with copper-titanium alloys [Bang and Liu 1994]. Furthermore, it has been suggested that oxygen may be involved in the mechanism responsible for the wetting of bare alumina by molten aluminium (see Chapter 4, section 4.3) [Levi and Kaplan 2002].

Additional evidence that thermodynamic calculations based on simple exchange reactions may be overly simplistic for predicting interfacial reactions involving active brazes is provided by studies of pure copper wetted on silicon carbide. Studies have confirmed that the reaction product in this case is a previously unreported copper carbo-silicide [Wang and Wynblatt 1998]. Clearly, more work is required to establish a rigorous theoretical understanding of active brazing processes that can account for the detail of the empirical observations that are being made.

7.2.6 Other Examples of Active Brazing

An example of an active brazing process, where the constituents of the braze are seemingly inactive, is provided by the brazing of graphite using copper-and gold-base alloys. Under normal circumstances, none of the constituents of these brazing alloy families will wet graphite. However, by incorporating within the joint a shim of iron, ostensibly to help alleviate thermal expansion mismatch stress, excellent wetting and bonding to the graphite is achieved. Figure 7.25 shows the resulting joint strength as

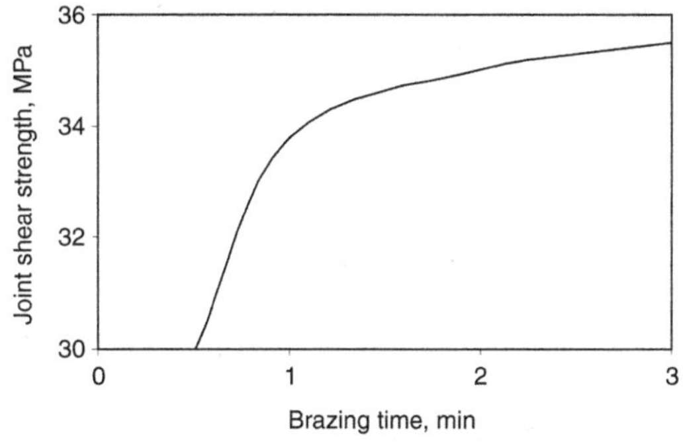

Fig. 7.25 Effect of holding time at the process temperature on the shear strength of joints between 0.11% carbon steel and graphite made with Au-62Cu braze at 1398 °C (2057 °F), incorporating a 120 μm (4.7 mils) thick foil of iron

a function of the brazing time. Bonding to the graphite occurs because the brazes wet the iron interposer, partially dissolving it in the process. Iron is metallurgically compatible with carbon, iron-carbon alloys being the basis of steel technology, so that the iron-enriched braze is then able to wet the graphite, and the solidified braze contains both iron and carbon (cementite, Fe_3C). The brazing cycle needs to be longer than normal because of the two-stage wetting process that is involved [Ohmura et al. 1994].

A version of active brazing that is much used for metallizing alumina and (preoxidized) aluminum nitride is the "direct copper bonding" (DCB) process. It utilizes the eutectic reaction between copper and cupric oxide at 1066 °C (1951 °F), as shown in Fig. 7.26. In this process, thin foils of copper are provided with a miniscule layer of cuprous oxide (CuO) and then pressed against a plate of alumina, or aluminum nitride, which has been subjected to a controlled oxidation of its surface to give an oxide thickness of 1 to 2 µm (40–80 µin.). When heated to the process temperature of 1070 °C (1960 °F) in an atmosphere with a controlled partial pressure of oxygen, the eutectic melts while most of the copper sheet remains solid. The copper oxide surface layer on the copper foil reacts with the ceramic to form a submicroscopic interfacial layer of spinel, according to the reaction:

$$CuO + Al_2O_3 = CuAl_2O_4$$

On cooling, the copper foils will be strongly bonded to the ceramic. The copper layers are typically of the order of 0.3 mm (12 mils) thick. The presence of the copper claddings improves the fracture toughness of the ceramic substrates, permitting thinner sheets to be used than would be possible if the ceramics were left bare. The copper also boosts thermal performance by virtue of the excellent heat spreading provided by the metal. Because the low expansion coefficients of the ceramics (~7.1 ppm/°C for Al_2O_3 and 4.1 ppm/°C for AlN) are not appreciably increased by the addition of the thin copper claddings, these DBC plates are widely used as substrates for bare silicon die in a variety of power

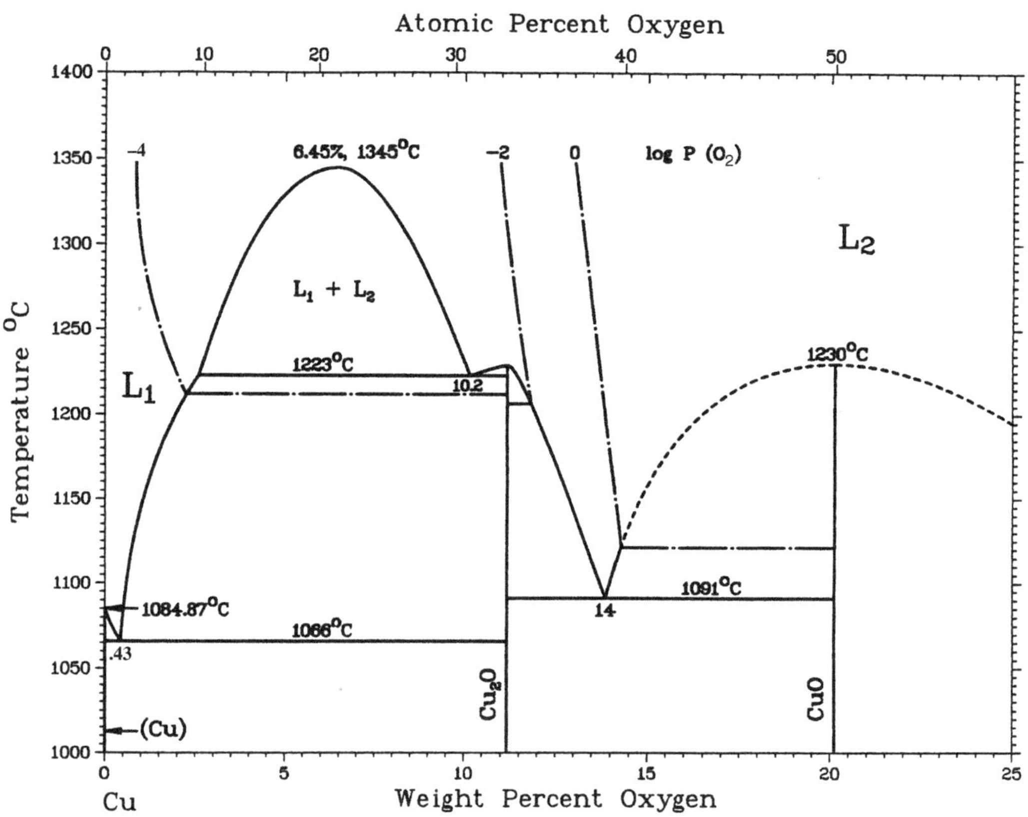

Fig. 7.26 Copper-oxygen phase diagram

electronic applications [Schulz-Harder and Maier 1996, Schulz-Harder 2003].

7.2.7 Hybrid Processes of Active Brazing with Diffusion Brazing

A hybrid process of active brazing with diffusion brazing is mentioned in Chapter 6, section 6.2, for joining Kovar components to alumina, using trifoils of titanium clad on both sides with nickel. A similar approach has been successfully used for joining parts of silicon nitride, employing foils of copper, 100 μm (4 mils) thick clad on both sides with a 2 to 3 μm (80–120 μin.) layer of titanium [Paulasto, Ceccone, and Peteves 1997]. Bonding was carried out in a vacuum of 0.2 mPa (3×10^{-8} psi), under a 50 kPa (7.25 psi) compressive loading, at a temperature of 950 °C (1740 °F) for just 10 min. Average joint strengths of 203 MPa (4.2 lb/ft^2) were achieved using this method, with failure in tensile testing occurring mostly within the ceramic. Several phases formed in the region of the ceramic-metal interface. Those identified by Paulasto and his coworkers comprised TiN, immediately adjacent to the silicon nitride, then a layer of Ti-Si-Cu-N compound, followed by a Ti-Cu or Ti-Cu-Si intermetallic. Increasing the thickness of the titanium layer increased only the thickness of the Ti-Cu and Ti-Cu-Si phases, without any benefit to the mechanical properties.

A different combination and configuration has been used to produce robust, actively diffusion-brazed joints to alumina components [Marks et al. 2000]. The materials, on this occasion, comprised a cleaned foil of niobium, 127 μm (5 mils) thick, as the active metal, and plates of alumina, each coated with an evaporated 3 μm (120 μin.) layer of copper. The bonding operation was carried out on a graphite hot press in a vacuum better than 13 mPa (2×10^{-6} psi) under a compressive load of 2 MPa (290 psi) and peak temperatures of 1150 °C (2100 °F) and 1400 °C (2550 °F) were used. Assemblies joined by this method achieved strengths of up to 240 MPa (5.0 lb/ft^2) and retained their mechanical integrity at elevated temperatures, with the strength dropping to 190 MPa (4.0 lb/ft^2) at 900 °C (1650 °F). By comparison, joints to ceramics, including alumina, with copper-silver-titanium brazes such as Cusil ABA (Ag-35.25Cu-1.75Ti), lose their mechanical integrity above about 500 °C (930 °F) (Fig. 7.27). Detailed metallographic analysis of fractured surfaces revealed that, during the joining operation, the copper film melts, alloys with some of the niobium, and then separates into copper-rich droplets, enabling direct bonding between the niobium and the alumina. The niobium-alumina bonding mechanism involves limited dissolution of the alumina, with the aluminum atoms entering into the niobium lattice substitutionally and the oxygen interstitially. The molten copper phase performs several functions in the joining process. As in other diffusion-brazing processes, the liquid fills voids along the interface and also provides a fast diffusion path for niobium to reach the alumina interface. It then enhances the growth of the contact region between the alumina and the niobium. The solidifying copper-rich particles

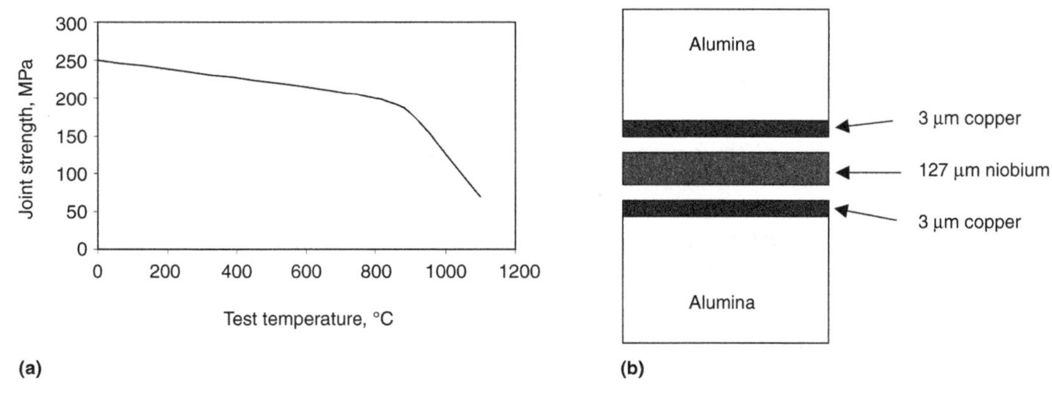

Fig. 7.27 Alumina assemblies joined by active diffusion brazing. (a) Plot of fracture strength in four-point bend tests of beams cut from assemblies. Adapted from Marks et al. [2000]. Each plate was coated with copper and joined at 1400 °C (2550 °F), in a vacuum better than 13 mPa (2×10^{-6} psi), with a niobium foil pressed between them. Note that joint strength is largely maintained up to 900 °C (1650 °F), which is about 400 °C (750 °F) higher than would be expected for an active copper-base brazed joint. (b) Schematic illustration of the interlayer structure used to produce the active diffusion brazed joints

adhere strongly to both the niobium and alumina surfaces and serve to reinforce the strength of the niobium-alumina bond. Indeed, reducing the thickness of the copper coating below 3 μm (120 μin.) results in a decrease in the joint strength.

7.3 Materials and Process Considerations

Key materials and process issues relating to joining of nonmetals using active brazing alloys will be reviewed. Emphasis will be placed on the differences in brazing to metals by established methods, as described in detail in Chapter 1, section 1.3.2.

Form of the Filler Metal

Active brazing alloys can be produced in four forms:

- *Mixed pastes:* Many stock brazing alloys are available in the form of pastes. Because the volume proportion of the active constituent in active brazes is relatively low, it is possible to convert the stock brazes into active brazes by adding titanium hydride powder without greatly altering the rheological properties of the paste. Titanium hydride decomposes into metallic titanium at about 500 °C (930 °F) so the active metal is effectively protected against degradation until the process atmosphere has been established.
- *Applied coatings:* The active constituent of many active brazes may be applied as a coating directly to the surface of the nonmetal and the brazing operation then conducted using a conventional filler metal. The coating process is similar to other metallizing schemes except that, in this case, the coating should be fully consumed by reaction with the nonmetal and dissolution in the braze. Suitable deposition methods include vapor deposition, powder-loaded paints, and mechanically cladding of thin foils. The general drawback of this approach is that the active metal will be exposed to the atmosphere prior to brazing and will therefore be covered with a stable oxide or nitride, which can impede wetting by the brazing alloy. In order to reduce this problem, it is important to use a highly inert atmosphere, with the combined oxygen and water vapor content as low as possible. A vacuum of 0.2 mPa (3×10^{-8} psi) has been found to be suitable for this purpose.
- *Braze claddings:* Titanium cored wires and foils are available commercially. Other active brazing alloy combinations can be produced as trifoils, with the active constituent generally in the form of a foil, clad or coated with the lower-melting-point brazing alloy (see section 7.2.7 of this chapter). These clad alloys have the advantage of being simple to manufacture because the ductility of the brazing alloy is unaffected by the titanium core, and the titanium is protected against atmospheric corrosion until the brazing alloy melts. The principal limitation of this form of active brazes is that the titanium is not able to initiate wetting of the nonmetal until after the braze cladding melts. This means the initial wetting of the nonmetal part is poor or extremely sluggish, and, in extreme cases, the braze can fail to wet the nonmetal, particularly when the other faying surface is of a readily wettable metal such as copper.
- *Homogeneous alloys:* True active brazing alloys (often referred to by an ABA designation) tend not to have favorable properties for mechanical working and are therefore prepared either as powders and thence pastes, or as foil or wire by rapid solidification casting technology (see Chapter 2, section 2.2.2 and section 7.2 of this chapter).

Preparation of the Components

A critical aspect of successful brazing of nonmetals using active brazes is the initial surface preparation of the nonmetal. The method used to shape nonmetal components prior to joining has a profound influence on the joint strength. Any machining process tends to generate subsurface cracks and other flaws that degrade the intrinsic strength of the material. Accordingly, it is recommended that the ceramics are either resintered following machining, if this is possible, or the damaged surfaces, which may be 50 to 100 μm (2–4 mils) in depth, are removed by gentle lapping or chemical etching prior to brazing. The benefits of this preparation are illustrated in Fig. 7.28 [Mizuhara and Mally 1985]. Active brazes benefit from smooth faying surfaces on the nonmetal ($R_a < 1$ μm, or 40 μin.) because their inherently slow rate of spreading will be further impeded by a gross surface topography. If the active constituent is exposed on the surface, it should be freshly cleaned before use (see Chapter 1, section 1.3.2.7 for effective cleaning procedures).

Joint Design

Joint design for brazing nonmetals is considered in the following section (7.4). A vital point that needs to be recognized is that where there is thermal expansion mismatch between the abutting components, joints made with active brazes are likely to require interlayer structures to relieve the resulting stress. Because joint stress is such a critical parameter in achieving mechanical integrity, care needs to be taken to ensure that any metal components in an assembly are always in a known metallurgical condition. This is preferably one in which some stress can be absorbed by elastic and plastic deformation. It is usually recommended that metal components are annealed prior to brazing to soften them, especially if they have been subject to cold work or rapid thermal excursions (e.g., welding) as part of their fabrication process.

Process Atmosphere

The active constituent in active brazes will react readily with oxygen and water vapor and sometimes even nitrogen. Consumption of the active ingredients by these reactions depletes the amount that remains for reacting with the nonmetal and may also give rise to dross formation that will impede wetting and spreading. Consequently, active brazing is often conducted in vacuum, argon, helium, or hydrogen atmospheres. The quality of the atmosphere in the process chamber needs to be monitored, or at least checked regularly, to ensure that the combined oxygen and water vapor content is below 10 ppm. If the active constituent is exposed at the surface, the combined value should be lower than half this figure, as mentioned previously. The sensitivity of active brazes to the brazing atmosphere is reflected by the data given in Fig. 7.20, which shows the strength of ceramic-metal assemblies brazed using silver-copper-titanium alloy in different atmospheres. In this example, the highest and most consistent joint strengths have been achieved in atmospheres of high-purity nitrogen and argon [Weise, Malikowski, and Krappitz 1989].

The importance of low-oxygen content in the atmosphere is further illustrated by the data given in Table 7.4, which shows the wetting angle of molten copper, at 1150 °C (2100 °F), in vacuum on some different oxides of titanium [Li 1993]. The higher the fraction of the metallic element in the oxide, the better is the wetting (contact angle) and also the resulting adhesion between the copper and the oxide.

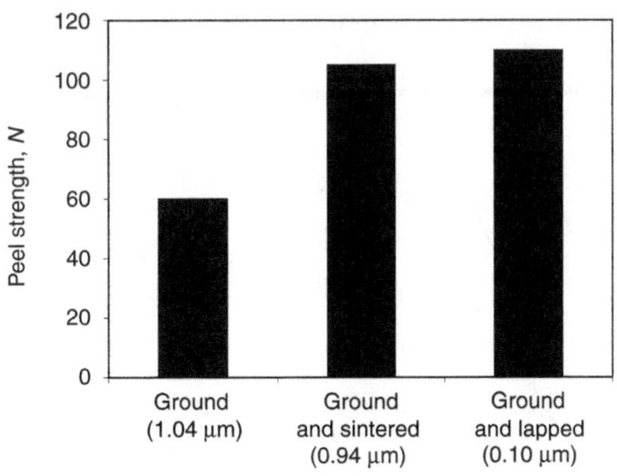

Fig. 7.28 Correlation between the surface preparation technique and peel strength (more correctly, peel failure stress) of joints between alumina and nickel components using a copper-silver-titanium braze

Heating and Cooling Rate

As with all filler metal joining processes, rapid heating is generally desirable, just as shortening of the time between preparation of the components and wetting by the braze minimizes the propensity for surface contamination to occur through oxidation, handling, or through other forms of exposure. However, because metals and nonmetals have very different specific heat capacities, thermal conductivities, and surface emissivities, there is a risk of developing adverse temperature gradients between and across the components. Good practice for active brazing is to heat rapidly to just below the solidus temperature of the braze, and, after a short dwell, to allow thermal equilibrium to become

Table 7.4 Contact angle and bond strength of molten copper on oxides of titanium

The higher the metallic content of the oxide, the better is the resulting wetting and bonding by the metal.

Oxide	Contact angle, degree	Bond strength
Ti_2O_3	113	Poor
$TiO_{1.14}$	82	Fair
TiO	75	Fair
$TiO_{0.86}$	72	Good

established, then proceed to heat to the brazing temperature. Cooling should be rapid to the solidus temperature of the braze, but thereafter will depend on the properties of the components and the difference in thermal expansion and the elastic modulus between them. Slower cooling from the solidus temperature usually assists in stress relief.

Brazing Temperature

Recommended brazing practice for of metal-to-metal brazing is that the peak process temperature should be 20 to 50 °C (35–90 °F) above the liquidus temperature of the braze. In active brazing, 50 °C (90 °F) is normally considered to be a minimum superheat; raising the process temperature increases the activity of the braze and reduces its viscosity, both of which benefit wetting and spreading. However, excessive temperatures will increase the thickness of the reaction layer formed in a given time, which is usually disadvantageous and may also exacerbate erosion of the metal component in metal/nonmetal joints. A minimum superheat is necessary to ensure all parts of the joint exceed the brazing temperature for the minimum time during the brazing cycle.

Process Duration

The brazing time, that is, the period for which the faying surfaces are above the liquidus temperature of the virgin braze, is usually dictated by the need to form a continuous reaction layer of specific thickness, and this may take as long as 15 to 20 minutes, or even longer, as indicated in Fig. 7.1. As discussed in section 7.1 of this chapter, brazed joints to nonmetals made using active brazes often have optimal mechanical properties when the reaction layer is thin and continuous and rarely more than 2 μm (80 μin.) thick. It should also have a textured interface.

Component Fixturing

Furnace fixturing intended for use with active brazes requires more thought in design than it does for conventional brazes. Active brazes will wet most materials so that accidental braze spillage can result in components being firmly bonded to fixturing and interior surfaces of the furnace. Needless to say, such joints will tend to be highly adherent. Jigging used in active brazing processes is commonly of graphite.

As mentioned in section 7.2.1, the fixturing should be designed so as to apply a light load (1–5 kPa, or 20–100 lb/ft^2) to the braze preform during the heating cycle, unless the active brazing process is coupled with diffusion brazing, as described in section 7.2.7 of this chapter, when higher compressive loadings should be applied. The effect of applied load on the resulting joint strength is illustrated for joints made to silicon nitride with copper-titanium braze in Fig. 7.29 [El Sawy et al. 1993]. Application of a light load substantially increases the joint strength. A modest loading will overcome any tendency for the filler to dewet from the nonmetal component before the reaction layer becomes established, but excessive pressure will tend to overcome the hydrostatic pressure of the braze and result in filler metal being expelled from the joint gap. Because many silver-copper-titanium brazes exhibit liquid immiscibility, such displacement of molten alloy can give rise to gross compositional changes in the filler metal and detrimentally affect wetting by the braze and, in turn, the mechanical properties of the resulting joint.

Process Economics

The requirements for more scrupulous surface preparation of the nonmetal, a high-quality process atmosphere, and a price premium for the braze can mean that an active brazing process is economically unfavorable compared with metallizing the nonmetal component and using a conventional braze. Thus, the most elegant technical solution may not be the most appropriate one from an economic perspective, and each case must be carefully examined in terms of the overall requirements and the respective merits of the processes that are available.

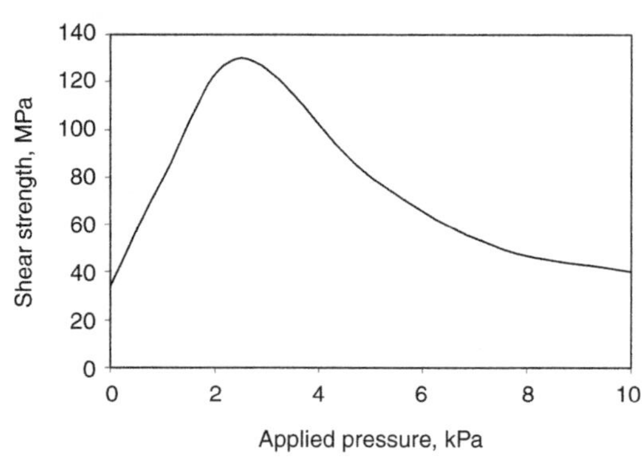

Fig. 7.29 Effect of applied load during the brazing cycle on the shear strength of silicon nitride assemblies joined using a copper-titanium braze

7.4 Design and Properties of Metal/Nonmetal Joints

Metal/nonmetal brazed joints can be difficult to exploit in industrial applications, even when the braze wets the nonmetal satisfactorily, joint filling is good, and acceptable fillets are formed. The mechanical properties of brazed joints in metal/nonmetal assemblies depend on a complex variety of factors, including the elastic properties of the metal and nonmetal, the plastic behavior of the metal, the thickness and mechanical properties of the interfacial layer, the geometry of the assembly, the service temperature and temperature range, and also the mode of loading applied to it in service. In addition, residual stress due to thermal expansion and elastic modulus mismatch, and temperature gradients play a significant role and must generally be minimized in order to obtain high and repeatable joint strengths.

In general, engineering ceramics and most nonmetals have lower coefficients of thermal expansion (CTEs) than do metals. During cooling from the solidus temperature of the braze, the metal part will contract more than the nonmetal part. One might therefore presume that this would give rise to the desirable condition of a compressive stress in the ceramic and a tensile stress in the nonmetal. The assembly should then have favorable mechanical properties because metals resist tensile forces well and compressive stress will suppress crack opening in the ceramic part. In practice, however, the differential contraction will tend to set up a bending moment that gives rise to high tensile forces acting in the ceramic at the joint edge and also within the body of the joint just below the joint interface, as shown in Fig. 7.30 [Suga and Elssner 1989]. This situation will be conducive to the formation and propagation of cracks in the assembly, the susceptibility depending on the mode of stressing. The cracks will propagate along the path with the lowest fracture toughness, which is usually close to the ceramic surface where the reaction products have formed.

Most brazed joints to nonmetals are of simple planar geometry. Finite element analysis, based on elastic-plastic models of component behavior, can be used to highlight in quantitative terms the importance of the relative coefficients of thermal expansion of the components, their elastic and yield properties, the size and geometry of the joints, and the response of the bonded assembly to different loading configurations. Detailed modeling, however, is currently hindered by a lack of accurate knowledge of the composition of the interfacial layers and uncertainty regarding the values to assign to their respective properties. Nevertheless, three important features may be observed [Howe 1993b]:

- The magnitude of the residual stress is determined by the thermoelastic parameter $\Delta\alpha$-$\Delta E \Delta T$, where $\Delta\alpha$ (coefficient of thermal expansivity differential) $= \alpha_{metal} - \alpha_{non\text{-}metal}$, ΔE (elastic modulus differential) $= E_{metal} - E_{non\text{-}metal}$ and ΔT (temperature differential) $= T_{braze\ solidus} - T_{service\ minimum}$. Thermal expansion mismatch induces tensile stresses in the component of lower CTE, which is usually the nonmetal. These stresses are transmitted from the joint interface, where the mismatch takes effect. A mismatch in elastic modulus generates interfacial tensile stresses at the edge of the joint, regardless of the sign of the mismatch and therefore always increases the propensity to fracture. The temperature differential between the braze solidus and the minimum temperature of service or operation of the assembly is usually set by other considerations, so there is not usually scope to make large changes to this term.

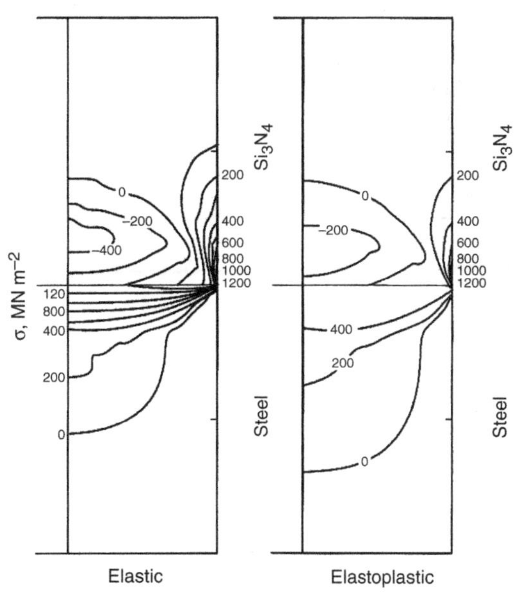

Fig. 7.30 Distribution of principal stresses (shown in contour units of MN/m^2) in a planar joint between silicon nitride and steel, according to elastic and elastoplastic finite element modeling. Adapted from Suga and Elssner [1989]

The effect of thermal expansion mismatch on the strength of brazed joints to silicon nitride is shown in Fig. 7.31. Failure of such assemblies generally occurs through the near-surface layer of the nonmetal, with fracture being initiated by small defects in the material. By applying fracture mechanics modeling, it is possible to calculate the minimum size of a defect inside the nonmetal that will cause spontaneous failure of such a brazed ceramic-metal joint as a function of the CTE mismatch. The calculated curve given in Fig. 7.32 shows the extreme sensitivity of critical defect size to CTE mismatch and highlights the need to minimize stress from this source and eliminate flaws in the ceramic component.

- The peak mismatch stress is related to the thickness of the metal component, with the dependence related to the assembly geometry and combination of materials used.
- Plastic deformation of the metal component may reduce the residual strain in the assembly considerably.

It follows that a metal/nonmetal joint will have least residual stress if the metal component is thin, soft, and possesses a CTE that is closely matched to the nonmetal, with the CTE of the metal slightly higher than that of the nonmetal over the widest possible temperature range. Because nonmetals have much lower CTE values than most metals, the metal in the joined couple should also possess a low CTE. Unfortunately, low-expansivity metals, notably molybdenum and tungsten, have high elastic moduli (for metals) so it is not possible to achieve all three requirements, mentioned previously, simultaneously. Three approaches are generally used to make robust metal/nonmetal joints. All involve use of interlayers to buffer the nonmetal from the metal component. The interlayer may take the form of compliant structures, in which case they are either soft and thin or low expansivity and thin, as described in sections 7.4.1 and 7.4.2 in this chapter. Alternatively, in some instances, the properties of the reaction products themselves can be exploited to provide all or part of this function, as described in section 7.4.3 of this chapter.

Increasing the braze layer thickness to reduce the mismatch strain is seldom effective because the mechanical properties of the assembly rapidly become dominated by those of the filler metal, which are mostly inferior to the metal and nonmetal parts. This is shown in Fig. 7.33 for alumina/aluminum/alumina joints made with a layer of braze 40 μm (1.6 mils) and 750 μm (30 mils) thick, respectively [Nicholas 1993].

7.4.1 Low-Modulus Interlayers

Low-modulus interlayers are either intrinsically soft or have a structural form that artificially provides the same benefit. The interlayer

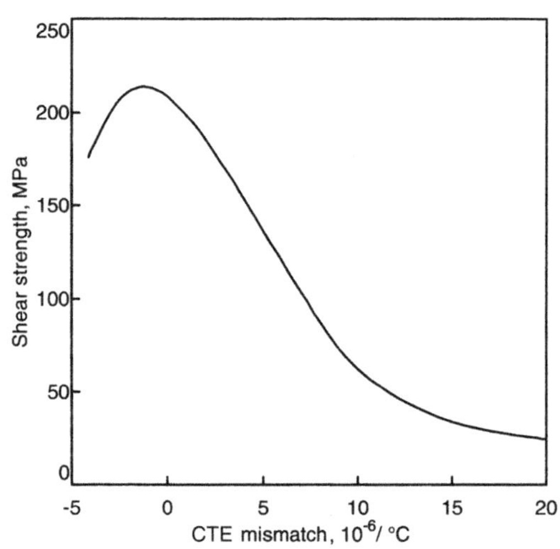

Fig. 7.31 Effect of coefficient of thermal expansion (CTE) mismatch, relative to that of the ceramic, on the shear strength of silicon nitride/metal brazed joints. Adapted from Naka, Kubo, and Okamoto [1989]

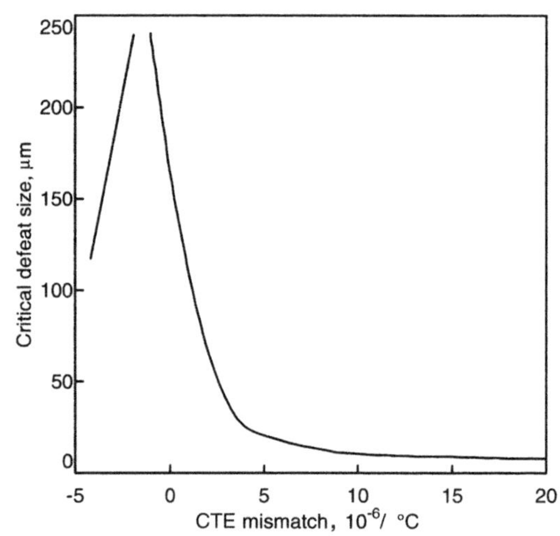

Fig. 7.32 Relationship between coefficient of thermal expansion (CTE) mismatch relative to that of the ceramic and the critical defect size that will cause failure of the ceramic due to imposed stress. Adapted from Akselsen [1992]

will therefore deform readily when subject to stress and thereby reduce the residual stress in the assembly. Copper sheet is a good first choice as a monolithic interlayer for most applications because it is soft (in an annealed condition), has high electrical and thermal conductivity, is compatible with many low-melting-point brazes, and is a relatively low-cost metal. Nickel is commonly used with higher-melting-point brazes and in products intended for use in environments that are fairly aggressive chemically. Further details on this approach are given in Chapter 4, section 4.2.2. Interlayers with a low effective elastic modulus may also be achieved by using compliant structures. Some examples and a discussion of the merits and limitations of this approach for minimizing residual stress arising from CTE and elastic modulus mismatch are described in Chapter 4, section 4.2.3.

The brazing alloy itself, once it is largely denuded of the active constituent through reaction with the nonmetal component, diffusion into the metal component, and losses to the furnace atmosphere, will be relatively soft and therefore can contribute to stress relief. For example, silver-copper-indium braze (without titanium) has a Young's modulus of 75 GPa (11 Mpsi), which compares with nearly 400 GPa (60 Mpsi) for a typical engineering ceramic. Furthermore, being a relatively low-melting-point alloy, this braze does not possess significant strength at elevated temperature. Figure 7.34 shows the yield stress of bulk samples of silver-copper-indium braze as a function of temperature [Levy 1991]. Above 300 °C (570 °F), the alloy will yield readily.

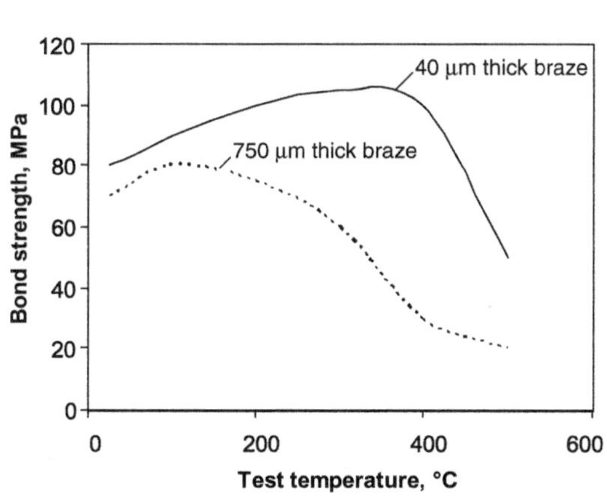

Fig. 7.33 Effect of test temperature on the tensile strength of alumina/aluminum/alumina samples vacuum brazed at 1000 °C (1830 °F). Thin joints are more resilient to elevated-temperature service than thick joints where the braze is unconstrained by the ceramic components.

While this is beneficial in minimizing mismatch stress, it also explains why thick brazed joints cannot be used where the component is required to sustain significant loads at elevated temperature in service (see Fig. 7.33). Therefore, interlayers must be used.

In addition to possessing a low elastic modulus, the interlayer must be stable against reaction with the brazing alloy during the heating cycle used for the brazing operation as well as during the lifetime of the product. A quantitative ranking of candidate interlayer metals, taking into account a host of factors including melting point, activation energies for diffusion and recrystallization, reveals niobium as possessing many favorable characteristics [Chularis 1994]. The disadvantages of niobium are its relatively low thermal conductivity (54 W/mK); relatively high cost, especially in sheet form; and a limited knowledge-base of niobium metallurgy, compared with the available information on other more commonly used candidate metals.

7.4.2 Low-Expansion Interlayers

Low-expansion interlayers are not intended so much to reduce the total residual stress in the assembly but rather to redistribute it to locations where it can be safely accommodated. The basis of the approach is to braze the nonmetal component to a metal that is reasonably closely matched in its CTE so the major proportion of the CTE mismatch is transferred to the metal/metal brazed joint between the interlayer and the metal component of the assembly. Sometimes different brazes are used to make these two joints. As mentioned earlier, low-expansion metals, with a few notable exceptions, have high

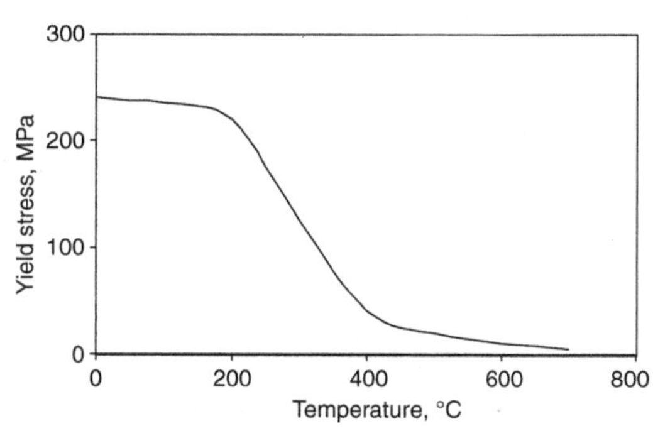

Fig. 7.34 Yield stress of silver-copper-indium braze, measured on bulk samples as a function of temperature. The braze melting range is 625 °C (1160 °F) to 755 °C (1390 °F).

elastic modulus so that they do not often reduce the total stress through significant elastic or plastic deformation. Further details on low-expansion materials and their use as interlayers are given in Chapter 4, section 4.2.2.

When considering the thermal expansivity of possible metals to use as interlayers, it is not the thermal expansivity (CTE) per se that is important, rather, it is the total expansion over the temperature range between the solidus temperature of the braze and the lowest service temperature of the assembly. In Chapter 4, section 4.2.1.1, mention was made of low-expansion iron-nickel alloys, which possess CTE values that are low over a limited range of temperature and composition. These alloys also have low elastic modulus and can be prepared in the form of thin foils. Therefore, when used as interlayers, they will relieve mismatch stress at high temperature in the manner described in the preceding section. However, below the Curie point of the alloy in question, where its CTE is relatively low, any further reduction in temperature results in only a small increase in stress between the ceramic and the interlayer because the thermal expansion coefficients of these two parts can be closely matched. Therefore, the net effect is to decrease the temperature interval over which the CTE mismatch develops, and this accounts for a doubling of both the bending strength and fracture toughness of brazed joints made between silicon nitride and molybdenum components with a nickel-iron interposer, which has been reported [Mutoh et al. 1993].

In general, a soft buffer layer will be more effective than one that is expansion-matched to the nonmetal component, owing to the problem of transient thermal gradients that arise when other properties, such as the specific heat capacity, of the metal and nonmetal also differ. This subject is discussed further in Chapter 4, section 4.2.4.

7.4.3 Mechanical Properties of Reaction Products

The strength of copper-alumina assemblies joined using the copper-copper oxide eutectic process is shown in Fig. 7.35 as a function of the thickness of the reaction layer. As the layer forms and grows, the joint strength increases progressively up to a maximum value for a reaction layer 5 μm (200 μin.) thick in this particular case. Further growth of the layer causes the strength to decline as cracks and voids develop within it. This characteristic is typical for metal-ceramic assemblies, although the optimum strength is mostly obtained for thinner reaction layers (see Fig. 7.36 [Hongqi, Zhihao and Xiaotian 1994]). At the optimum thickness of the reaction layer, the brazed joints can achieve mechanical properties that are comparable to those of the monolithic ceramic parts, as shown in section 7.1.2.2 of this chapter.

In joints to ceramics, occasionally the presence of the interfacial reaction layer often have a beneficial role in reducing thermal mismatch stresses. The bridging compounds frequently have coefficients of thermal expansivity that are

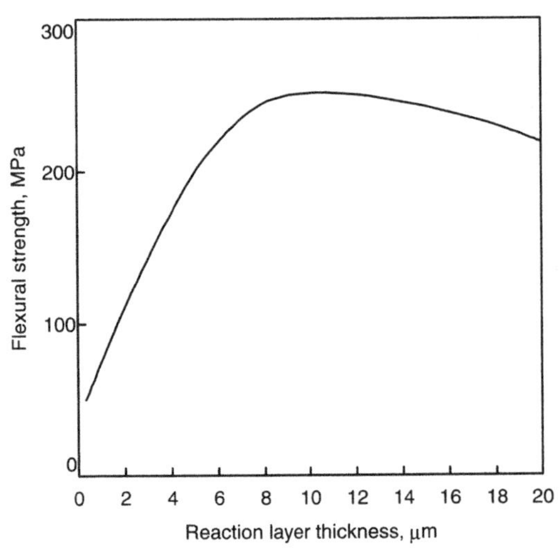

Fig. 7.35 Strength of copper-alumina joints made using the Cu/CuO_2 eutectic braze as a function of the thickness of the $AlCuO_2$ interfacial reaction layer formed. Adapted from Kim and Kim [1992]

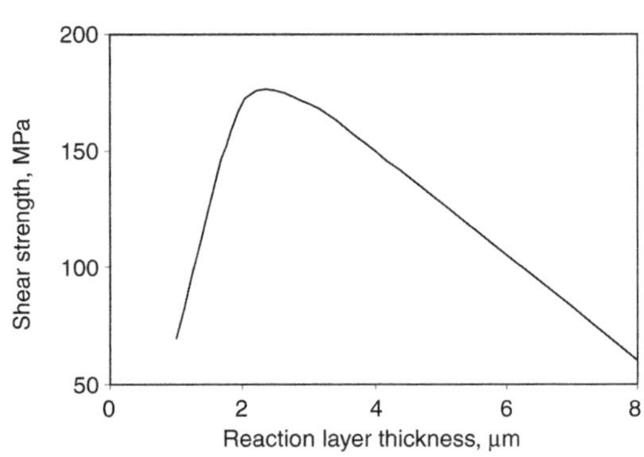

Fig. 7.36 Relationship between joint strength and reaction layer thickness for alumina components joined with silver-copper titanium active braze

Table 7.5 Bending strength of Si_3N_4/steel joints made with brazes containing three different active constituents

Typical process conditions used, suggesting some correlation between joint strength and the coefficient of thermal expansion (CTE) of the interfacial compounds formed by reaction

Active constituent	Bending strength, MPa	CTE of the interfacial compound formed by reaction, 10^{-6}/°C(a)
Titanium	84	9.3
Zirconium	69	7.2
Hafnium	148	6.9

(a) CTE of Si_3N_4, 3.2 × 10^{-6}/°C; Cr/Ni steel, 18 10^{-6}/°C

Table 7.6 Coefficients of thermal expansion (CTE) of selected transition metal carbides and nitrides at room temperature

Reaction product	CTE, 10^{-6}/°C
Si_3N_4	3.2
SiC	4.1
ZrC	6.7
HfC	6.7
HfN	6.9
ZrN	7.2
TiC	7.7
TiN	9.3

Adapted from Lugscheider and Tillmann [1990]

intermediate between those of the ceramic and metal components in the brazed assembly. For example, the $CuAl_2O_4$ layer (α = 11 ppm/°C) formed in a copper-alumina joint is equivalent to having a thin intermediate plate that reduces the stress concentration between the abutting components (copper α = 17 ppm/°C, alumina α = 7 ppm/°C). Table 7.5 shows the bending strength of joints made between silicon nitride ceramic and a chromium/nickel steel using silver-copper braze activated with titanium, zirconium, or hafnium. There is a trend linking the mechanical properties of the assemblies to the CTE of the intermetallic compounds formed by reaction in each case. Table 7.6 lists coefficients of thermal expansion for various transition metal carbides and nitrides. They are all intermediate between most metals, including brazing alloys and nonmetals.

7.4.4 Measurement of Residual Stress

Measurement of the residual stress in metal/nonmetal assemblies can be accomplished by using x-rays to determine the deviation of the lattice constants of the component materials from their relaxed condition [Lancu and Munz 1990]. An alternative method is to use scanning acoustic microscopy (SAM) because the acoustic velocity in materials varies with externally applied compression or bending. This has been demonstrated for joints between nickel and silicon nitride, formed by brazing with a chromium-nickel-titanium alloy [Narita, Ishikawa, and Ishikawa 1992].

REFERENCES

- Akselsen, O.M., 1992. Review—Advances in Brazing of Ceramics, *J. Mater. Sci.,* Vol 27, p 1989–2000
- Ashby, M.F., 1997. Material Property Charts, *Material Selection and Design,* Vol 20, *ASM Handbook,* ASM International, 1997
- Asthana, R. and Sobczak, N., 2000. Wettability, Spreading, and Interfacial Phenomena in High-Temperature Coatings, *JOM-e,* Vol 52 (No. 1), p 1–18
- Bang, K.-S. and Liu, S., 1994. Interfacial Reaction between Alumina and Cu-Ti Filler Metal during Reactive Metal Brazing, *Weld. J. Res. Suppl.,* Vol 73 (No. 3), p 54s–60s
- Camilo de Camargo, P.R., 1995. "Role of Oxygen in the Cu-O-Ti/Sapphire Interfacial Region Formation," Ph.D. dissertation, T-4726, Colorado School of Mines
- Chidambaram, P.R., Edwards, G.R., and Olson, D.L., 1991. A Thermodynamic Criterion for Predicting Wettability at Metal-Alumina Interfaces, *Metall. Trans. B,* Vol 23B, p 215–222
- Chularis, A.A., 1994. A Calculation-Graphical Method for Evaluating the Criterion of Selecting a Refractory Metal of the Barrier Metal When Joining Active Metals to Other Metals and Alloys, *Weld. Intl.,* Vol 8 (No. 3), p 236–238
- Crispin, R.M. and Nicholas, M., 1976. The Wetting and Bonding Behaviour of Some Nickel Alloy-Alumina Systems, *J. Mater. Sci.,* Vol 11 (No. 1), p 17–21
- Drevet, B., Kalogeropoulou, S., and Eustathopoulos, N., 1993. Wettability and Interfacial Bonding in the Au-Si/SiC System, *Acta Metall. Mater.,* Vol 41 (No. 11), p 3119–3126
- El Sawy, A.H. et al., 1993. Direct Brazing of Silicon Nitride Ceramics to Copper Using Active Filler Metal, *Proc. Conf. Trends in Welding Research,* June 1–5 (Gatlinburg,

TN), David, S.A. and Vitek, J.M., Ed., p 1119–1127

- Eustathopoulos, N., Nicholas, M.G., and Drevet, B., 1999. *Wettability at High Temperatures,* Pergamon Press
- Finnis, M.W., 1996. The Theory of Metal-Ceramic Interfaces, *J. Phys., Condens. Matter,* Vol 8 (No. 32), p 5811–5836
- Frage, N., Froumin, N., and Dariel, M.P., 2002. Wetting of TiC by Non-Reactive Liquid Metals, *Acta Mater.,* Vol 50 (No. 2), p 237–245
- Froumin, N. et al., 2000. Wetting Phenomena in the TiC/(Cu-Al) System, *Acta Metall. Mater.,* Vol 48 (No. 7), p 1435–1441
- Granta Design Limited, 2004. Cambridge Engineering Selector v4.5, Cambridge, UK
- Hongqi, H., Zhihao, J., and Xiaotian, W., 1994. The Influence of Brazing Conditions on Joint Strength in Al_2O_3/Al_2O_3 Bonding, *J. Mater. Sci.,* Vol 29, p 5041–5046
- Howe, J.M., 1993a. Bonding, Structure and Properties of Metal/Ceramic Interfaces, Part 1: Chemical Bonding, Chemical Reaction and Interfacial Structure, *Intl. Mater. Rev.,* Vol 38 (No. 5), p 233–256
- Howe, J.M., 1993b. Bonding, Structure and Properties of Metal/Ceramic Interfaces, Part 2: Interfacial Fracture Behaviour and Property Measurement, *Intl. Mater. Rev.,* Vol 38 (No. 5), p 257–271
- Kim, S.T. and Kim, C.H., 1992. Interfacial Reaction Product and its Effect on the Strength of Copper to Alumina Eutectic, *J. Mater. Sci.,* Vol 27, p 2061–2066
- Kubaschewski, O., 1988. Silver-Copper-Titanium, in Ternary Alloys. A Comprehensive Compendium of Evaluated Constitutional Data and Phase Diagrams, Vol 2, Petzow, G. and Effenberg, G., Ed., VCH, p 55–59
- Kuzumaki, T., Ariga, T., and Miyamoto, Y., 1990. Effect of Additional Elements in Ag-Cu Based Filler Metal on Brazing of Aluminium Nitride to Metals, *ISIJ Int.,* Vol 30 (No. 12), p 1135–1141
- Lancu, O.T. and Munz, D., 1990. Residual Stress State of Brazed Ceramic/Metal Compounds, Determined by Analytical Methods and X-Ray Residual Stress Measurements, *J. Am. Ceram. Soc.,* Vol 73 (No. 5), p 1144–1149
- Landry, K. and Eustathopoulos, N., 1996. Dynamics of Wetting in Reactive Metal/Ceramic Systems: Linear spreading, *Acta Mater.,* Vol 44 (No. 10), p 3923–3932
- Levi, C.G., Abbaschian, G.J, and Mehrabian, R., 1978. Interface Interactions during Fabrication of Aluminum Alloy-Alumina Fibre Composites, *Metall. Trans. A,* Vol 9A (No. 5), p 697–702
- Levi, G. and Kaplan., W.D., 2002. Oxygen Induced Interfacial Phenomena During Wetting of Alumina by Liquid Aluminum, *Acta Mater.,* Vol 50 (No. 1), p 75–88
- Levy, A., 1991. Thermal Residual Stresses in Ceramic-to-Metal Brazed Joints, *J. Am. Ceram. Soc.,* Vol 74 (No. 9), p 2141–2147
- Li, J.G., 1993. Understanding of Wetting, Adhesion and Interfacial Bonding of Metals and Ceramic Oxides, *Proc. Conf 2nd European Colloquium Designing Ceramic Interfaces II: Understanding and Tailoring Interfaces for Coating, Composite and Joining Applications,* Nov 11–13, 1993 (Petten, Belgium), p 481–498
- Loehman, R.E., 1988. Joining and Bonding Mechanisms in Nitrogen Ceramics, *Proc. Conf. Intl. Meeting on Advanced Materials,* Vol 8, *Metal-Ceramic Joints,* June 2–3 (Tokyo, Japan), p 49–59
- Loehman, R.E., 1994. Wetting and Joining of Mullite Ceramics by Active-Metal Braze Alloys, *J. Am. Ceram. Soc.,* Vol 77 (No. 1), p 271–274
- Lugscheider, E., Bürger, W., and Broich, U., 1997. Development and Characterisation of Joining Techniques for Dispersion-Strengthened Alumina, *Weld. J. Res. Suppl.,* Vol 76 (No. 9), p 349s–355s
- Lugscheider, E. and Tillmann, W., 1990. Development of New Active Filler Metals in a Ag-Cu-Hf System, *Weld. J. Res. Suppl.,* Vol 69 (No. 11), p 416s–421s
- Lugscheider, E. and Tillmann, W., 1991. Development of New Active Filler Metals for Joining Silicon-Carbide and -Nitride, *Proc. Conf. British Association for Brazing and Soldering, 6th International Conference,* Sept 3–5 (Stratford-upon-Avon), Paper 11
- Lugscheider, E. and Tillmann, W., 1993. Interfacial Reactions between New Active Filler Metals and Nonoxide Ceramics, *Proc. Conf 2nd European Colloquium Designing Ceramic Interfaces II, Understanding and Tailoring Interfaces for Coating, Composite and Joining Applications,* Nov 11–13 (Petten, Belgium), p 499–517
- Marks, R.A. et al., 2000. Joining of Alumina via Copper/Niobium/Copper Interlayers, *Acta Mater.,* Vol 48 (No. 18/19), p 4425–4438

- Meier, A., Javernick, D.A., and Edwards, G.R., 1999. Ceramic-Metal Interfaces and the Spreading of Reactive Liquids, *JOM,* Vol 51 (No. 2), p 44–47
- Minegishi, T., Sakurai, T., and Morozumi, S., 1991. Electric Field-Assisted and Field-Depressed Segregation of Reactive Metals to the Bond Interface in Braze Alloy Joining, *J. Mater. Sci.,* Vol 26, p 5473–5480
- Mizuhara, H. and Mally, K., 1985. Ceramic-to-Metal Joining with Active Brazing Filler Metal, *Weld. J. Res. Suppl.,* Vol 64 (No. 10), p 27s–32s
- Moorhead, A.J., Henson, H.M., and Henson, T.J., 1987. In *Ceramic Microstructures '86: Role of Interfaces,* Pask, J.A., and Evans, A.G., Ed., Plenum Press
- Morrell, R., 1985. *Handbook of Properties of Technical and Engineering Ceramics,* Her Majesty's Stationary Office
- Mutoh, Y. et al. 1993. Strength and Fracture Toughness of Si_3N_4-Metal Joints with Phase Transforming Interlayer Metal, *Proc. Conf. 9th Biennial Conference on Fracture,* Sept 21–25, 1992 (Varna, Bulgaria), Pub., *Reliability and Structural Integrity of Advanced Materials,* Sedmak, S., Sedmak, A., and Ruzic, D., Ed., Vol 2, p 1043–1048
- Naka, M., Kubo, M., and Okamoto, I., 1989. Brazing of Si_3N_4 to Metals with Al Filler (Report II), *Trans. JWRI,* Vol 2 (No. 18), p 33–36
- Nakao, Y., Nishimoto, K., and Saida, K., 1989. Bonding of Si_3N_4 to Metals with Active Filler Metals, *Trans. Jpn. Weld. Soc.,* Vol 20 (No. 1), p 66–76
- Naidich, Y.V., Zhuravlev, V., Krasovskaya, N., 1998. The Wettability of Silicon Carbide by Au–Si Alloys, *Mater. Sci. Eng.,* Vol A245, p 293–299
- Narita, T., Ishikawa, T., and Ishikawa, I., 1992. Simultaneous Measurements of Stress and Distortion of the Ceramic Metal Joints by Scanning Acoustic Microscopy, *Nondestruct. Test. Eval.,* Vol 8-9, p 709–716
- Nicholas, M.G., 1988. Reactive Brazing of Ceramics, *Proc. Conf. Intl. Meet. Adv. Mater.,* Vol 8, *Metal-Ceramic Joints,* June 2–3 (Tokyo, Japan), Materials Research Society, p 49–59
- Nicholas, M.G., 1989a. Reactive Metal Brazing, *Proc. Conf. 3rd International Conference Joining Ceramics, Glass and Metal,* April 26–28 (Bad Nauheim, Germany), Kraft, W., Ed., p 3–6
- Nicholas, M.G., 1989b. In *Surfaces and Interfaces of Ceramic Materials,* Kluwer Academic Press, Dufour, L.C. et al., Ed.
- Nicholas, M.G., 1990. *Joining of Ceramics,* The Institute of Ceramics/Chapman and Hall
- Nicholas, M.G., 1993. Fabrication of Ceramic Interfaces and Joints, *Proc. Conf 2nd European Colloquium Designing Ceramic Interfaces II, Understanding and Tailoring Interfaces for Coating, Composite and Joining Applications,* Nov 11–13 (Petten, Belgium), p 173–189
- Nicholas, M.G. and Crispin, R.M., 1986. Copper-Ceramic Materials, *INCRA Report* 383
- Nicholas, M.G. and Peteves, S.D., 1991. Reactive Joining of Silicon Nitride Ceramics, *Proc. Conf. British Association for Brazing and Soldering, 6th International Conference,* Sept 3–5 (Stratford-upon-Avon), Paper 10
- Nicholas, M.G. and Peteves, S.D., 1994. Reactive Joining; Chemical Effects on the Formation and Properties of Brazed and Diffusion Bonded Interfaces, *Scr. Metall. Mater.,* Vol 31 (No. 8), p 1091–1096
- Ohmura, H. et al., 1994. A Technique for Brazing Graphite/Graphite and Stainless Steel/High-Carbon Steel Joints, *Weld. J. Res. Suppl.,* Vol 73 (No. 10), p 249s–256s
- Paulasto, M., Ceccone, G., and Peteves, S.D., 1997. Joining of Silicon Nitride via a Transient Liquid, *Scr. Mater.,* Vol 26 (No. 10), p 1167–1173
- Paulasto, M. and Kivilahti, J., 1995. Formation of Interfacial Microstructure in Brazing of Si_3N_4 with Ti-Activated Ag-Cu Filler Alloys, *Scr. Metall. Mater.,* Vol 32 (No. 8), p 1209–1214
- Paulasto, M. and Kivilahti, J., 1998. Metallurgical Reactions Controlling the Brazing of Al_2O_3 with Ag-Cu-Ti Filler Alloy, *J. Mater. Res.,* Vol 13 (No. 2), p 343–352
- Peteves, S.D. et al., 1996. Joining Silicon Nitride to Itself and to Metals, *JOM,* Vol 48 (No. 1), p 48–77
- Saiz, E., Cannon, R.M., and Tomsia, A.P., 2001. Reactive Spreading in Ceramic/Metal Systems, *Oil Gas Sci. Technol.—Rev. IFP,* Vol 56 (No. 1), p 89–96
- Schulz-Harder, J. and Maier, P.H., 1996. DCB Substrates with Reduced Ceramic Thickness in Power Semiconductor Modules, *PCIM Europe,* No. 1, 5 p
- Schulz-Harder, J., 2003. Advantages and New Development of Direct Bonded Copper

Substrates, *Microelectron. Reliability,* Vol 43 (No. 3), p 359–365

- Schwartz, M.M., 1990. *Ceramic Joining,* American Society for Materials
- Scott, P.M., Nicholas, M., and Dewar, B., 1975. The Wetting and Bonding of Diamonds by Copper-Base Binary Alloys, *J. Mater. Sci.* Vol 10 (No. 11), p 1833–1840
- Selverian, J.H. and Kang S., 1992. Ceramic-to-Metal Joints, Part II: Performance and Strength Prediction, *Am. Ceram. Soc. Bull.,* Vol 71 (No. 10), p 1511–1520
- Selverian, J. H., O'Neil, D., and Kang, S., 1992. Ceramic-to-Metal Joints, Part I: Joint Design, *Am. Ceram. Soc. Bull.,* Vol 71 (No. 9) p 1403–1409
- Shiue, R.K., Wu, J.M.O., and Wang, J.Y., 2000. Microstructural Evolution at the Bonding Interface during the Early-Stage Infrared Active Brazing of Alumina, *Metall. Mater. Trans. A.,* Vol 31A (No. 10), p 2527–2536
- Suga, T. and Elssner, G., 1989. Structural Features to Relax Thermal Stress at Metal/Ceramic Joined Interfaces, *Proc. Conf. MRS International Meeting on Advanced Materials,* June 2–3 (Tokyo, Japan), Materials Research Society, Doyama, M. et al., Ed., *Metal-Ceramic Joints,* Vol 8
- Suganuma, K., 1990. Recent Advances in Joining Technology of Ceramics to Metals *ISIJ Int.,* Vol 30 (No. 12), p 1046–1058
- Torvund, T. et al., 1996. A Process Model for Active Brazing of Ceramics, Part 1: Growth of Reaction Layers, *J. Mater. Sci.,* Vol 31, p 6215–6222
- Villars, P., Prince, A., and Okamoto, H. 1995. *Handbook of Ternary Alloy Phase Diagrams,* Vol 3, American Society for Materials
- Wang, Z. and Wynblatt, P., 1998. Study of a Reaction at the Solid Cu/α-SiC Interface, *J. Mater. Sci.,* Vol 33, p 1177–1181
- Warren, R. and Andersson, C.H., 1984. Silicon Carbide Fibres and Their Potential for Use in Composite Materials II, *Composites,* Vol 15 (No. 2), p 101–111
- Waterman, N.A. and Ashby, M.F., 1997. *The Materials Selector,* 2nd ed., Chapman and Hall, p 46
- Weise, W., Malikowski, W., and Krappitz, H., 1989. Wetting and Strength Properties of Ceramic to Metal Joints Brazed with Active Filler Metals Depending on Brazing Conditions and Joint Geometry, *Proc. 3rd International Conference Joining Ceramics, Glass and Metal,* Kraft, W., Ed., April 26–28 (Bad Nauheim, Germany), DGM, p 33–42
- Xiong, H., Wan, C., and Zhou, Z., 1998. Development of a New CuNiTiB Brazing Alloy for Joining Si_3N_4 to Si_3N_4, *Metall. and Mater. Trans. A,* Vol 29A (No. 10), p 2591–2596
- Zhou, X.B. and De Hosson, J.Th.M., 1995. Reactive Wetting of Liquid Metals on Ceramic Substrates, *Acta Metall. Mater.*, Vol 44 (No. 2), p 421–426

Abbreviations and Symbols

A	atomic weight
A	area
ABA	active brazing alloy
c	crack length
C	concentration
CTE	coefficient of thermal (linear) expansion, see also α
D	rate of diffusion
$D_{\bar{\alpha}}$	average diffusivity
DIN	Deutsches Institut fur Normung (German Standardization Institute)
E	internal energy or Young's modulus
F	force
g	acceleration due to gravity
G	gauss (alternate cgs-emu system)
G	Gibbs free energy
h	height of braze cap
HAZ	heat-affected zone
HK	Knoop hardness
HV	Vickers hardness
k	Boltzman constant
K	dissolution rate constant
K	stress concentration factor
M	represents a metal in a reaction; chemical symbol for Mischmetal
MEMS	microelectromechanical system
MPD	melting-point depressant
ppm	parts per million
P	pressure
PTFE	polytetrafluoroethylene
Q	energy, heat
R	gas law constant (8.314 kJ/mol K)
R_a	surface smoothness (or roughness)
RE	chemical symbol for rare earth elements
RH	relative humidity
S	entropy
SAM	scanning acoustic miscroscopy
SHS	self-propagating high temperature synthesis heating method
SLID	solid-liquid interdiffusion
SPC	statistical process control
t	time
T	temperature
T_m	absolute melting temperature
TLP	transient liquid-phase (joining)
TLPB	transient liquid-phase bonding
TLV	threshold limit value
UHV	ultrahigh vacuum
V	volume
W	work
W_a	work of adhesion
α	coefficient of linear thermal expansion; see also CTE
ΔG_r	change in Gibbs free energy
δ	small change
η	viscosity
γ	surface tension
λ	wavelength
θ	contact angle
θ^*	effective contact angle
ρ	density
σ	strength
σ_s	yield stress

subscripts:

S	solid phase
V	vapor phase
L	liquid phase

Index

A

D

E

F

G

H

I

M

N

O

S

U

V

W

X

Y

Z

FSC
www.fsc.org
MIX
Papier | Fördert
gute Waldnutzung
FSC® C083411

Kolibri 360 GmbH
Ferdinand-Jühlke-Straße 7
99095 Erfurt, Deutschland
produktsicherheit@kolibri360.de